Tissue Culture of Epithelial Cells

Tissue Culture of Epithelial Cells

Edited by
Mary Taub
State University of New York at Buffalo
Buffalo, New York

Plenum Press • *New York and London*

Library of Congress Cataloging in Publication Data

Main entry under title:

Tissue culture of epithelial cells.
   Includes bibliographies and index.
   1. Tissue culture. 2. Epithelial cells. 3. Epithelium. I. Taub, Mary, date–
[DNLM: 1. Epithelium — cytology. 2. Tissue Culture. QS 532.5.E7 T616]
QM561.T56   1984   591′.07′24   84-13470
ISBN 0-306-41740-5

© 1985 Plenum Press, New York
A Division of Plenum Publishing Corporation
233 Spring Street, New York, N.Y. 10013

Printed in the United States of America

Contributors

DAVID BARNES   Department of Biological Sciences, University of Pittsburgh, Pittsburgh, Pennsylvania 15260

G. BEATY   Department of Health Sciences, Universidad Autonóma Metropolitana, Unidad Iztapalapa, México, D.F., Mexico

J. J. BOLÍVAR   Centro de Investigación y de Estudios Avanzados, Departamento de Fisiología y Biofísica, México, D.F. 07000, Mexico

L. BORBOA   Centro de Investigación y de Estudios Avanzados, Departamento de Fisiología y Biofísica, México, D.F. 07000, Mexico

M. CEREIJIDO   Centro de Investigación y de Estudios Avanzados, Departamento de Fisiología y Biofísica, México, D.F. 07000, Mexico

BARBARA EDDY   Department of Molecular Immunology, Roswell Park Memorial Institute, Buffalo, New York 14263

S. FERNÁNDEZ–CASTELO   Department of Electrical Engineering, Universidad Autónoma Metropolitana, Unidad Iztapalapa, México, D.F., Mexico

CLAIRE M. FRASER   Department of Molecular Immunology, Roswell Park Memorial Institute, Buffalo, New York 14263

L. GONZÁLEZ–MARISCAL   Centro de Investigación y de Estudios Avanzados, Departamento de Fisiologia y Biofísica, México, D.F. 07000, Mexico

PAUL A. INSEL   Department of Medicine, Division of Pharmacology, M-013 H, University of California, San Diego, La Jolla, California 92093

ARNOST KLEINZELLER   Department of Physiology, School of Medicine, University of Pennsylvania, Philadelphia, Pennsylvania 19104

JULIA E. LEVER   Department of Biochemistry and Molecular Biology, The University of Texas Medical School, Houston. Texas 77225

JOACHIM G. LIEHR   Department of Pharmacology, The University of Texas Medical School at Houston, Houston, Texas 77225

R. LÓPEZ–VANCELL   Department of Health Sciences, Universidad Autonóma Metropolitana, Unidad Iztapalapa, México, D.F., Mexico

KATHRYN E. MEIER   Department of Medicine, Division of Pharmacology, M-013 H, University of California, San Diego, La Jolla, California 92093

DAYTON S. MISFELDT   Palo Alto Veterans Medical Center, Palo Alto, California 94304; and Stanford University, Stanford, California 94305

JAMES M. MULLIN   Department of Human Genetics, Yale University School of Medicine, New Haven, Connecticut 06510

JEREMY D. PEARSON   ARC Institute of Animal Physiology, Babraham, Cambridge CB2 4AT, England. *Present address:* MRC Clinical Research Centre, Harrow HA1-3UJ, England

MILTON H. SAIER, Jr.   Department of Biology, University of California, San Diego, La Jolla, California 92093

MARTIN J. SANDERS   Palo Alto Veterans Medical Center, Palo Alto, California 94304; and Stanford University, Stanford, California 94305. *Present address:* Center for Ulcer Research and Education, VA Wadsworth Medical Center, Los Angeles, California 90073

URSINA SCHMIDT   Department of Molecular Immunology, Roswell Park Memorial Institute, Buffalo, New York 14263

GIORGIO SEMENZA   Laboratorium fur Biochemie der Eidgenossischen Technischen Hochschule ETH-Zentrum, CH-8092, Zurich, Switzerland

FRANCISCO V. SEPÚLVEDA   ARC Institute of Animal Physiology, Babraham, Cambridge CB2 4AT, England

DAVID A. SIRBASKU   Department of Biochemistry and Molecular Biology, The University of Texas Medical School at Houston, Houston, Texas 77225

MARY TAUB   Department of Biochemistry, State University of New York at Buffalo, School of Medicine, Buffalo, New York 14214

J. CRAIG VENTER   Department of Molecular Immunology, Roswell Park Memorial Institute, Buffalo, New York 14263

Preface

Epithelial cells are present in many different tissues in the body, and possess a diverse number of functional properties. However, all epithelial cells share some common characteristics. The cells possess a morphological polarity (an apical and basolateral surface), and are interconnected by tight junctions. The epithelial cells also possess the capacity to transport select solutes across the monolayer. Transport systems localized on either the apical or basolateral surface are responsible for this vectorial transport. Such characteristics of epithelial cells can be examined in the tissue culture situation. This volume discusses the use of cell culture techniques to study these fundamental properties of epithelial cells. Major questions concerning epithelia which may be examined in culture are addressed. The approaches which are taken to answer these questions are described in detail with regards to kidney cell cultures. Similar investigations may be done with epithelial cell cultures derived from other tissues, following the kidney cell culture paradigm.

The differentiated function of kidney tubule epithelial cells *in vivo* is transepithelial solute transport (Koushanpoor, 1976). When kidney tubule cells are put into the culture situation, they often retain the capacity to express this differentiated function. The microscopic observation of multicellular domes in cultured epithelial monolayers is an indicator of vectorial transport (Leighton *et al.*, 1970). Dome formation can be induced *in vitro* by hormones, and agents which cause differentiation (Lever, 1979). Transepithelial solute transport by cultured epithelial cells can also be detected *in vitro* by means of Ussing flux chambers (Misfeldt *et al.*, 1976; Cereijido *et al.*, 1978). Not only can the electrical properties of the cultured epithelial cells be quantitated by this classic physiologic technique, but the transport functions involved in the vectorial movement of solutes can be studied (Saier, 1981; Mullin *et al.*, 1980; Sepúlveda and Pearson, 1982). The dependence of transport on metabolism can also be evaluated using this technique.

Epithelial cell cultures are particularly amenable for biochemical and genetic studies. Genetic variants which are altered in their transport functions, or hormone responses may be isolated from the epithelial cell cultures. The transport systems and the hormone responses of normal and "mutant" epithelial cells can then be compared (Taub and Saier, 1981, Taub *et al.*, in press). A particular advantage of established epithelial cell cultures is cell cloning. Cloning permits cell populations originating from a single cell type to be obtained. Thus, the hormone responses and transport properties of particular epithelial cell types can be analyzed (Meier *et al.*, 1983). Hybridomas can be obtained which make monoclonal antibodies. Such monoclonal antibodies can be used to purify the hormone receptors, and the membrane transport proteins of the epithelial cells (Schmidt *et al.*, 1983). Many of these plasma membrane proteins are distributed in an asymmetric manner in the epithelial plasma membrane, either in the apical or basolateral domain. The mechanism of biogenesis of such plasma membrane asymmetry (or polarity) can readily be examined using epithelial cell cultures. Enveloped viruses have been convenient probes for *in vitro* studies of the polarity problem (Rodriguez-Boulan and Sabatini, 1978).

Cell culture studies which have a more direct bearing upon physiological conditions, are now possible due to the availability of hormonally defined serum-free culture media. Such defined media have recently been developed for many different types of animal cells (Sato, 1975; Sato and Reid, 1978; Rizzino *et al.*, 1979; Barnes and Sato, 1980). The *in vivo* environment of epithelial cells can be closely mimicked *in vitro* when using hormonally defined medium. *In vivo* the epithelial cells face two body compartments. The apical surface of the epithelial monolayer faces a lumenal compartment, whereas the basolateral surface (which is associated with a basement membrane) faces an interstitial compartment. Components from the blood diffuse through interstitial tissue, and finally reach the epithelial cells. In serum-free medium those components of the blood, and the basement membrane which are essential for epithelial cell growth can be identified, and routinely added to the culture medium. Unlike the case with serum, potentially cytotoxic components need not be present in the culture situation. Hormonally defined media have had a significant impact on the use of tissue culture to study epithelial cells. Precise studies of hormonal effects on epithelial cells are possible, and synergistic interactions between hormones can be observed. Many new types of primary epithelial cells can be maintained and grown in culture using hormonally defined medium. Several of the primary epithelial cell cultures have been shown to retain their differentiated functions *in vitro* (Taub and Sato, 1980; Chung *et al.*, 1982). Presumably, new epithelial cell lines may then be established. Such epithelial cell lines can then be compared both with the original tissue and with primary cell cultures.

In some cases cultured epithelial cells have now been taken back to the *in vivo* situation (Rindler *et al.,* 1979; Cameron, 1983). The growth and functional properties of the cells may then again be studied in the animal. Investigations with estrogen-dependent renal tumor cells have demonstrated the power of this combined *in vivo/in vitro* approach. (Sirbasku and Kirkland, 1976). Similar studies may be done concerning the hormone responses of epithelial cells in serum-free medium. The confirmation that *in vitro* results have a direct bearing on *in vivo* physiology will make tissue culture a more powerful tool in the study of epithelia.

Mary Taub

REFERENCES

Barnes, D. and Sato, G., 1980, Methods for growth of cultured cells in serum-free medium, *Anal. Biochem.* **102**:255–270.

Cameron, R., 1983, Personal communication.

Cereijido, M., Robbins, E. S., Dolen, W. S. Rotunno, C. A. and Sabatini, D. D., 1978, Polarized monolayers formed by epithelial cells on a permeable and translucent support, *J. Cell Bio.* **77**:853–880.

Chung, S. D., Alavi, N., Livingston, D., Hiller, S. and Taub, M., 1982, Characterization of primary rabbit kidney cultures that express proximal tubule functions in a hormonally defined medium, *J. Cell Biol.* **95**:118–126.

Koushanpoor, E., 1976, *Renal Physiology, Principles and Functions,* W. B. Saunders Co., Philadelphia.

Leighton, J. L., Estes, W., Mansukhari, J., and Brady, L., 1970, A cell line derived from normal dog kidney (MDCK) exhibiting properties of papillary adenocarcinoma and of renal tubular epithelium, *Cancer* **26**:7022–7028.

Lever, J. E., 1979, Inducers of mammalian cell differentiation stimulate dome formation in a differentiated kidney epithelial cell line (MDCK), *Proc. Natl. Acad. Sci. USA* **76**:1323–1327.

Meier, K. E., Snavely, M. D., Brown, S. L., Brown, S. H. and Insel, P. A., 1983, Alpha- and Beta$_2$-adrenergic receptor expression in the Madin-Darby Canine Kidney epithelial cell line, *J. Cell Biol.* **97**:405–415.

Misfeldt, D. S., Hamamoto, S. S., and Piklka, D. R., 1976, Transepithelial transport in culture, *Proc. Nat. Acad. Sci. USA,* **73**:1212–1216.

Mullin, J. M., Weibel, J., Diamond, L. and Kleinzeller, A., 1980, Sugar transport in the LLC-PK., Renal epithelial cell line: Similarity to mammalian kidney and the influence of cell density, *J. Cell Physiol.* **104**:375–389.

Rindler, M. J., Chuman, L. M., Shaether, L., and Saier, M. H., Jr., 1979, Retention of differentiated properties in an established dog kidney epithelial cell line (MDCK), *J. Cell Bio.* **81**:635–648.

Rizzino, A., Rizzino, H., and Sato, G., 1979, Defined media and the determination of nutritional and hormonal requirements of mammalian cells in culture, *Nutrition Reviews* **37**:369–378.

Rodriguez-Boulen, E. and Sabatini, D. D., 1978, Asymmetric budding of viruses in epithelial monolayers: A model for study of epithelial polarity, *Proc. Natl. Acad. Sci. USA* **75**:5071–5075.

Saier, M. H., Jr., 1981, Growth and differentiated properties of a kidney epithelial cell line (MDCK), *Am. J. Physiol.* **240**:C106–C109.

Sato, G. H., 1975, *The Role of Serum in Cell Culture in Biochemical Actions of Hormones* (G. Litwack, ed.), Academic Press, New York, pp. 391–396.

Sato, G. H. and Reid, L., 1978, The replacement of serum in cell culture by hormones, in: *Biochemistry and Mode of Action of Hormones, Volume II* (N. Rickenberg, ed.), University Park Press, Baltimore.

Sepulveda, F. V. and Pearson, S. D., 1982, Characterization of neutral amino acid uptake by cultured epithelial cells from pig kidney, *J. Cell Physiol.* **112**:182–188.

Schmidt, U. M., Eddy, B., Fraser, C. M., Venter, J. C. and Semenza, G., 1984, Purification of the Na^+,D-glucose cotransporter of rabbit intestinal brush border membranes using monoclonal antibodies, *FEBS Letters,* in press.

Sirbasku, D. A. and Kirkland, W. L., 1976, Control of cell growth IV growth properties of a new cell line established from an estrogen-dependent kidney tumor of the Syrian hamster, *Endocrinol.* **981**:1260–1277.

Taub, M. and Sato, G., 1980, Growth of functional primary cultures of kidney epithelial cells in defined medium, *J. Cell Physiol.* **105**:369–378.

Taub, M. and Saier, M. H., Jr., 1981, Amiloride resistant Madin–Darby Canine Kidney (MDCK) cells exhibit decreased cation transport, *J. Cell Physiol.* **106**:191–199.

Taub, M., Devis, P. E. and Grohol, S. H., 1984, PGE_1 Independent MDCK cells have elevated intracellular cyclic AMP, but retain the growth stimulatory effects of glucagon and epidermal growth factor in serum free medium, *J. Cell Physiol.,* in press.

Acknowledgments

The editor thanks the contributors for their excellent manuscripts. Each author has made a significant contribution to the area of epithelial cell culture. The editor's interest in the area of epithelial cell culture was started by her interactions with Dr. Milton H. Saier, Jr., and the concept of a book on this topic was initiated by Dr. Gordon H. Sato. Her investigations concerning membrane transport and animal cell cultures were initiated in the laboratory of Dr. Ellis Englesberg. The editor thanks Dr. Alexander Brownie, the Chairman of the Biochemistry Department of the State University of New York at Buffalo for making available the staff and facilities of the Biochemistry Department. Mrs. Esther Olczak, Mrs. Angela Grys, Ms. Gloria Viola, and Mrs. Diane Forrest have all participated significantly in the typing and other processing of the book. The members of my research laboratory, Dr. Soon Dong Chung, Ms. Patricia Devis, Dr. Sue Ford, Ms. Sue Hiller Grohol, Ms. Janny Seto and Dr. M. Anwar Wagar, are also thanked for continuing the investigation of kidney epithelial cell culture during the book's processing, and for their help in the reviewing of several manuscripts. The editor was funded by National Institutes of Health Grant 1R01 CA78111-04 and National Institutes of Health Research Career Development Award 1 KO4 CA/AM 00888-01.

Contents

Section I

Domes

Chapter 1

*Inducers of Dome Formation in Epithelial Cell Cultures including
Agents That Cause Differentiation*

JULIA E. LEVER

Section II
Electrophysiology and Ion Transport in MDCK Cells

Chapter 2

Electrical Properties of MDCK Cells

L. GONZÁLEZ-MARISCAL, L. BORBOA, R. LÓPEZ-VANCELL,
G. BEATY, AND M. CEREIJIDO

Chapter 3

Ion Transport in MDCK Cells

S. FERNÁNDEZ-CASTELO, J. J. BOLÍVAR, R. LÓPEZ-VANCELL, G.
BEATY, AND M. CEREIJIDO

Chapter 4

*Application of the Microbiological Approach to the Study of Passive
Monovalent Salt Transport in a Kidney Epithelial Cell Line, MDCK*

MILTON H. SAIER, JR.

Section III
Transport of Neutral Solutes by LLC–PK₁ Cells

Chapter 5

Sugar Transport in the Renal Epithelial Cell Culture

JAMES M. MULLIN AND ARNOST KLEINZELLER

Chapter 6

Neutral Amino Acid Transport in Cultured Kidney Tubule Cells

FRANCISCO V. SEPÚLVEDA AND JEREMY D. PEARSON

Chapter 7

Transepithelial Transport in Cell Culture: Mechanism and
Bioenergetics of Na$^+$, D-Glucose Cotransport

DAYTON S. MISFELDT AND MARTIN J. SANDERS

Section IV

Membrane Proteins and Hormone Receptors

Chapter 8

Viruses in the Study of the Polarity of Epithelial Membranes

MARY TAUB

Chapter 9

Hormone Receptors and Response in Cultured Renal Epithelial Cell Lines

KATHRYN E. MEIER AND PAUL A. INSEL

Chapter 10

Monoclonal Antibodies to Integral Membrane Transport and Receptor Proteins: Novel Reagents for Protein Purification

J. CRAIG VENTER, URSINA SCHMIDT, BARBARA EDDY, GIORGIO SEMENZA, AND CLAIRE M. FRASER

Section V

Epithelial Cell Cultures and Hormonally Defined Medium

Chapter 11

Estrogen-Dependent Kidney Tumors

JOACHIM G. LIEHR AND DAVID A. SIRBASKU

Chapter 12

Hormonally Defined, Serum-Free Media for Epithelial Cells in Culture

DAVID BARNES

Chapter 13

Importance of Hormonally Defined, Serum-Free Medium for In Vitro Studies Concerning Epithelial Transport

MARY TAUB

I

DOMES

Epithelial monolayers line body cavities, often separating a lumenal compartment from a compartment connected to the blood. The differentiated function of such epithelial monolayers *in vivo* is to transport fluids from one compartment to another. The expression of this differentiated function of epithelial cells is facilitated by the structural and functional polarity of the epithelial cells. The cells possess an apical domain within their plasma membrane (which faces the lumenal compartment) and a basolateral domain within the plasma membrane (which faces the blood side). Transepithelial transport occurs across the epithelial cell layer by means of specialized transport systems which are localized in particular membrane domains. Occluding junctions between adjacent cells also regulate vectorial solute movements. Epithelial cells similarly express their differentiated functions *in vitro*. The occurrence of vectorial fluid transport *in vitro* can be visualized microscopically as dome formation, the subject of this section.

1

Inducers of Dome Formation in Epithelial Cell Cultures including Agents That Cause Differentiation

JULIA E. LEVER

1. INTRODUCTION

Epithelia are cell sheets that line body cavities such as the lumina of the intestine, kidney tubules, and salivary glands. The differentiated phenotype of epithelia involves structural and functional specializations of the plasma membrane which enable these cells to transport ions and fluids across the cell layer. The plasma membrane of this cell type is polarized into distinct apical and basolateral regions which differ functionally, morphologically, and biochemically. Occluding junctions form between adjacent cells at the apical/basolateral boundary and can regulate the passage of ions across the cell layer. In addition, occluding junctions have been proposed to maintain cell polarity by serving as a barrier to lateral diffusion within the membrane of intrinsic proteins from basolateral and apical domains. The Na^+/K^+ ATPase of epithelia is restricted to the basolateral surface. Na^+ pump polarity is believed to create the driving force for vectorial transport of Na^+ across the cell layer, setting up an osmotic gradient which drives an accompanying net water flux in the basolateral direction.

Confluent cultures of certain established epithelial cell lines express this differentiated phenotype. Transepithelial fluid transport in cell culture is visible

JULIA E. LEVER • Department of Biochemistry and Molecular Biology, The University of Texas Medical School, Houston, Texas 77225. Work supported by National Institutes of Health grant AM 27400-01. Recipient of National Institutes of Health Research Career Development Award AM 00768.

by phase contrast microscopy as dome formation (Leighton *et al.*, 1969). Dome formation, expressed both by the Madin–Darby canine kidney (MDCK) cell line (proposed to originate from distal tubule or collecting duct of the kidney) and by the LLC–PK$_1$ cell line (proposed to originate from proximal tubule of kidney) (Hull *et al.*, 1976), is a lifting of the epithelial cell monolayer owing to active transport of ions and water *across* the cell monolayer in an apical to basolateral direction, trapping a bubble of fluid between the cell layer and the culture dish (Fig. 1). Domes do not occur in cultures grown on porous supports such as collagen-coated mesh or nitrocellulose filters since fluid accumulation on the basolateral side is prevented.

Dome formation in epithelial cell cultures is acceptable presumptive evidence of the following processes that are required for its expression: functional plasma membrane polarization, formation of occluding junctions, and vectorial transepithelial active ion transport. Each of these properties has been directly demonstrated in kidney epithelial cell culture. Evidence for plasma membrane polarization has come from morphological studies demonstrating microvilli on the apical surface (Cereijido *et al.*, 1978), polarized budding of enveloped viruses (Rodriguez–Boulan and Sabatini, 1978), restriction of ^{3}H-ouabain-binding

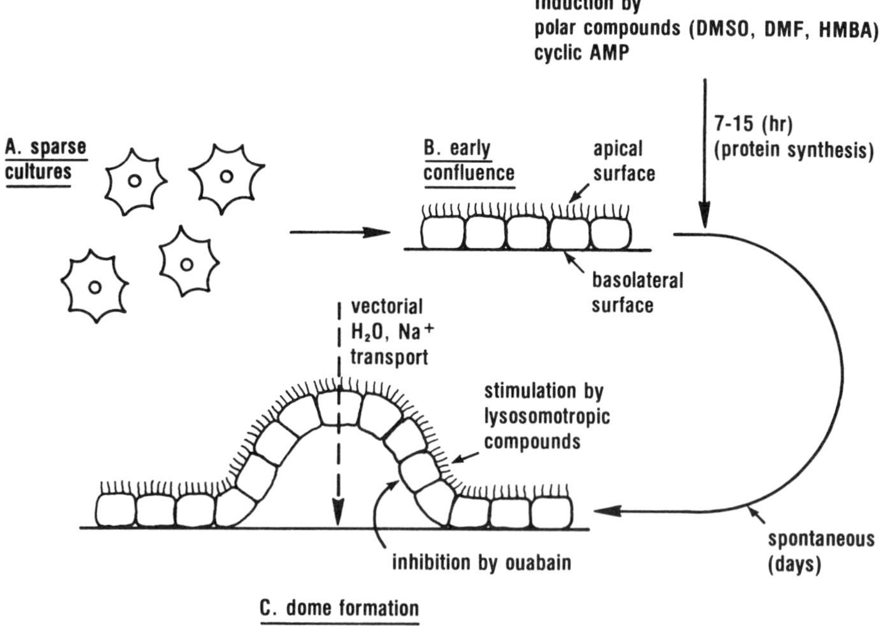

Figure 1. Diagrammatic explanation of dome formation. The apical (microvillar) cell surface is shown in contact with the medium; the basolateral surface is shown facing the plastic culture dish.

activity, a probe to localize Na^+/K^+ ATPase activity to the basolateral surface (Mills *et al.*, 1979; Cereijido *et al.*, 1981a; Lamb *et al.*, 1981), polarity of lectin binding (Dragsten *et al.*, 1981), and polarity of specific transport activities (Rabito, 1981; Rabito and Karish, 1982). A monoclonal antibody prepared against MDCK cells was localized on the basolateral membrane. Interestingly, when tested on kidney frozen sections, this antibody specifically recognized antigens on cells of the thick ascending limb of Henle's loop and distal convoluted tubule (Herzinger *et al.*, 1982). The presence of occluding junctions in these cultures is evidenced by morphological studies, measurements of transepithelial electrical resistance, a measure of ion flux mediated by occluding junctions (Cereijido *et al.*, 1980; Misfeldt *et al.*, 1976), and measurement of transepithelial permeability to ions and other solutes (Cereijido *et al.*, 1978, 1981a; Misfeldt *et al.*, 1976). Vectorial active transport of ions and water has been directly measured using cell culture monolayers grown on filters and mounted in Ussing chambers (Misfeldt *et al.*, 1976).

2. STIMULATION OF DOME FORMATION BY INDUCERS OF DIFFERENTIATION

An analogy between the regulation of numbers of domes in epithelial cell cultures and the regulation of expression of differentiated properties studied in cultures of a variety of other cell types became evident from a series of unexpected observations.

A number of compounds that had already been identified as potent inducers of differentiation of Friend erythroleukemia cells in culture (Marks and Rifkind, 1978) caused a dramatic increase in the number of domes when added to confluent MDCK kidney epithelial cell cultures (Lever, 1979a). Inducers included nonphysiological polar compounds such as dimethylsulfoxide, dimethylformamide (DMF), hexamethylene bisacetamide (HMBA), and compounds of physiological occurrence such as butyrate, hypoxanthine, and adenosine. Stimulation of dome formation was also observed after addition of agents that elevate intracellular cyclic AMP levels (Lever, 1979a,b,c; Valentich *et al.*, 1979), but induction could not be explained solely on the basis of cyclic AMP mediation (Lever, 1979c; Thomas *et al.*, 1982). The effects of these inducers on dome formation were not restricted to kidney epithelial cells in culture; the same group of inducers triggered a similar pattern of induction of dome formation in mammary epithelial cell cultures (Lever, 1979b,c).

2.1. Kinetics of Induction

The number of domes per unit area of the cell monolayer increased as a function of inducer concentration reaching a characteristic plateau level (Fig. 2).

By contrast, the size of domes was a characteristic of the particular inducer and did not differ significantly as a function of inducer concentration. Kinetics of the time course of induction indicated a lag phase of about 7 hr followed by a gradual increase in numbers of domes reaching a maximum after 3 days. Induction was reversible after removal of inducer, requiring 7–10 hr for complete reversal of the morphological response.

2.2. Inhibition Characteristics

Results from inhibitor studies indicated that distinct cellular functions were required for expression of the dome-forming phenotype. Continuous protein synthesis was required for induction and maintenance of domes. Induction was prevented by cycloheximide; also, addition of cycloheximide to cultures after induction caused disappearance of domes (Lever, 1979a). Interestingly, once cell confluence was achieved, inhibition of DNA synthesis did not affect dome formation, indicating that dome formation was regulated independently of cell profileration (Lever, 1979a,b,c). Cells in domes are active mitotically (Das *et al.*, 1974; Lever, 1981). A role of cytoskeletal elements in the dome-forming

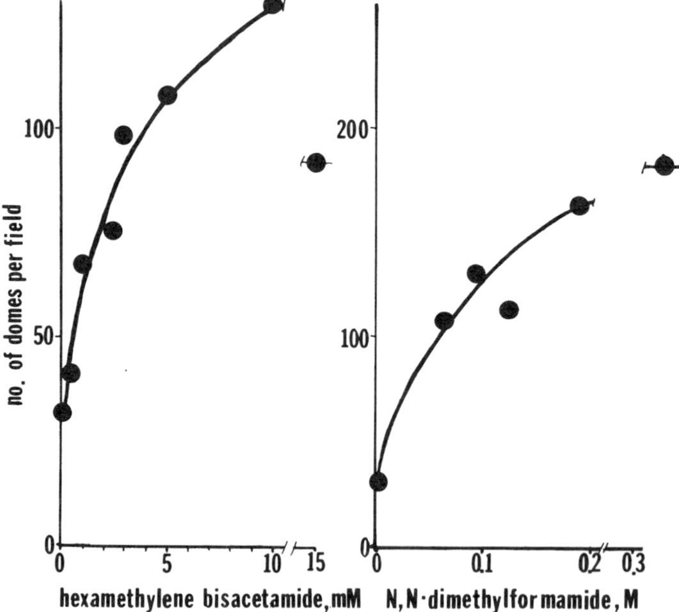

Figure 2. Induction of domes in MDCK cells as a function of inducer concentration.

response was indicated by observations that dome formation was blocked by either cytocholasin B, vinblastine, or colcemid (Lever, 1979a,b,c).

Tunicamycin, an inhibitor of N-glycosydically linked glycosylation, did not block dome formation over a range of concentrations that inhibited glycosylation of cellular glycoproteins and cell proliferation (Lever and Sari, 1983). Inhibition of dome formation was observed only at high concentrations of tunicamycin which also caused partial inhibition of protein synthesis. This result suggested that glycosylation was not an important determinant of polarized membrane properties required for expression of dome formation. Meiss *et al.* (1982) reached a similar conclusion based on the observation that lectin-resistant MDCK cell mutants with genetic alterations in N-linked oligosaccharides exhibited dome formation and morphological polarity. Ouabain (Abaza *et al.*, 1974) and vanadate (Mullin *et al.*, 1980a) caused inhibition of dome formation presumably by inhibiting the Na^+/K^+ ATPase required for vectorial ion and fluid transport.

2.3. Analogies with Other Cell Culture Model Systems for Differentiation

Several striking parallels were noted (summarized in Table I) between the induction of dome formation in epithelial cultures and the induction of differentiation of Friend erythroleukemia cells in culture (Lever, 1982a).

2.3.1. Spectrum of Inducers and Relative Potencies

The same broad spectrum of chemically diverse compounds triggered differentiation when added to different cell types in culture. Inducers differed from one another in optimal concentrations and potencies, but with few exceptions, the same inducer acted at a similar concentration with a similar relative response whether tested for induction of erythroid differentiation in Friend erythroleukemia cells or for induction of dome formation in either MDCK kidney epithelial cell cultures or Rama 25 mammary epithelial cell cultures (Lever, 1982a). The observation that the same group of compounds can induce differentiation in profoundly different cell types suggests that similar mechanisms regulate differentiation in each case. Among polar solvent inducers, dimethylsulfoxide, dimethylformamide, and the model compound hexamethylene bisacetamide (HMBA), synthesized to incorporate structural features that maximize biological activity (Reuben *et al.*, 1978), caused the largest biological response. Compounds such as tetramethylurea, dimethylacetamide, *N*-methylacetamide, diethylene glycol, and 1-methyl-2-pyrrolidone were also inducers of dome formation and differentiation (Lever, 1979b). n-Butyrate was almost as potent an

Table I.   Comparison between the Characteristics of Induced Epithelial Cell Dome Formation and Those of Friend Cell Differentiation

	Dome formation	MELC differentiation
1. Inducers		
Polar solvents (e.g., DMSO, HMBA)	Yes	Yes
Butyrate	Yes	Yes
Purines	Yes	Yes
Ouabain	No	Yes
Actinomycin D	No	Yes
2. Commitment process		
Reversible	Yes	Yes (at early stages)
cAMP increase	Yes	Yes
Inhibition by phorbol esters	Yes	Yes
Latent period	Yes	Yes
Dependence on inducer concentration	Yes	Yes
Expression in increased fraction of cell populations	Yes	Yes
Terminal cell division	No	Yes
3. Expression (biochemical, morphogenetic)		
Reversibility	Yes	No
Dependence on inducer concentration	No	No
Continued proliferation	Yes	No

inducer as HMBA on a concentration basis and in terms of maximal response. By contrast, purines and agents that elevate cyclic AMP levels were more potent than HMBA on a molar basis but elicited a weaker response.

2.3.2. *Effect of Tumor Promoters*

Another similarity between induction of dome formation and differentiation concerned effects of compounds known as tumor promoters. Phorbol esters such as 12-0-tetradecanoylphorbol 13-acetate (TPA) stimulate tumor cell proliferation but inhibit the acquisition of differentiated functions in cell culture (Weinstein *et al.*, 1977). Dome formation was inhibited after the addition of TPA to mammary epithelial cells (Dulbecco *et al.*, 1980) or LLC–PK$_1$ kidney epithelial cells (Amsler and Cook, 1982). TPA abolished transepithelial electrical resistance properties (a measure of occluding junction permeability) when added to MDCK cell monolayers mounted in an Ussing chamber for analysis (Ojakian, 1981). Thus, inhibition of dome formation by TPA could be explained by opening of occluding junctions and the resulting collapse of the transepithelial ion gradient required for vectorial fluid transport. Amsler and Cook (1982) showed that TPA blocked the appearance of a differentiated transport property, Na-$^+$stimulated glucose uptake in LLC–PK$_1$ cells. The mechanism of TPA's effects on growth and differentiation of a broad range of cell types is unknown, but there is evidence that TPA interacts with specific plasma membrane receptors (Lee and Weinstein, 1978).

2.3.3. Inducers Trigger an Increased Fraction of Cells in the Population to Express the Phenotype

One problem in measurement of differentiation in cell culture concerns how to define it experimentally. Most studies with various epithelial cell types such as hepatocytes or mammary epithelial cells have defined differentiation by quantitation of specific differentiated markers (such as casein production in the case of mammary cells) in the total cell population rather than in individual cells (Bissell, 1981). This approach does not distinguish whether a response occurs uniformly over the cell population or is restricted to a subpopulation of responsive cells. Analysis of erythroid differentiation on a single cell basis in Friend cells (Marks and Rifkind, 1978) has revealed that inducers trigger heme production in an increased fraction of cells in the population, but the amount of heme per committed cell remains the same as the uninduced level. Thus the process of differentiation in cell culture is perhaps best definable as the appearance of an increased fraction of cells in the population which expresses the differentiated phenotype.

Since dome formation in epithelial cell cultures requires the participation of more than one cell, it was not possible to demonstrate this criterion of differentiation on a single-cell basis. Instead, we made use of isolated colonies, each theoretically derived from a single cell. At a certain size (about 0.5 cm diameter), each colony becomes a "mini" confluent epithelial monolayer which expresses dome formation over the colony area. In a large number of colonies derived without selective pressure from MDCK cells, about 16% of the colonies expressed dome formation spontaneously (Lever, 1979a). If inducer was added to preexisting colonies, then 50–60% of the colonies exhibited domes. This experiment suggested that inducers recruit an increased fraction of cells in the population to express dome formation (Lever, 1979a,b,c).

2.3.4. Similarity to Proposed Committment Step in Differentiation

Erythroid differentiation in Friend cells has been divided into two stages: the commitment step, which represents the cell's decision to subsequently express the differentiated phenotype; and the expression stage, during which the cell begins to exhibit its differentiated biochemical and morphological changes (Rifkind *et al.*, 1978).

Table I indicates that induction of dome formation in epithelial cells exhibits several of the properties associated with the commitment stage of Friend cell erythroid differentiation. As discussed in preceding paragraphs, each type of biological response increases quantitatively as a function of inducer concentration. Induction in each case refers to response by an increased fraction of the cell population rather than an increased response by a constant fraction of cells in the population. A similar latent period of at least 9 hr is required after addition of inducer. Other parallels between the two cell systems include inhibitory effects of phorbol esters and stimulatory effects of elevation of cyclic AMP levels. Both dome formation and early stages of Friend cell commitment are reversible after removal of inducer and are not blocked after inhibition of DNA synthesis.

A fundamental difference between induction of domes and induction of Friend cell differentiation is the requirement for a terminal cell division in the case of erythroid differentiation preceding progression from the commitment stage to the expression stage. Thus, inhibition of DNA synthesis at late stages of commitment prevents progression to the expression stage (Rifkind *et al.*, 1978). Also, after progression to the expression stage, Friend cell differentiation becomes irreversible after removal of inducer (Marks and Rifkind, 1978).

Other differences between the two cell systems concern response to certain hormones and inhibitors. The differences presumably reflect biochemical differences between each type of fully differentiated cell. For example, actinomycin D is an inducer of Friend cell differentiation (Terada *et al.*, 1978) but inhibits dome formation (Lever, 1979a). Similarly, ouabain is an inducer of Friend cell differentiation (Bernstein *et al.*, 1976). Hydrocortisone inhibits Friend cell differentiation (Scher *et al.*, 1978) but is required for dome formation (Lever, unpublished observation) and is a component of the defined, hormone-supplemented medium for MDCK cells (Taub *et al.*, 1979).

2.4. Inducer-Resistant Variants

Subpopulations of MDCK cells exhibited heritable differences in response to inducers. Variant cell lines were isolated without selection by single-cell plating from the MDCK parental cell line (Lever, 1981). Colonies obtained were grown to mass culture then screened for inducibility. Cell lines were obtained that did not form domes either spontaneously or after addition of inducers, yet retained epithelial morphology. Another group of variant MDCK cell clones responded to induction of domes by HMBA but not DMF. This property suggests that HMBA and DMF act by different mechanisms. The "wild-type" MDCK cell line used in our studies is MDCK cl 4, obtained by single-cell plating of the parental MDCK cell line. MDCK cl 4 exhibits a much higher frequency of spontaneous dome formation compared with the parental cell line, which appears to be a mixture of several cell types. These variant cell lines are further evidence for an interaction of inducers with specific cellular targets and will provide useful tools to dissect the mode of action of various inducers.

3. EXPRESSION OF A SPECIFIC, INDUCIBLE DIFFERENTIATED TRANSPORT FUNCTION IN KIDNEY EPITHELIAL CELL CULTURES

3.1. *Inducibility of Na^+-Stimulated Glucose Transport Activity in LLC–PK_1 Kidney Cell Cultures*

The Na^+-stimulated glucose transport system is a differentiated function of kidney proximal tubule. This transport activity is localized on the apical membrane of confluent cultures of the LLC–PK_1 kidney epithelial cell line (Rabito, 1981) but is almost undetectable in subconfluent cultures (Mullin *et al.*, 1980b). The relationship of this active sugar transport activity to a coordinate program of cell differentiation in culture was underscored by the observation that addition of either HMBA or agents causing elevated cyclic AMP levels to LLC–PK_1 cultures accelerated the development of concentrative sugar transport activity (Amsler and Cook, 1982). Addition of the phorbol ester TPA to LLC–PK_1 cells prevented the appearance of Na^+-stimulated glucose transport (Amsler and Cook, 1982). TPA is known to prevent differentiation in several cell systems, as noted in Section 2.3.2.

Na^+-stimulated active sugar transport, measured using the analog α-methyl-D-glucopyranoside (αMeGlc), became expressed at the time domes first appeared in the culture and increased steadily in specific activity thereafter. By contrast, Na^+-stimulated amino acid transport activity, which shares the same electrochemical Na^+ gradient driving force utilized by active sugar transport, showed a reciprocal regulation. It was maximal in subconfluent cultures and declined accompanying the onset of dome formation and appearance of active sugar transport (Lever, 1982b).

Since Na^+-stimulated glucose transport activity represents a specific, quantifiable differentiated renal function that is regulated in parallel with dome formation, we characterized the properties of this transport mechanism expressed in isolated membranes from confluent LLC–PK_1 cells.

3.2. *Characteristics of Na^+-Stimulated Glucose Transport Activity Expressed in Apical Membrane Vesicles from LLC–PK_1 Cells*

Active, Na^+-stimulated αMeGlc transport activity is expressed in vesicles derived from apical membranes of LLC–PK_1 cultures (Lever, 1982b). The αMeGlc transport activity in vesicles from LLC–PK_1 cells exhibits the same sugar and inhibitor specificity as the corresponding system in kidney proximal tubule and is electrogenic, stimulated by an interior-negative membrane potential. However, a stoichiometry of 2 Na^+:1 glucose was determined for the transport mechanism in LLC–PK_1 vesicles (Lever, 1982b) in contrast with the

1:1 stoichiometry established using renal brush border vesicles (Beck and Sacktor, 1978). This observation of a 2:1 stoichiometry for Na^+-coupled hexose transport in vesicles confirms that determined by short-circuit current measurements in intact LLC–PK$_1$ cells (Sanders and Misfeldt, 1981) and may be a unique property of the LLC–PK$_1$ cell line.

4. MECHANISMS OF ACTION OF INDUCERS ON TARGET CELLS

The broad spectrum of low-molecular-weight compounds known to trigger cellular differentiation in diverse cell types encompasses widely different structural and functional characteristics. Thus it has proven difficult to identify the cellular target of their action in attempts to explain mechanisms of induction of differentiation.

4.1. Role of Cyclic AMP Levels

Several but not all of the inducers of dome formation caused an elevation in intracellular cyclic AMP levels in MDCK cells (Thomas *et al.*, 1982). Confluent, uninduced cultures exhibited 14-fold elevated intracellular levels of cyclic AMP compared to subconfluent uninduced cultures. When HMBA was added to confluent cultures, a 2.5-fold further elevation in cyclic AMP levels was observed. Similar effects of HMBA on cyclic AMP levels have been reported for Friend erythroleukemia cells (Rifkind *et al.*, 1978), mammary epithelial cells (Lever, 1979c), and the LLC–PK$_1$ pig kidney epithelial cell line (Amsler and Cook, 1982). DMSO, another inducer of differentiation and dome formation (Lever, 1979a,b,c), elevated cyclic AMP levels 4.5-fold and has been reported to stimulate cyclic AMP levels in Friend erythroleukemia cells (Rifkind *et al.*, 1978). n-Butyrate, which induces dome formation (Lever, 1979a,b,c) and is also known as a Friend cell differentiation inducer (Rifkind *et al.*, 1978), elevated cyclic AMP levels in MDCK cells, but isobutyrate neither induced dome formation nor affected cyclic AMP levels.

In contrast, DMF was as potent an inducer of dome formation as HMBA but did not affect cyclic AMP levels (Thomas *et al.*, 1982). This result suggested that elevation of cyclic AMP levels was not required to induce domes.

Amsler and Cook (1982) concluded that although elevation of cyclic AMP levels in LLC–PK$_1$ cells was an early response in the development of Na^+-stimulated hexose transport, high intracellular cyclic AMP was not a sufficient condition.

The isolation of MDCK cell variants, which could be induced to form domes by HMBA but not DMF (Lever, 1981), taken together with the previous observations suggests that induction of dome formation by inducers of differentiation may involve multiple and separate mechanisms rather than a direct and obligatory action of cyclic-AMP-mediated protein phosphorylation.

4.2. Role of Ca^{2+}

Extracellular Ca^{2+} is required for sealing of occluding junctions in MDCK monolayers, but these junctions open when the concentration of Ca^{2+} in the cytoplasm was increased after the addition of a calcium ionophore (Cereijido *et al.*, 1981a,b). Cereijido and his colleagues observed a drop in transepithelial resistance of MDCK monolayers after removal of Ca^{2+} by a calcium chelator, which could be restored upon readdition of extracellular Ca^{2+} to the bathing medium.

In addition to its direct effect on occluding junctions and Na^+ transport processes (Taub and Saier, 1979) in MDCK cells, Ca^{2+} by interaction with its intracellular receptor calmodulin has been proposed as an intracellular mediator of hormone action (Means and Dedman, 1980) which would intersect the pathways of cellular regulation mediated by cyclic AMP (Rasmussen *et al.*, 1972). Levenson *et al.* (1980) have proposed a role for increased intracellular Ca^{2+} in regulating the commitment stage of Friend cell differentiation.

Addition of Ca^{2+} antagonists such as trifluoperazine, verapamil, or lanthanum chloride to MDCK cells resulted in a stimulation of dome formation, although less pronounced than that triggered by HMBA (Lever, unpublished observations). These results suggest that intracellular Ca^{2+} levels may play a role in cellular responses to inducers of differentiation.

4.3. Cellular Targets of Inducers

4.3.1. Polar Solvent Inducers

Certain inducers such as DMSO and DMF are well-known cryoprotective agents. These polar compounds would be expected to compete with water for hydrogen bonding and would exert solvent effects on macromolecular conformation. However, the effects of this category of inducer are not easily explainable on this basis. Thus, the measured ability of these compounds to unfold globular proteins (Herskovits *et al.*, 1977) does not correlate with their potency as an inducer of differentiation. Also, more stringent structural features were required for optimal inductive activity than compatible with simple solvent effects. In a series of methylene bisacetamides, optimal biological activity was observed when acetamides were separated by six methylene residues (Reuben *et al.*, 1978), suggesting that HMBA interacts with specific cellular targets. The finding of variant MDCK cell lines (Lever, 1981) which respond to HMBA but not DMF provides further evidence that inducers interact with specific targets. The plasma membrane (Lyman *et al.*, 1976) and the methylated state of chromatin (Sheffery *et al.*, 1982) have been proposed as possible cellular targets of these polar solvents. As noted previously, a surface membrane event associated with many but not all of the inducers is elevation of cyclic AMP levels. Intracellular pathways of metabolism of these compounds have not been characterized; it is possible that a metabolite of these compounds may be the active inducer.

4.3.2. Fatty Acid Inducers

Butyrate is a potent inducer of dome formation in both mammary and kidney epithelial cells (Lever, 1979c). Isobutyrate was without inductive activity. When other saturated and unsaturated fatty acids of chain length up to C-18 were tested, only myristic acid showed inductive activity.

The effects of butyrate on gene expression and cellular morphology of mammalian cells in culture are complex. Butyrate stimulates differentiation of Friend erythroleukemia cells (Leder and Leder, 1975). Butyrate causes elevation of cyclic AMP levels in a variety of cell types including MDCK (Thomas *et al.*, 1982), LLC–PK$_1$ kidney epithelial cells (Amsler and Cook, 1982), and mammary epithelial cells (Lever, 1979c). Studies with other cell types have shown that butyrate inhibits histone deacetylation (Boffa *et al.*, 1978) and has effects on cellular morphology and expression of cell surface glycoproteins (Simmons *et al.*, 1975; Storrie *et al.*, 1978).

5. CELLULAR FUNCTIONS THAT CONTRIBUTE TO REGULATION OF DOME FORMATION

The mechanism of stimulation of dome formation after the addition of differentiation-inducing chemicals is not understood. Three important contributions must be considered. First, ionic pathways across the plasma membrane could respond to induction. Ouabain (Abaza *et al.*, 1974) or furosemide (Lever, unpublished observations) caused collapse of domes. A second force to consider is the strength of adhesion of the cell monolayer to the substratum. Manipulation of adhesiveness of MDCK cells by variation of the type of substratum caused corresponding changes in dome formation (Rabito *et al.*, 1980). Third, occluding junctions that regulate paracellular flux of ions could play a role.

5.1. Ion Transport Activities

The Na$^+$, K$^+$ ATPase provides a driving force for transepithelial ion and fluid transport, a prerequisite for dome formation. Thus the finding of reduced Na$^+$ pump activity under conditions of maximal expression of dome formation in MDCK cells was unexpected. Na^+,K^+,ATPase activity of MDCK cells, measured by ouabain-sensitive $^{86}Rb^+$ uptake, was reduced up to 50% after 2 days' exposure of the cells to inducers of differentiation (Lever, 1982c; Kennedy and Lever, 1984). Similar findings were reported by Mager and Bernstein (1978) in the case of Friend cell differentiation.

A model has been proposed to explain the possible role of inhibition of Na^+,K^+,ATPase activity in regulating commitment of Friend cells to erythroid differentiation (Smith *et al.*, 1982). This model is based on that proposed to explain the cardiotonic actions of digitalis (Sweadner and Goldin, 1980). A Ca^{2+}/Na^+ antiport secondary active transport system in the plasma membrane catalyzes expulsion of Ca^{2+} from the cytoplasm coupled to Na^+ gradients maintained by Na^+,K^+ATPase activity. Dissipation of transmembrane Na^+ gradients by inhibition of the Na^+ pump could result in increased levels of cytoplasmic Ca^{2+}, which would presumably act as the signal for commitment to differentiation. Although some evidence for the validity of this model has been obtained in the Friend cell system, the existence of Na^+/Ca^{2+} antiport has not yet been demonstrated in MDCK cells (Taub and Saier, 1979).

5.2. Occluding Junctions

Cereijido *et al.* (1981b) found evidence which suggested that, in MDCK cells, occluding junctions were sealed in domes and open in nondome areas of the cell monolayer. Contrasting evidence that no differences in occluding junction structure or function exist in domes has also been presented (Rabito *et al.*, 1978; Pickett *et al.*, 1975).

Passage of an electrical current across the cell monolayer proceeds largely by the paracellular route limited at the apical side by occluding junctions. The magnitude of transepithelial resistance (r_t) properties of MDCK cell monolayers (grown on a Millipore filter and mounted in an Ussing chamber for electrical measurement) can be used to estimate functional changes associated with occluding junctions averaged over the cell monolayer. Such monolayers exhibited r_t values of 300–400 ohms cm^2.

HMBA triggered a significant increase in both cyclic AMP levels and transepithelial resistance values, whereas DMF, an equally potent inducer of dome formation, did not (Thomas *et al.*, 1982). Ojakian (1981) has reported that elevation of cyclic AMP levels does not affect transepithelial resistance of MDCK cell monolayers. Thus elevation in r_t, an indication of increased sealing of occluding junctions, can be dissociated from an obligatory role in induction of domes.

5.3. Adhesion to the Substratum

Adhesion to a solid substratum is a prerequisite for cell growth and continued viability of the kidney epithelial cultures used in this study. MDCK cells are viable for only a short time when maintained in suspension culture (Valentich *et al.*, 1979). The observation by Rabito *et al.* (1980) that the frequency of dome formation in MDCK cells can be modulated by plating on various substrata that affect the strength of cell adhesion suggests that a possible target of inducers of dome formation may be cell surface components involved in adhesion. We cannot demonstrate differences in MDCK cell adhesiveness to plastic dishes after treatment with chemical inducers, partly because of the difficulty of setting up a valid cell adhesion assay method. However, several of our variant MDCK cell clones which are not inducible by DMF also exhibit a significant resistance to cell detachment from plastic by trypsin treatment (Lever, in preparation). This observation suggests that these variants exhibit a cell surface alteration affecting cell adhesiveness which modifies response to inducers.

5.4. Cell Polarity

Functional membrane polarization is a prerequisite for transepithelial fluid transport. Therefore, regulation of the unknown processes by which proteins sort out to either the apical or basolateral surface may be involved in controlling the onset of vectorial transport activities. Occluding junctions are thought to serve as a barrier to lateral diffusion of integral membrane proteins from the apical side to the basolateral side of the cell (U *et al.*, 1979, 1980; Dragsten *et al.*, 1981). Also, suspension of epithelial cells in collagen gels has been reported to influence cell polarity (Chambard *et al.*, 1981; Greenberg and Hay, 1982). Recent studies suggest that carbohydrate residues of proteins are not important determinants of their localization on either apical or basolateral surfaces in MDCK cells (Roth *et al.*, 1979; Meiss *et al.*, 1982; Lever and Sari, 1983).

5.5. Cytoskeletal Framework

Our hypothesis that dome formation is associated with cell differentiation in culture predicts that cells in domes may differ functionally and biochemically from cells in the surrounding monolayer. The first direct evidence for this hypothesis was provided by fluorescence histochemistry studies which showed that LLC–PK$_1$ cells in domes exhibited morphological differences associated with the actin cytoskeleton when compared with cells in the surrounding monolayer (Lever, 1982a). Cereijido *et al.* (1981b) provided evidence for the hypothesis that occluding junctions surrounding cells in domes are closed but junctions surrounding cells in the monolayer are open in MDCK cells. By contrast, occluding junctions of cells grown on filters were uniformly closed over the entire cell layer. Meza *et al.* (1980) observed that addition of cytochalasins, which inhibit actin polymerization, modulated the permeability of occluding junctions in MDCK cells. Claude and Goodenough (1973) and Claude (1978) correlated transepithelial resistance with the number of occluding junction strands viewed by freeze–fracture electron microscopy.

Taken together, these studies suggest that epithelial cells in domes or grown on filters may have undergone a cytoskeletal rearrangement which causes occluding junctions to be sealed more tightly than those of cells attached to plastic.

6. HYPOTHESIS: DETERMINANTS OF EPITHELIAL CELL DIFFERENTIATION

Cell differentiation represents a complex, unknown sequence of events leading to coordinated changes in multiple biochemical and morphological parameters representing the new differentiated state. Several conditions have been observed to affect the spontaneous or chemically induced development of a functional, polarized epithelium in cell culture. Subconfluent, actively proliferating epithelial cell cultures can be viewed as relatively undifferentiated epithelial cells, which have not yet expressed occluding junctions, microvilli, or certain characteristic functions, e.g., Na^+-stimulated transport in the case of LLC–PK$_1$ cells. Accompanying cell confluence and decreased cell proliferation, several differentiated properties such as the development of occluding junctions, microvilli, dome formation, and other characteristic functions begin to appear.

Among factors known to influence epithelial cell differentiation in culture, the nature of the cell substratum has received the most attention, as documented in reports too numerous for review here. For example, extracellular matrix and floating collagen gels promote differentiated properties of mammary epithelial cells (Emerman *et al.*, 1979; Yang *et al.*, 1979; Wicha *et al.*, 1982), hepatocytes (Reid *et al.*, 1981; Sirica *et al.*, 1979), and kidney epithelial cells (Valentich, 1981). Superimposed on substratum effects is regulation by serum and hormones, which may itself be altered by substratum-induced cellular alterations (Emerman *et al.*, 1977). The substratum may exert direct effects based on its chemical constitution and organization. Also, enhanced nutrient accessibility to the basal cell surface may be a factor in effects of floating collagen gels and filter cell supports.

Sanders and Dickau (1981) observed that domes formed from chick embryo epiblast tissue were lined by a basal lamina and possessed apical microvilli, whereas nondome, plastic-adherent cells lacked these structures. Interestingly, Valentich (1982) found that MDCK kidney epithelial cells elaborate a basal lamina when grown on hydrated collagen gels but not on plastic. Furthermore, when MDCK cell cultures were grown on Nucleopore filters, basal lamina formation was observed only on the regions of the monolayer over the pores of the filter. This suggested that cell monolayer detachment *per se* as well as enhanced basal nutrient accessibility may promote additional modes of differentiated phenotypic expression.

Based on these diverse observations with several different epithelial cell systems, it is tempting to speculate that the locally detached regions of the cell monolayer found in domes may exhibit a variety of differentiated phenotypic changes analogous to those observed over the whole cell layer on floating collagen gels. Indeed, Sirica *et al.* (1979) observed a pronounced buildup of actin microfilaments in collagen–gel cultures of hepatocytes similar to those which we observed in domes of kidney epithelial cell cultures (Lever, 1982a). Cytoskeletal components interact with the inner surface of the cell membrane near regions of cell attachment to the substratum (reviewed by Korn, 1976), and there may be a reciprocal regulatory interaction between cell attachment–detachment and cytoskeletal configuration. The role of chemical inducers in triggering cytoskeletal changes (Lever, 1982a) remains to be explored.

REFERENCES

Abaza, N. A., Leighton, J., and Schultz, S. G., 1974, Effects of ouabain on the function and structure of a cell line (MDCK) derived from canine kidney. I. Light microscopic observations of monolayer growth, *In Vitro* **10**:172–183.

Amsler, K., and Cook, J. S., 1982, Development of Na^+-dependent hexose transport in a cultured line of procine kidney cells, *Am. J. Physiol.* **242**:C94–C101.

Beck, J. C., and Sacktor, B., 1978, The sodium electrochemical potential-mediated uphill transport of D-glucose in renal brush border membrane vesicles, *J. Biol. Chem.* **253**:5531–5535.

Bernstein, A., Hunt, D. M., Crichley, V., and Mak, T. W., 1976, Induction by ouabain of hemoglobin synthesis in cultured Friend erythroleukemic cells, *Cell* **9**:375.

Bissell, M. J., 1981, The differentiated state of normal and malignant cells or how to define a normal cell in culture, *Int. Rev. Cytol.* **70**:27–100.

Boffa, L. C., Vidali, G., Manor, R. S., and Allfrey, V. G., 1978, Suppression of histone deacetylation *in vivo* and *in vitro* by sodium butyrate, *J. Biol. Chem.* **253**:3364–3366.

Cereijido, M., Rotunno, C. A., Robbins, E. S., and Sabatini, D. D., 1978, Polarized epithelial membranes produced *in vitro,* in: *Membrane Transport Processes* (J. F. Hoffman, ed.), Raven Press, New York, p. 433.

Cereijido, M., Stefani, E., and Palomo, A. M., 1980, Occluding junctions in a cultured transporting epithelium: Structural and functional heterogeneity, *J. Membr. Biol.* **53**:19–32.

Cereijido, M., Meza, I., and Martinez–Palomo, A., 1981a, Occluding junctions in cultured epithelial monolayers, *Am. J. Physiol.* **240**:C96–C102.

Cereijido, M., Ehrenfeld, J., Fernandez–Castelo, S., and Meza I., 1981b, Fluxes, junctions and blisters in cultured monolayers of epitheliod cells (MDCK), *Ann. N.Y. Acad. Sci.* **372**:422–441.

Chambard, M., Gabrion, J., and Mauchamp, J., 1981, Influence of collagen gel on the orientation of epithelial cell polarity: Follicle formation from isolated thyroid cells and from preformed monolayers, *J. Cell Biol.* **91**:157–166.

Claud, P., 1978, Morphological factors influencing transepithelial permeability: A model for the resistance of the zonula occludentes, *J. Membrane Biol.* **39**:219–232.

Claude, P., and Goodenough, D. A., 1973, Fracture faces of zonulae occludentes from "tight" and "leaky" epithelia, *J. Cell Biol.* **58**:390.

Das, N. K., Hosick, H. L., and Nandi, S., 1974, Influence of seeding density on multicellular organization and nuclear events in cultures of normal and neoplastic mouse mammary epithelium, *J. Natl. Cancer Inst.* **52:**849–855.

Dragsten, P. R., Blumenthal, R., and Handler, J. S., 1981, Membrane asymmetry in epithelia: Is the tight junction a barrier to diffusion in the plasma membrane? *Nature* **294:**718–722.

Dulbecco, R., Bologna, M., and Unger, M., 1980, Control of differentiation of a mammary cell line by lipids, *Proc. Natl. Acad. Sci. USA* **77:**1551–1555.

Emerman, J. T., Enami, J., Pitelka, D. R., and Nandi, S., 1977, Hormonal effects on intracellular and secreted casein in cultures of mouse mammary epithelial cells on floating collagen membranes, *Proc. Natl. Acad. Sci. USA* **74:**4466–4470.

Emerman, J. T., Burwen, S. J., and Pitelka, D. R., 1979, Substrate properties influencing ultrastructural differentiation of mammary epithelial cells in culture, *Tissue Cell.* **11:**109–119.

Greenberg, G., and Hay, E. D., 1982, Epithelia suspended in collagen gels can lose polarity and express characteristics of migrating mesenchymal cells, *J. Cell Biol.* **95:**333–339.

Herskovits, T. T., Behrens, C. F., Suita, P. B., and Pandolfelli, E. R., 1977, Solvent denaturation of globular proteins. Unfolding by the monoalkyl and dialkyl-substituted formamides and ureas, *Biochim. Biophys. Acta* **490:**192–199.

Herzinger, D. A., Easton, T. G., and Ojakian, G. K., 1982, The MDCK cell line expresses a cell surface antigen of the kidney distal tubule, *J. Cell Biol.* **93:**269–277.

Hull, R. N., Cherry, W. R., and Weaver, G. W., 1976, The origin and characteristics of a pig kidney cell strain LLC-PK$_1$, *In Vitro* **12:**670–677.

Kennedy, B. G. and Lever, J. E., 1984, Regulation of Na$^+$, K$^+$, ATPase activity in MDCK kidney epithelial cell cultures: Role of growth state, cyclic AMP and chemical inducers of dome formation and differentiation, *J. Cell Physiol.*, in press.

Korn, E. D., 1976, Introductory workshop: Membranes and their association with contractile proteins, in: *Cell Motility*, Volume 3, Cold Spring Harbor Conferences on Cell Proliferation (R. Goldman, T. Pollard, and J. Rosenbaum, eds.), Cold Spring Harbor Press, New York, pp. 623–629.

Lamb, J. F., Ogden, P., and Simmons, N. L., 1981, Autoradiographic localization of [^{3}H]ouabain bound to cultured epithelial cell monolayers of MDCK cells, *Biochim. Biophys. Acta* **644:**333–340.

Leder, A., and Leder, P., 1975, Butyric acid, a potent inducer of erythroid differentiation in cultured erythroleukemic cells, *Cell* **5:**319–322.

Lee, L. S., and Weinstein, I. B., 1978, Tumor-promoting phorbol esters inhibit binding of epidermal growth factor to cellular receptors, *Science* **202:**313–315.

Leighton, J., Brada, Z., Estes, L. W., and Justh, G., Secretory activity and oncogenicity of a cell line (MDCK) derived from canine kidney, *Science* **163:**472–473.

Levenson, R., Housman, D., and Cantley, L., 1980, Amiloride inhibits murine erythroleukemia cell differentiation: Evidence for a Ca^{2+} requirement for commitment, *Proc. Natl. Acad. Sci. USA* **77:**5948–5952.

Lever, J. E., 1979a, Inducers of mammalian cell differentiation stimulate dome formation in a differentiated kidney epithelial cell line (MDCK), *Proc. Natl. Acad. Sci. USA* **76:**1323–1327.

Lever, J. E., 1979b, Cyclic AMP and inducers of mammalian cell differentiation stimulate dome formation in mammary and renal epithelial cell cultures, in: *Hormones and Cell Culture*, Volume 6 (G. Sato and R. Ross, eds.), Cold Spring Harbor Conferences on Cell Proliferation, Cold Spring Harbor Press, New York, p. 727.

Lever, J. E., 1979c, Regulation of dome formation in differentiated epithelial cell cultures, *J. Supramol. Struct.* **12:**259–272.

Lever, J. E., 1981, Regulation of dome formation in kidney epithelial cell cultures, *Ann. N.Y. Acad. Sci.* **372:**371–383.

Lever, J. E., 1982a, Cell differentiation and dome formation in polarized epithelial cell monolayers, in: *Cell Growth in Hormonally Defined Media,* Volume 9 (G. Sato, A. B. Pardee, and D. Sirbasku, eds.), Cold Spring Harbor Conferences on Cell Proliferation, Cold Spring Harbor Press, New York, p. 541.

Lever, J. E., 1982b, Expression of a differentiated transport function in apical membrane vesicles isolated from an established kidney epithelial cell line: Sodium electrochemical potential-mediated active sugar transport, *J. Biol. Chem.* **257**:8680–8686.

Lever, J. E., 1982c, Transepithelial ion transport and differentiation in epithelial cell cultures, in: *Ions, Cell Proliferation and Cancer* (A. L. Boynton, W. L. McKeehan, and J. F. Whitfield, eds.), Academic Press, New York, pp. 187–203.

Lever, J. E., and Sari, C. E., 1983, Effect of tunicamycin on polarized membrane functions of an established kidney epithelial cell line, *Biochim. Biophys. Acta,* **762**:265–271.

Lyman, G. H., Preisler, H. D., and Papahadjopoulos, D., 1976, Membrane action of DMSO and other chemical inducers of Friend leukaemic cell differentiation, *Nature* **262**:360–363.

Mager, D., and Bernstein, A., 1978, Early transport changes during erythroid differentiation of Friend leukemic cells, *J. Cell Physiol.* **94**:275–286.

Marks, P. A., and Rifkind, R. A., 1978, Erythroleukemic differentiation, *Ann. Rev. Biochem.* **47**:419–448.

Means, A. R., and Dedman, J. R., 1980, Calmodulin—an intracellular calcium receptor, *Nature* **285**:73–77.

Meiss, H. K., Green, R. F., and Rodriguez–Boulan, E. J., 1982, Lectin-resistant mutants of polarized epithelial cells, *Mol. Cell. Biol.* **2**:1287–1294.

Meza, I., Ibarra, G., Sabanero, M., Martinez–Palomo, A., and Cereijido, M., 1980, Occluding junctions and cytoskeletal components in a cultured transporting epithelium, *J. Cell Biol.* **87**:746–754.

Mills, J. W., MacKnight, A. D. C., Dayer, J. M., and Ausiello, D. A. 1979, Localization of ^{3}H-ouabain-sensitive Na^+ pump sites in cultured pig kidney cells, *Am. J. Physiol.* **236**:C157–C162.

Misfeldt, D. S., Hamamoto, S. T., and Pitelka, D. R., 1976, Transepithelial transport in cell culture, *Proc. Natl. Acad. Sci. USA* **73**:1212–1216.

Mullin, J. M., Diamond, L., and Kleinzeller, A., 1980a, Effects of ouabain and ortho-vanadate on transport-related properties of the $LLC–PK_1$ renal epithelial cell line, *J. Cell. Physiol.* **105**:1–6.

Mullin, J. M., Weibel, J., Diamond, L., and Kleinzeller, A. J., 1980b, Sugar transport in the $LLC–PK_1$ renal epithelial cell line: Similarity to mammalian kidney and the influence of cell density, *J. Cell. Physiol.* **104**:375–389.

Ojakian, G. K., 1981, Tumor promoter-induced changes in the permeability of epithelial cell tight junctions, *Cell* **23**:95–103.

Pickett, P. B., Pitelka, D. R., Hamamoto, S. T., and Misfeldt, D. S., 1975, Occluding junctions and cell behavior in primary cultures of normal and neoplastic mammary gland cells, *J. Cell Biol.* **66**:316–332.

Rabito, C. A., 1981, Localization of the Na^+-sugar cotransport system in a kidney epithelial cell line $(LLC\ PK_1)$, *Biochim. Biophys. Acta* **649**:286–290.

Rabito, C. A., and Karish, M. V., 1982, Polarized amino acid transport by an epithelial cell line of renal origin $(LLC–PK_1)$: The basolateral systems, *J. Biol. Chem.* **257**:6802–6808.

Rabito, C. A., Tchao, R., Valentich, J., and Leighton, J., 1978, Distribution and characteristics of the occluding junctions in a monolayer of a cell line (MDCK) derived from canine kidney, *J. Membr. Biol.* **43**:351–365.

Rabito, C. A., Tchao, R., Valentich, J., and Leighton, J., 1980, Effect of cell-substratum interaction on hemicyst formation by MDCK cells, *In Vitro* **16**:461.

Rasmussen, H., Goodman, D. B. P., and Tenenhouse, A., 1972, The role of cyclic AMP and calcium in cell activation, *CRC Crit. Rev. Biochem.* **1:**95–148.

Reid, L., Morrow, B., Jubinsky, P., Schwartz, E., and Gatmaitan, Z., 1981, Regulation of growth and differentiation of epithelial cells by hormones, growth factors, and substrates of extracellular matrix, *Ann. N.Y. Acad. Sci.* **372:**354–370.

Reuben, R. C., Khanna, P. L., Gazitt, Y., Breslow, R., Rifkind, R., and Marks, P. A., 1978, Inducers of erythroleukemic differentiation. Relationship of structure to activity among planar-polar compounds, *J. Biol. Chem.* **253:**4214–4218.

Rifkind, R. A., Fibach, E., Reuben, R. C., Gazitt, Y., Yamasaki, H., Weinstein, I. B., Nudel, V., Shumida, I., Terada, M., and Marks, P. A., 1978, Erythroleukemia cells: Commitment to differentiate and the role of the cell surface, in: *Differentiation of Normal and Neoplastic Hematopoietic Cells,* Volume 5 (B. Clarkson, P. A. Marks, and J. E. Till, eds.), Cold Spring Harbor Conferences on Cell Proliferation, Cold Spring Harbor Press, New York, p. 209.

Rodriguez–Boulan, E., and Sabatini, D. D., 1978, Asymmetric budding of viruses in epithelial monolayers: A model system for study of epithelial polarity, *Proc. Natl. Acad. Sci. USA* **75:**5071–5075.

Roth, M. G., Fitzpatrick, J. P., and Compans, R. W., 1979, Polarity of influenza and vesicular stomatitis virus maturation in MDCK cells: Lack of a requirement for glycosylation of viral glycoproteins, *Proc. Natl. Acad. Sci. USA* **75:**6430–6434.

Sanders, E. J., and Dickau, J. E., 1981, Morphological differentiation of an embryonic epithelium in culture, *Cell Tissue Res.* **220:**539–548.

Sanders, M. J., and Misfeldt, D. S., 1981, Transepithelial transport in cell culture: Different stoichiometries of Na^+:phlorizin binding and Na^+:D-glucose cotransport, *J. Cell Biol.* **91:**424 (abstract).

Scher, W., Tsuei, D., Sassa, S., Price, P., Gabelman, N., and Friend, C., 1978, Inhibition of dimethylsulfoxide Friend cell erythrodifferentiation by hydrocortisone and other steroids, *Proc. Natl. Acad. Sci. USA* **75:**3851–3855.

Sheffery, M., Rifkind, R. A., and Marks, P. A., 1982, Murine erythroleukemia cell differentiation: DNAse I hypersensitivity and DNA methylation near the globin genes, *Proc. Natl. Acad. Sci. USA* **79:**1180–1184.

Simmons, J. L., Fishman, P. H., Freese, E., and Brady, R. O., 1975, Morphological alterations and ganglioside sialytransferase activity induced by small fatty acids in HeLa cells, *J. Cell Biol.* **66:**414–424.

Sirica, A. E., Richards, W., Tsukada, Y., Sattler, C. A., and Pitot, H. C., 1979, Fetal phenotypic expression by adult rat hepatocytes on collagen gel/nylon meshes, *Proc. Natl. Acad. Sci. USA* **76:**283–287.

Smith, R. L., Macara, I. G., Levenson, R., Housman, D., and Cantley, L., 1982, Evidence that a Na^+/Ca^{2+} antiport system regulates murine erythroleukemia cell differentiation, *J. Biol. Chem.* **257:**773–780.

Storrie, B., Puck, T. T., and Wigner, L., 1978, The role of butyrate in the reverse transformation reaction in mammalian cells, *J. Cell Physiol.* **94:**69–76.

Sweadner, K. J., and Goldin, S. M., 1980, Active transport of sodium and potassium ions, *New Engl. J. Med.* **302:**777–783.

Taub, M., and Saier, M. H., 1979, Regulation of ^{22}Na transport by calcium in an established kidney epithelial cell line, *J. Biol. Chem.* **254:**11440–11444.

Taub, M., Chuman, L., Saier, M. H., and Sato, G., 1979, Growth of Madin-Darby canine kidney epithelial cell (MDCK) line in hormone-supplemented, serum-free medium, *Proc. Natl. Acad. Sci. USA* **76:**3338–3342.

Terada, M., Epner, E., Nudel, U., Salmon, J., Fibach, E., Rifkind, R. A., and Marks, P. A., 1978,

Induction of murine erythroleukemia differentiation by actinomycin D, *Proc. Natl. Acad. Sci. USA* **75**:2795–2799.

Thomas, S. R., Schultz, S. G., and Lever, J. E., 1982, Stimulation of dome formation in MDCK kidney epithelial cell cultures by inducers of differentiation: Dissociation from effects on transepithelial resistance and cyclic AMP levels, *J. Cell Physiol.* **113**:427–432.

U, H. S., Saier, M. H., Jr., and Ellisman, M. H., 1979, Tight junction formation is closely linked to the polar redistribution of intramembranous particles in aggregating MDCK epithelia, *Exp. Cell Res.* **122**:384–392.

U, H. S., Saier, M. H., Jr., and Ellisman, M.H., 1980, Tight junction formation in the establishment of intramembranous particle polarity in aggregating MDCK cells, *Exp. Cell Res.* **128**: 223–237.

Valentich, J. D., 1981, Morphological similarities between the dog kidney cell line MDCK and the mammalian cortical collecting tubule, *Ann. N.Y. Acad. Sci.* **372**:384–404.

Valentich, J. D., 1982, Basal lamina assembly by the dog kidney epithelial cell line MDCK, in: *Growth of Cells in Hormonally-Defined Media,* Volume 9 (G. Sato, A. B. Pardee, and D. Sirbasku, eds.), Cold Spring Harbor Press, New York, pp. 567–579.

Valentich, J. D., Tchao, R., and Leighton, J., 1979, Hemicyst formation stimulated by cyclic AMP in dog kidney cell line MDCK, *J. Cell. Physiol.* **100**:291–304.

Weinstein, I. B., Wigler, M., and Pietropaolo, C., 1977, The action of tumor promoting agents in cell culture, in: *Origins of Human Cancer* (H. H. Heath, J. D. Watson, and J. A. Wingston, eds.), Cold Spring Harbor Press, New York, pp. 751–772.

Wicha, M. S., Lowrie, G., Kohn, E., Bagavardoss, P., and Mahn, T., 1982, Extracellular matrix promotes mammary epithelial growth and differentiation *in vitro, Proc. Natl. Acad. Sci. USA* **79**:3213–3217.

Yang, J., Richards, J., Bowman, P., Guzman, R., Enami, J., McCormick, K. Hamamoto, S., Pitelka, D., and Nandi, S., 1979, Sustained growth and three-dimensional organization of primary mammary tumor epithelial cells embedded in collagen gels, *Proc. Natl. Acad. Sci. USA* **76**:3401–3405.

II

ELECTROPHYSIOLOGY AND ION TRANSPORT IN MDCK CELLS

The MDCK cell line was established from the kidney of a normal cocker spaniel in 1958. Remarkably, this dog kidney epithelial cell line retains many of the differentiated properties typical of kidney tubule cells in the loop of Henle or in the distal tubule. MDCK cells have been shown to possess morphological polarity; transepithelial salt and water transport was demonstrated by means of an Ussing flux chamber. Both electrophysiological and microbiological techniques have been used to study the transport systems and the junctions of MDCK cells. MDCK was shown to possess an amiloride-sensitive Na^+ transport system, as well as a furosemide-sensitive chloride transport system. Factors (hormones, calcium) that regulate transport have been identified, and hormonal control of cyclic AMP synthesis has been studied. MDCK cells respond to arginine–vasopressin by making cyclic AMP; this response is typical of cells in the distal tubule and the loop of Henle. Arginine vasopressin also stimulates ion transport. Variant (or ''mutant'') MDCK cells have been isolated and used in the study of membrane transport in these cultured kidney cells.

2

Electrical Properties of MDCK Cells

L. GONZÁLEZ–MARISCAL, L. BORBOA,
R. LÓPEZ–VANCELL, G. BEATY, and M. CEREIJIDO

1. INTRODUCTION

The electrical characterization of a transporting epithelium requires information three main aspects: (1) the overall electrical properties across the whole cell layer; (2) the transcellular permeation route; and (3) the paracellular pathway. In the present article we describe assumptions, techniques, efforts, and information collected on these aspects of the monolayer of MDCK cells, which under certain culturing conditions behave as a natural transporting epithelium (Misfeldt *et al.*, 1976; Cereijido *et al.*, 1978a,b; Valentich *et al.*, 1979). Furthermore, we also describe our efforts to relate the electrical properties to different structural features (e.g., the strands of the occluding junctions) and to intracellular organelles (e.g., the cytoskeleton), both under steady-state conditions and under transient conditions such as the establishment of the junctions, the removal of Ca^{2+}, and the addition of test substances such as cytochalasin and colchicine.

2. ELECTRICAL PROPERTIES OF THE PLASMA MEMBRANE OF MDCK CELLS

We have developed a procedure that uses an intracellular microelectrode of very fine tip (electrical resistance ca. 100 MΩ) (Stefani and Cereijido, 1983).

L. GONZÁLEZ-MARISCAL, L. BORBOA, and M. CEREIJIDO • Centro de Investigación y de Estudios Avanzados, Departmento de Fisiología y Biofísica, México, D. F. 07000, Mexico. R. LÓPEZ-VANCELL and G. BEATY • Department of Health Sciences, Universidad Autonóma Metropolitana, Unidad Iztapalapa, México, D. F., Mexico.   Work supported by research grants from the National Research Council of Mexico (CONACYT) and by grant AM 26481 from the U.S. National Institute of Health.

With this procedure it is possible to deliver a square pulse of current and, at the same time, record the voltage variation produced (Fig. 1). The time course of this variation permits in turn measurement of the electrical capacity. The electrical potential found was around -50 mV, the electrical resistance was 61.6 MΩ, and the capacity 45.1 pF. This value appears somewhat higher than expected for a cell the size of an MDCK diameter 14 μm, height 5 μm). Yet, in mature monolayers (plated at confluence for more than one day), it is not due to an intercellular electric coupling, because the cell responds as a single homogeneous compartment, and no intercellular coupling was detected in 20 pairs of neighboring cells explored. The high capacity is instead related to the fact that the high degree of infolding of the lateral plasma membrane, and the micro-villi of the apical regions, afford a larger amount of membrane than one would predict on the basis of the size and shape of the MDCK cell.

A fraction of this plasma membrane will have a higher resistance than the whole plasma membrane of an MDCK cell (R_τ). Therefore, if in order to traverse the transcellular route a current should cross two fractions membrane (once to enter through the apical and then to exit through the basolateral) the electrical resistance through a single MDCK (R) cell would be given by

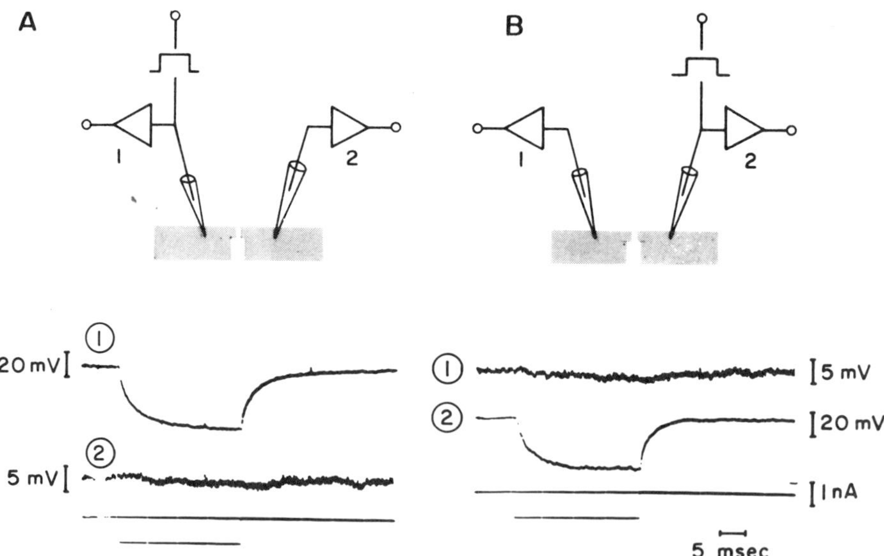

Figure 1. Impalement of two adjacent MDCK cells. The upper schemes illustrate the experimental arrangement. Records 1 and 2 were obtained with microelectrodes 1 and 2, respectively. The resistance of the microelectrodes was about 90 MΩ in both of them. Notice the large gain in records A2 and B1. Cells were of passage 73 and were plated 4 days before. (Taken with kind permission from Stefani and Cereijido, 1983.)

$$R = \frac{R_\tau}{F_a + F_b} \tag{1}$$

where F_a is the fraction occupied by the apical and F_b the fraction occupied by the basolateral. Furthermore, since $F_a + F_b$ is equal to 1, one may transform equation 1 into

$$R = \frac{R_\tau}{F_a - F_a^2} \tag{2}$$

This is a parabola whose minimum would be given when the occluding junction divides the plasma membrane into 50% apical and 50% basolateral, i.e., when F_a is 0.5. If we take this value of F_a and the value of the resistance R_τ found experimentally (61.6 MΩ), we conclude that the transcellular route through a single MDCK cell should be, at least, 250 MΩ. Since a typical monolayer contains some $2-5 \times 10^5$ of such cells per square centimeter, the transcellular route of the monolayer would have a minimal resistance of 500–1300 Ωcm^2. We shall return to this value once we discuss the electrical properties of the whole monolayer in Section 3.

Other interesting electrical properties of the MDCK cell are the following: (1) they present an asymmetric voltage/current relationship, and (2) under certain circumstances (e.g., low Na$^+$, addition of amiloride) with large depolarizing current pulses the voltage response becomes time dependent (Fig. 2); i.e., the membrane of the MDCK cell responds like that of an excitable cell.

3. ELECTRICAL PROPERTIES OF THE WHOLE MONOLAYER OF MDCK CELLS

These monolayers can be prepared on a permeable support. We use a disk of a nylon cloth 1.3 cm in diameter, coated with collagen. This preparation can be

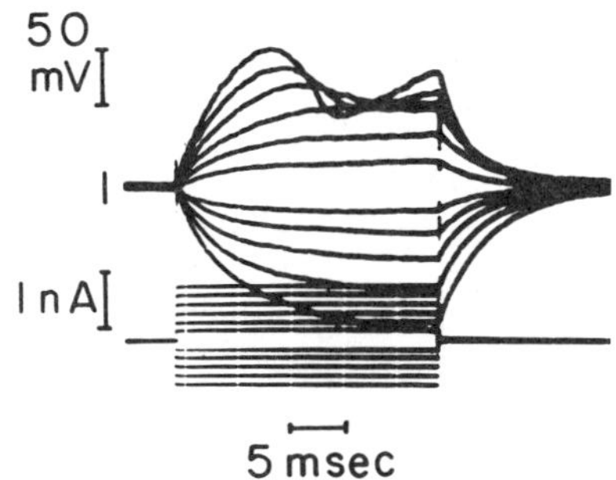

Figure 2. Time-dependent resistance obtained during large positive pulses in the presence of 10^{-4} M amiloride. Notice that even before reaching the time-dependent responses the voltage deflections produced by the pulses of current were asymmetric. (Taken with kind permission from Stefani and Cereijido, 1983.)

mounted as a flat sheet between two lucite chambers as if it were a frog skin or a gallbladder. The electrical resistance across varies from 80–300 Ωcm^2 obtained in most laboratories (Misfeldt *et al.*, 1976; Cereijido *et al.*, 1978b; Rabito *et al.*, 1978) to more than 4000 Ωcm^2 obtained by Simmons (1981).

According to Valentich (1981), typical batches of MDCK cells contain subpopulations that differ widely in their electrical properties, and culturing conditions would favor the predominance of one or the other, thus accounting for the large difference obtained by different authors.

In our hands the monolayer of MDCK cells has some 100–300 Ωcm^2 and presents a linear conductance/Na-concentration relationship; an asymmetric instantaneous current/voltage relationship; a 9:1 Na^+/Cl^- discrimination; a decrease in this ability to discriminate when the pH is lowered from 7.4 to 3.8, suggesting that cation-specific channels, which exclude Cl^-, contain acidic groups dissociated at neutral pH; and a characteristic pattern of ionic selectivity (Eisenman's series VI), which suggests that the negatively charged sites are highly hydrated and of medium field strength (Cereijido *et al.*, 1978a,b).

4. THE PARACELLULAR PATHWAY

As discussed in Section 2, on the basis of a transcellular route, the monolayer should have a minimal resistance of 500–1300 Ωcm^2. The fact that it has 100–300 Ωcm^2 indicates clearly that it contains another permeation route that avoids the cytoplasm of the cells. In order to localize this route, we passed pulses of current of 20–50 $\mu A\ cm^{-2}$ and a duration of 5 msec, first in the positive and then in the negative direction to avoid polarization of the electrodes, and the voltage variation elicited was scanned with a microelectrode on the apical surface of the monolayer. The amplitude of the signal at the point where the microelectrode is placed is proportional to the electrical conductance, and this permits detection of the points where current flows through the monolayer (Cereijido *et al.*, 1980). This indicated (1) that the monolayer of MDCK cells has in fact a paracellular permeation route, and that the high conductance is not due to incomplete growth or faulty sealing; and (2) that the paracellular path has large variations along the perimeter of a given cell, and the intercellular space appears to be actually *tight,* except for some conducting spots studded along the intercellular space.

The same population of monolayers used for voltage scanning was analyzed with freeze–fracture electron microscopy, and it was observed that the number of strands varies abruptly from 1 to 10 within a few nanometers. We wondered whether natural leaky epithelia would also have a heterogeneous distribution of conductance and, in particular, if this heterogeneity was associated with the chaotic distribution of strands. Therefore, we applied the same voltage scanning technique to the *Necturus* gallbladder (Cereijido *et al.*, 1982) and found that in this preparation both conductance and number of strands are homogenously distributed along the interspace. Since this was clearly different from the monolayer of MDCK cells, the comparison suggests that the MDCK monolayer has mixed characteristics of *leaky* and *tight* epithelia. In this respect it might be pertinent to take into account the observations of Simmons (1981) and of Valentich (1981), mentioned previously, indicating that culturing conditions would determine the resistance of the monolayer, probably through a shift in the proportion between sealing and conductive elements.

5. DO STRANDS DETERMINE THE RESISTANCE OF THE OCCLUDING JUNCTION?

The discussion of Section 4 might give the impression that the junctions of the monolayer of MDCK cells have an irregularly distributed conductance because it has an irregular pattern of strands. Actually there is considerable evidence that under certain circumstances the number of strands and the conductance of natural epithelia are associated (see Bentzel *et al.*, 1980; Meldolesi *et al.*, 1978). However, there is a serious controversy on this point (see Martínez–Palomo and Erlij, 1975). Discrepancies come mainly from the comparison of different epithelia and different animal species. Hence, in order to analyze two situations where the same monolayer of MDCK cells would have a drastic difference in resistance and/or structure, we resorted to the effect of temperature.

Table I shows the reversible effect that temperature has on electrical transepithelial resistance in spite of the constancy of the structural parameters of the occluding junction (González-Mariscal *et al.*, 1984). The fact that a change of 306% in the electrical resistance is not accompanied by detectable modifications in the arrangement of the strands (Fig. 3) suggests that the irregularly distributed

Table I.   Occluding Junctions of MDCK Cells at 3° and 37°C

	3°C[a]	37°C[a]
Number of strands	4.8 ± 0.2 (129)	5.2 ± 0.2 (153)
Junctional width (μm)[b]	0.23 ± 0.02 (129)	0.25 ± 0.01 (153)
Junctional density[c]	18 (129)	21 (153)
Conductance (mmho/cm^2)	5.19 ± 0.24 (21)	16.03 ± 0.40 (28)

[a]Values are expressed as mean ± standard error (number of observations).
[b]Distance between uppermost and lowermost strand.
[c]Number of strands divided by junctional width.

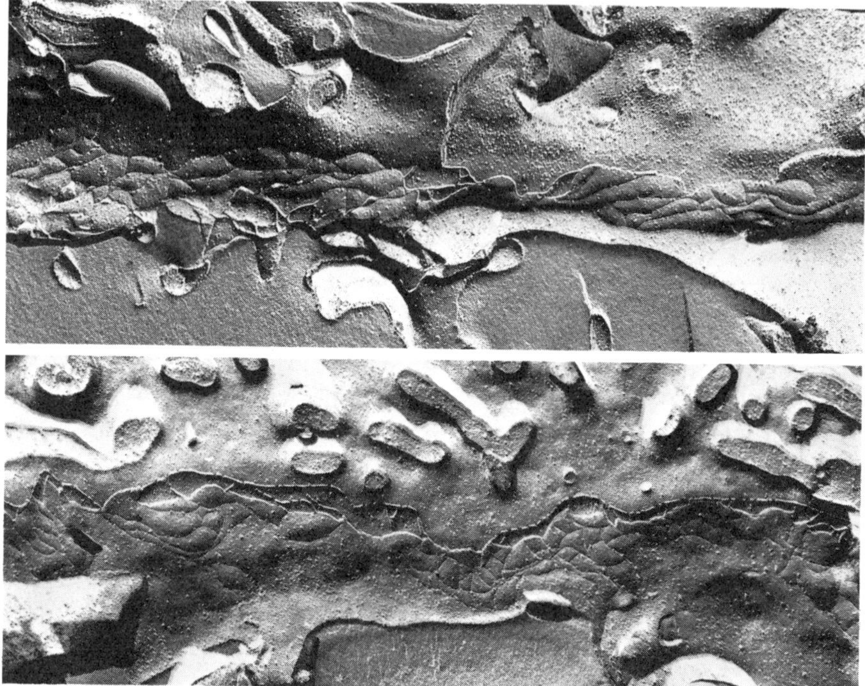

Figure 3. Freeze–fracture replicas of occluding junctions of MDCK cells fixed at 3° (upper) and 37°C (lower) (X10,500). (Taken with kind permission from González-Mariscal *et al.*, 1984.)

number of strands and the heterogeneously distributed resistance described previously might be just coincidental.

6. *DE NOVO* FORMATION OF OCCLUDING JUNCTIONS

When MDCK cells are trypsinized and suspended, they lose their occluding junctions and polarity. After they are reseeded at confluence on the collagen-coated disk, they reestablish both junctions and polarity. Junction formation may be followed by gauging the electrical resistance across the monolayer (Fig. 4) (Cereijido *et al.*, 1978a,b). There is an initial period when the addition of inhibitors of the protein synthesis would affect the formation of junctions and a later period when inhibitors are no longer effective despite the fact that the electrical resistance is still low. The observation that inhibitors of the synthesis of proteins impair the formation of the occluding junction does not necessarily mean that the junctions themselves are constituted by proteins.

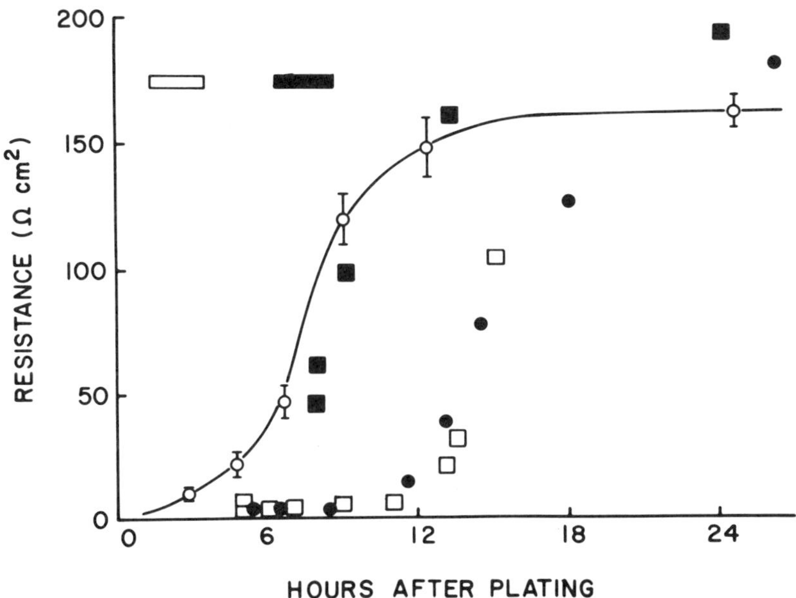

Figure 4. Development of electrical resistance across monolayers of MDCK cells plated at confluence at time 0. Although the monolayer is completed in less than 0.5 hr, electrical resistance develops slowly as junctions are synthesized and sealed (open circles). Cycloheximide (6 μg/ml), present during the period marked by open block, delays development of junctions (open squares). Puromycin (10^{-4} M) has the same effect (filled circles). If cycloheximide is present instead during the period marked by filled block, it does not prevent the rise of electrical resistance (filled squares). (Taken with kind permission from Stefani and Cereijido, 1983.)

During the first 24-h period the junctions are fully sealed, yet their ability to discriminate between cations and anions is very low, suggesting that sealing and ion-permeating components of the junctions are not only different, but are also installed with different time courses.

7. THE SEALING OF OCCLUDING JUNCTION

The sealing of the occluding junction requires Ca^{2+}. If one plates the monolayer for 90 min and then removes Ca^{2+} from the bathing medium, the electrical resistance does not develop (Fig. 5). Yet, if on the following day one adds Ca^{2+}, the electrical resistance rises with a faster time course than when the junction has to be synthesized *de novo* (Borboa and Cereijido, unpublished observations). Moreover, this sealing effect of Ca^{2+} can be observed in spite of

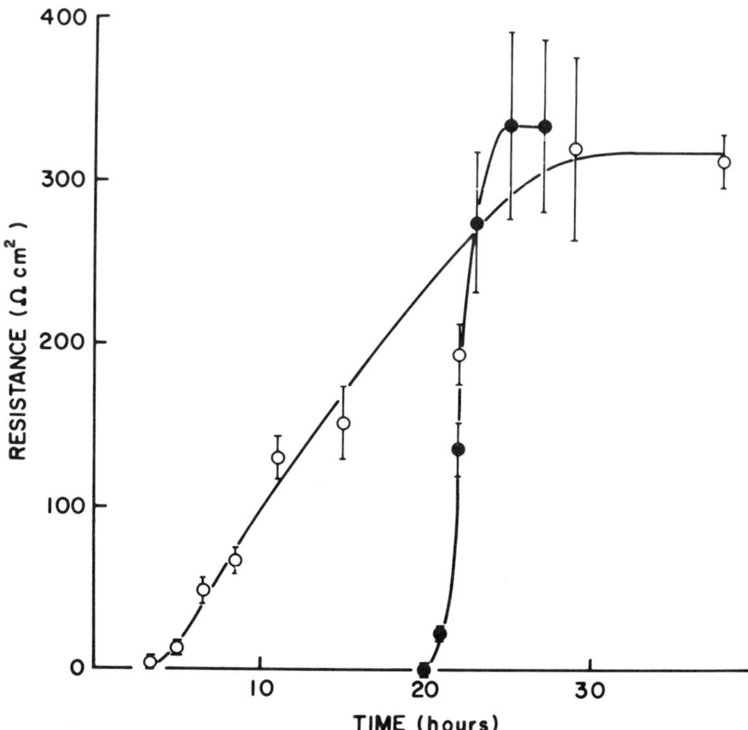

Figure 5. Time course of the electrical resistance across monolayers of MDCK cells. Open circles, disks are plated in CMEM for 90 min and then removed to MEM without cells. Filled circles, disks that have been plated for 90 min and then transferred to Ca^{2+}-free MEM overnight. At the twentieth hour they were bathed in CMEM with Ca^{2+}. (Taken with kind permission from Borboa, L., González-Robles, A., and Cereijido, H., unpublished observations.)

the presence of inhibitors of the synthesis of proteins. Before sealing, freeze–fracture replicas fail to detect any functional strand.

   Once the junction is formed, it can be reopened by removing Ca^{2+} (Fig. 6) and resealed by restoration of this ion. Barium and magnesium would not substitute Ca^{2+} in this resealing (Cereijido *et al.*, 1981; Martínez-Palomo *et al.*, 1980).

8. MICROTUBULES, MICROFILAMENTS, AND OCCLUDING JUNCTIONS

   We have prepared rabbit antibodies against actin and tubulin of MDCK cells and treated these cells with them. In order to visualize the bound antibodies, we

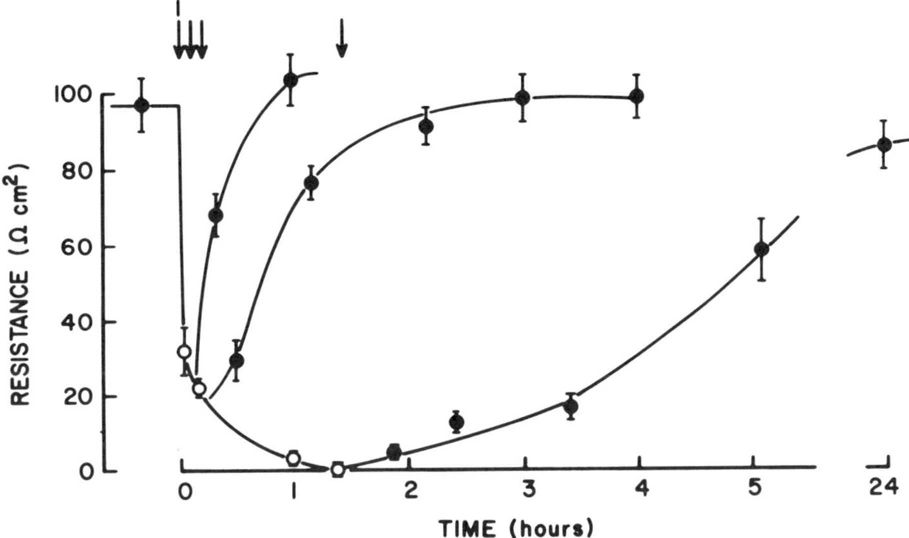

Figure 6. Recovery of the electrical resistance across MDCK monolayers. At arrow 1, all disks were put into Petri dishes containing Ca^{2+}-free MEM with 2.5 mM EGTA. Ca^{2+} was restored at the times marked by the arrows. Curves were drawn by eye. (Reprinted with kind permission from Martínez-Palomo *et al.*, 1980.)

used a second antibody (goat-anti-rabbit) which was fluoresceinated. We observed that, whereas microtubules concentrated mainly around the nucleus, microfilaments formed a characteristic ring in close contact with the cellular borders, suggesting that microfilaments may be involved in junctional events (Meza *et al.*, 1980).

Colchicine, a powerful disruptor of microtubules, has no detectable effect on the steady-state resistance of the monolayer or on the opening and resealing elicited by removal and restoration of Ca^{2+}. On the contrary, cytochalasin B, a drug that disrupts microfilaments, produces a gradual opening of resting occluding junctions and impairs resealing of junctions that had been opened by the removal of Ca^{2+} (Meza *et al.*, 1980).

Hoi Sang *et al.* (1980) have obtained indications that cytochalasin B may block junction formation in monolayers of just-trypsinized cells.

Since the information discussed previously suggested a close association between microfilaments and junctions, we fixed MDCK cells, labeled the surface membrane proteins with ^{131}I using a lactoperoxidase method, and then disolved the membrane with triton. The only iodinated proteins were those of the cellular membrane which remained attached to the cytoskeleton. We then solubilized the cytoskeleton with the proteins attached using 1% SDS and ran electrophoretic analysis in polyacrylamide gels. We expected that if occluding junctions *were* or *contained* proteins, we would be able to identify them in a given band. Although we found that many of the membrane proteins were iodinated and attached to the cytoskeleton, we found no difference in the patterns of proteins obtained under several of the conditions that would profoundly modify the function of the junction (i.e., with or without Ca^{2+}, with cytochalasin B, with colchicine, trypsinized). This, of course, does not refute the possibility that junctions are or contain proteins; it may just indicate that the amount of protein is too small to be detected with the method used, or that the different treatments that affect the sealing capacity of the junction would just modify the state of sealing or ion-translocating elements.

9. THE OCCLUDING JUNCTION AS A FUNCTION OF TIME

Occluding junctions appear to be highly dynamic structures. This is evident from the fact that a plated monolayer does not maintain its cellular density constant, and growing and addition of new cells denote reaccommodation and

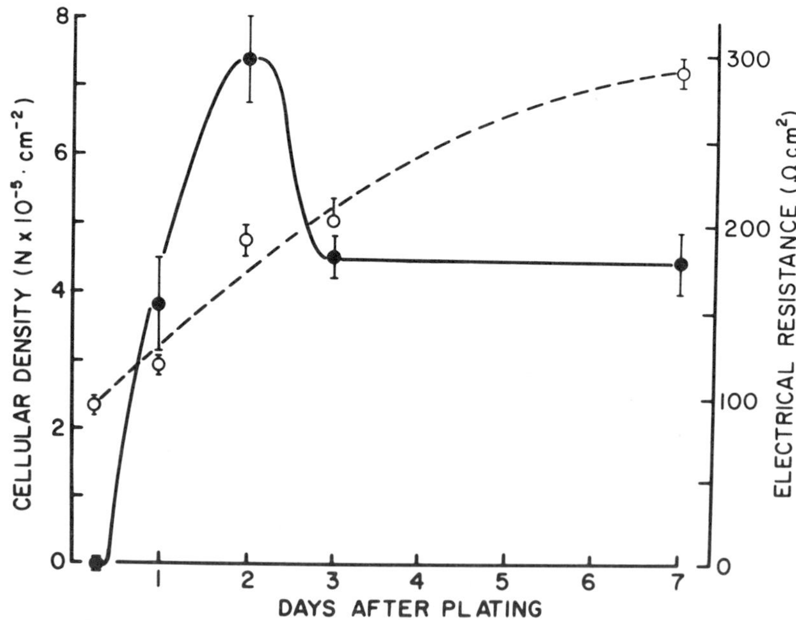

Figure 7. Cellular density (open circles) and electrical resistance across monolayers (filled circles) of MDCK cells as a function of time after plating. Cellular density was counted in the same disks used for the electrical measurements. (Taken with kind permission from Borboa, L., González-Robles, A., and Cereijido, H., unpublished observations.)

require a permissive intercellular attachment. Moreover, since epithelia have cells of different types that are at different stages of their cycle and are dequamated without producing a breach in the diffusion barrier presented by the epithelium, one must conclude that junctions have a plasticity that permits such rearrangement of cells. Occluding junctions appear as dynamic structures whose properties depend on the age of the epithelium.

We have studied the electrical resistance across monolayers of different ages, and with different cellular densities, which in turn connoted a different amount of occluding junctions per unit area. Figure 7 shows that, although monolayers of 32 and 72 hr respond as expected (i.e., with a higher conductance as the number of cells increases), younger monolayers have just the opposite behavior (Borboa and Cereijido, unpublished observations). Of course, as these monolayers increase their cell density, they have more conducting elements working in series. Yet the electrical properties of the cells themselves do not seem to be different at 1 or 5 days after plating. This suggests that the properties of the occluding junctions vary as a function of the age of the monolayer.

ACKNOWLEDGMENTS.   We wish to thank Amparo Lazaro and Roberto Carmona for their efficient technical assistance.

REFERENCES

Bentzel, C. J., Hainau, B., Ho, S., Hui, S. W., Edelman, A., Anagstopoulus, T., and Benedetti, E. L., 1980, Cytoplasmic regulation of tight junction permeability: Effect of plant cytokinins, *Am. J. Physiol.* **239**:C75–C89.

Cereijido, M., Robbins, E. S., Dolan, W. J., Rotunno, C. A., and Sabatini, D. D., 1978a, Polarized monolayers formed by epithelial cells on a permeable and transluscent support. *J. Cell Biol.* **77**:853–880.

Cereijido, M., Rotunno, C. A., Robbins, E. S., and Sabatini, D. D., 1978b, Polarized epithelial membranes produced *in vitro*, in: *Membrane Transport Processes I* (J. F. Hoffman, ed.), Raven Press, New York.

Cereijido, M., Stefani, E., and Martínez–Palomo, A., 1980, Occluding junctions in a cultured transporting epithelium: Structural and functional heterogeneity, *J. Membr. Biol.* **53**:19–32.

Cereijido, M., Meza, I., and Martínez–Paloma, A., 1981, Occluding junctions in cultured epithelial monolayers, *Am. J. Physiol.* **240**:C96–C102.

Cereijido, M., Stefani, E., and Chávez de Ramírez, B., 1982, Occluding junctions of the *Necturus* gallbladder, *J. Membr. Biol.* **70**:15–25.

González–Mariscal, L., Chávez de Ramírez, B., and Cereijido, H., 1984, The effect of temperature on the occluding junctions of monolayers of epithelioid cells (MDCR), *J. Membr. Biol.* **79**:175–184.

Hoi Sang, V., Saier, M. H., Jr., and Ellisman, M. H., 1980, Tight junction formation in the establishment of intramembranous particle polarity in aggregating MDCK cells. Effect of drug treatment, *Exp. Cell Res.* **128**:223–235.

Martínez–Palomo, A., and Erlij, D., 1975, Structure of tight junctions in epithelia with different permeability, *Proc. Natl. Acad. Sci. U.S.A.* **72**(11):4487–4491.

Martínez–Palomo, A., Meza, I., Beaty, G., and Cereijido, M., 1980, Experimental modulation of occluding junctions in a cultured transporting epithelium, *J. Cell. Biol.* **87:**736–745.

Meldolesi, J., Castiglioni, G., Parma, R., Nassivera, N., and De Camilli, P., 1978, Ca^{2+} dependent desassembly and reassembly of occluding junctions in guinea pig pancreatic acinar cells. Effect of drugs, *J. Cell. Biol.* **79:**156–172.

Meza, I., Ibarra, G., Sabanero, M., Martínez–Palomo, A., and Cereijido, M., 1980, Occluding junctions and cytoskeletal components in a cultured transporting epithelium, *J. Cell. Biol.* **87:**746–754.

Misfeldt, D. S., Hammamoto, S. T., and Pitelka, D. K., 1976, Transepithelial transport in cell culture, *Proc. Natl. Acad. Sci. U.S.A.* **73:**1212–1216.

Rabito, C., Tchao, R., Valentich, J. D., and Leighton, J., 1978, Distribution and characteristics of the occluding junctions in a monolayer of a cell line (MDCK) derived from canine kidney, *J. Membr. Biol.* **43:**351–365.

Simmons, N. L., 1981, Ion transport in "tight" epithelial monolayers of MDCK cells, *J. Membr. Biol.* **59:**105–114.

Stefani, E., and Cereijido, M., 1983, Electrical properties of cultured epithelioid cells (MDCK), *J. Membr. Biol.* **73:**177–184.

Valentich, J. D., 1981, Morphological similarities between the dog kidney cell line MDCK and the mammalian cortical collecting tubule, *Ann. N.Y. Acad. Sci.* **372:**384–405.

Valentich, J. D., Tchao, R., and Leighton, J., 1979, Hemicyst formation stimulated by cyclic AMP in dog kidney cell line (MDCK), *J. Cell Physiol.* **100:** 291–304.

3

Ion Transport in MDCK Cells

S. FERNÁNDEZ–CASTELO, J. J. BOLÍVAR,
R. LÓPEZ-VANCELL, G. BEATY, and M. CEREIJIDO

1. INTRODUCTION

The use of monolayers of MDCK cells as a model system for natural transporting epithelia created the necessity of learning about their ion-translocating mechanisms. These mechanisms have been analyzed directly by measuring unidirectional fluxes of tracers, and indirectly through their consequences on electrical parameters and blister-forming activity of the monolayer. This chapter reviews briefly the information available on the mechanisms and also describes the studies performed to determine whether they respond to agents like serum, ADH, changes in sodium concentration, and viral infection, which are known to affect transport parameters in other biological preparations.

2. BLISTERS

Much of the evidence that MDCK cells retain the property of transporting fluid vectorially from the apical to the basolateral side came from the studies on

J. J. BOLÍVAR and M. CEREIJIDO • Centro de Investigación y Estudios Avanzados, Departamento de Fisiología y Biofisica, México, D.F. 07000, Mexico.   S. FERNÁNDEZ–CASTELO • Department of Electrical Engineering, Universidad Autónoma, Metropolitana, Unidad Iztapalapa, México, D.F., Mexico.   R. LÓPEZ–VANCELL and G. BEATY • Department of Health Sciences, Universidad Autónoma, Metropolitana, Unidad Iztapalapa, México, D.F., Mexico.   Work supported by research grants from the National Research Council of Mexico (CONACyT) and by grant AM 26481 from the U.S. National Institutes of Health.

blister formation in monolayers. Leighton and his co-workers (1969) observed that blisters are dynamic structures in continuous formation and collapse. Presumably, fluid was transported across the MDCK monolayer. The transported fluid, which had accumulated between the monolayer and the impermeable support, resulted in detachment of the cells from their support and in formation of a dome. A dome would only form if both the pumping force and the resistance of the occluding junctions prevail over the firmness of the attachment of MDCK cells to the dish. The resistance of the occluding junctions prevents escape of the fluid that has been pumped to the basolateral surface. However, blistering and nonblistering regions in the monolayer seem to have identical properties. If one weakens experimentally the attachment of the cells to the substratum, the size and number of blisters increase significantly (Rabito *et al.*, 1978, 1980). The fact that at a given moment only a fraction of the total area (10–20%) is blistering has been attributed to the prevalence in nonblistering areas of the pumping and attachment forces over the resistance of the junctions. As a consequence, the junctions burst in such nonblistering areas (Cereijido *et al.*, 1981). Misfeldt *et al.* (1984) have performed a detailed mathematical analysis of the size and shape of the blisters as a function of the intervening forces.

In agreement with the hypothesis that blister formation is due to fluid transport is the observation that the blisters disappear after the monolayer is treated with ouabain (Abaza *et al.*, 1974). The number of blisters is also affected by agents such as cAMP, theophylline, 1-methyl-3-isobutylxanthine, papaverine, prostaglandin E_1, and cholera toxin, which act on cellular transporting mechanisms (Valentich *et al.*, 1979; Lever, 1979a,b,c; Lever, 1981; Rindler *et al.*, 1979a). The induction of blister formation and vectorial transport has been proposed to be due to the capacity of MDCK cells to differentiate. Presumably, MDCK cells have been retaining this potential since their removal from the kidney of a normal dog (Madin and Darby, 1958) and their adaptation to cell culture conditions. This hypothesis is supported by the observation that inducers of mammalian cell differentiation such as dimethylsulfoxide, dimethylformamide, hexamethylene bisacetamide (polar solvents), hypoxanthine, inosine, adenosine (purines), and n-butyrate (fatty acid) stimulate dome formation (Lever, 1979a,b,c).

3. IONIC FLUXES ACROSS THE WHOLE MONOLAYER

MDCK cells can be cultured in monolayers on permeable supports and mounted as flat sheets between two Lucite chambers, so that both electrical studies and flux studies may be performed (Misfeldt *et al.*, 1976; Cereijido *et al.*, 1978a). Judging from the position of the occluding junctions, the location of microvilli, the distribution of intramembrane particles, and the labeling with ^{3}H-ouabain, these membranes have a considerable degree of asymmetry (Misfeldt *et al.*, 1976; Cereijido *et al.*, 1978a,b; Cereijido *et al.*, 1980; Lamb *et al.*, 1981). As explained in Chapter 2 of this book, these monolayers may have a transepithelial resistance ranging from 80–4000 cm^2 depending on the culture conditions. In low-resistance monolayers the transepithelial flux of sodium is around 20 μmole h^{-1} cm^{-2} (Cereijido *et al.*, 1980). The permeability of sodium, calculated with the results of tracer studies and with diffusion potentials (in the latter case the Goldman-Hodkin-Katz equation is used), affords essentially the same value (1.71×10^{-5} cm sec^{-1}). Thus phenomena like "exchange diffusion" (Ussing, 1954) and "single-file diffusion" (Hodgkin and Keynes, 1955) are either absent or play a negligible role (Cereijido *et al.*, 1978b).

High-resistance monolayers of MDCK cells do not exhibit a significant net sodium flux (Simmons, 1981a) but have a net basolateral-to-apical secretion of Cl^-. In these high-resistance monolayers Cl^- secretion may be stimulated by exogenous ATP and inhibited by the addition of furosemide (10^{-4} M) to the basolateral side (Simmons, 1981b). Although inhibition by furosemide is observed only when the diuretic is present on the basolateral side, phloretin inhibits from either side. These observations suggest the existence of two different mechanisms for translocation of Cl^-. The Na^+ dependence of Cl^- transport is also consistent with the existence of two transport systems. The removal of Na^+ from the apical side does not affect Cl^- movement, whereas the removal of Na^+ from the basolateral side does inhibit Cl^- secretion (Simmons *et al.*, 1984). Cl^- and K^+ movements have also been observed to be related. Furosemide not only inhibits Cl^- flux, but also decreases the passive movement of K^+. Thus at least a fraction of the passive movement of K^+ is through the basolateral membrane and is in association with Cl^- (Simmons, 1981b). Finally, Cl^- transport has been observed to be hormonally regulated. Adrenaline produces a 20- to 30-fold increase in the short-circuit current (due to Cl^- secretion) and a proportional decrease in the electrical conductance (Simmons *et al.*, 1984).

4. IONIC FLUXES ACROSS THE PLASMA MEMBRANE

Unidirectional ionic fluxes have been measured in confluent MDCK monolayers cultured in Petri dishes. Under these conditions, the MDCK cells are bathed on their apical side with the desired test solution. Nevertheless, some of these fluxes (e.g., the active K flux) do not proceed through the apical, but through the basolateral side. Furthermore, these fluxes do not seem to be damped by the necessary diffusion through the occluding junctions. This latter observation was in part the basis for the suggestion that the junctions must be open in monolayers cultured on nonpermeable supports, except within the blisters, which occupy less than 20% of the area of the monolayer (Cereijido *et al.*, 1981).

In these monolayers Na^+ moves through an active mechanism that may be inhibited with ouabain, through a channel that is blocked by amiloride, and through a Na–Na exchange diffusion mechanism (Rindler *et al.*, 1979a,b; Cereijido *et al.*, 1980). Sodium uptake is stimulated by K^+ and Rb^+ and is inhibited by Li^+. These results suggest the presence of a Na/K cotransport mechanism in MDCK cells (Rindler *et al.*, 1979b). Bicarbonate, isopropionate, and acetate are stimulatory to Na^+ uptake. Replacement of Cl^- does not affect the rate of ^{22}Na uptake. Nevertheless, furosemide and bumetanid, which inhibit Cl^--translocating mechanisms, also inhibit the influx of Na^+ (Rindler *et al.*, 1982).

McRoberts *et al.* (1982) have shown that these inhibitors decrease ^{36}Cl uptake provided Na^+ and K^+ are present in the bathing solution simultaneously. These inhibitors also decrease the operation of the Na/K cotransport system. This information indicates that the cotransport system of MDCK cells may be of the type "$Na^+ + K^+ + Cl^-$" described by Field (1978) and Frizzel *et al.* (1979) for natural epithelia. The transport system of Field and Frizzel was similarly sensitive to furosemide and bumetanide. Rindler *et al.* (1982) have shown that this $Na^+ + K^+ + Cl^-$ transport system in MDCK cells depends on the presence (but not the consumption) of ATP.

An increase in the extracellular pH stimulated the influx of ^{22}Na and decreased the outflux of Na^+. Thus the presence of a Na/H antiport system was suggested. This system would be a neutral exchanger, as it is not affected by valinomycin. Valinomycin is a compound that increases the electrical conductance across cell membranes and thus abolishes the transmembrane electrical potential (Rindler *et al.*, 1979b; Rindler and Saier, 1981).

Taub and Saier (1979) have shown that calcium is inhibitory when added to the extracellular bathing solution, and that other divalent cations, like Ba^{2+}, Mg^{2+}, and Mn^{2+}, have a similar effect. In contrast, intracellular Ca^{2+} results in a stimulation of the sodium flux. This stimulation of Na entry by intracellular Ca^{2+} does not seem to be due to a Na/Ca exchange system, but to an allosteric regulation of Na^+ transport by the Na channel.

5. EFFECT OF SODIUM CONCENTRATION ON IONIC MOVEMENTS

In studies where the cytoplasmic concentration of Na^+ in MDCK cells was varied experimentally (from 14–50 mEq cm^{-2}), the rate of active pumping of Na was observed to increase by some 300% (Cereijido *et al.*, 1981). The rate constant for the Na/Na exchange diffusion was observed to have the opposite tendency. This phenomenon is known to exist also in the frog skin (Cereijido *et al.*, 1964) and has been proposed to be a homeostatic resource: when the cell is offered a high Na load, it reacts by lowering the Na^+ permeability of the apical membrane (so the penetration of Na^+ into the cell decreases) and by increasing the active extrusion of Na^+ through the basolateral membrane via the Na^+-K^+ ATPase. In this manner the cell regains its normally low Na content.

The effect of the cytoplasmic Na^+ concentration on the rate of K^+ influx into MDCK cells also was studied. In these studies the cytoplasmic Na^+ concentration was gradually increased. During this time period the outflux of Na^+ and influx of K^+ were being measured simultaneously. Initially, when the Na content of the cells was increased, the pumping of both the Na^+ and K^+ was observed to increase in a 1:1 ratio up to 30 nEq cm^{-2}. Yet as the Na content was further increased, the active uptake of K^+ was observed to decrease. However, the active Na efflux continues to increase (although with a tendency to saturate). At a Na^+ content of around 53 nEq cm^{-2} the pump exchanges two Na^+ per one K^+, constituting a so-called "electrogenic pump" (Cereijido et al., 1981). These variations in the active movement of K^+ and Na^+ are not accompanied by a change in the number of active sites involved, as the labeling with ^{3}H-ouabain remains constant: 3.06 ± 0.14 (4) and 3.15 ± 0.17 (4) sites $\times 10^{11}$ cm^{-2} at cellular sodium contents of 14 or 49 nEq cm^{-2} (Cereijido et al., 1981).

6. EFFECT OF SERUM

Serum is known to produce a significant increase in the permeability of cultured cells to ions (Cunningham and Pardee, 1969; Rozengurt and Heppel, 1975; Smith and Rozengurt, 1978; Villereal, 1981). In monolayers the number of blisters per unit area is also sensitive to the serum content of the bathing media (Lever, 1981). Yet, as mentioned previously, the blister formation depends not only on the ability to translocate fluid, but also on (1) the tightness of the occluding junctions as well as (2) the degree of attachment of the cells to their substratum. Therefore, serum may not act specifically on transport. We have investigated the effect of serum on the movement of ions and have found (Fig. 1) that the total outflux of Na in cells grown in Dulbecco's minimum essential medium (DMEM) containing 10% of calf serum was considerably higher than in cells maintained in DMEM without serum.

If cells were not plated at confluence, but were seeded at about one half of the confluence density and then grown to confluence in low (0.5%) or high (10%) serum, the difference in Na^+ outflux was even larger (Fernández-Castelo and Cereijido, unpublished results). In fibroblasts the increase of the Na^+ fluxes produced by serum occurs within a few minutes (Rozengurt and Heppel, 1975). Yet in MDCK cells the increase in the rate of Na^+ outflux produced by a switch from low to high serum is not detectable for up to 12 hr.

Table I shows that the effect of serum on the efflux of sodium is abolished if the cells are treated either with ouabain alone or with ouabain in the absence of

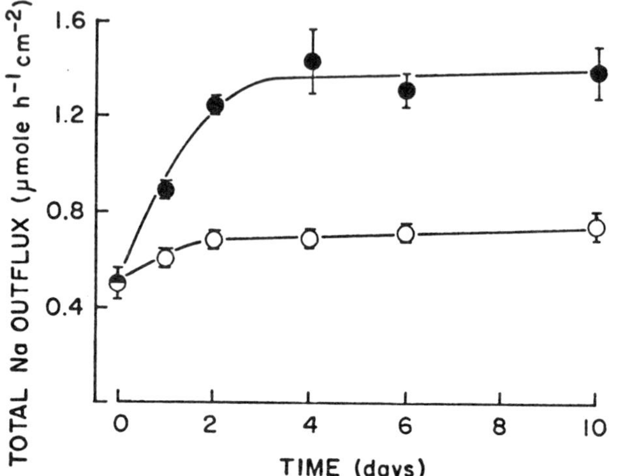

Figure 1. Effect of serum on the total outflux of sodium. The outflux was calculated on the basis of the washout of ^{22}Na measured in confluent monolayers of MDCK cells cultured in 35-mm plastic Petri dishes as described by Cereijido *et al.* (1980). Cells were grown to confluence in DMEM containing 10% of calf serum (CS). At day 0, monolayers were divided into two groups: one continued in DMEM with 10% CS (●) and the other in DMEM without serum (○). Media were changed every other day and 4 hr prior to the measurement of fluxes. Each point represents 4–10 measurements.

Na^+. These observations indicate that serum acts on the active extrusion of Na^+. The amount of Na^+ actively extruded depends on two basic factors: the number of pumps, and the functional capacity of each pump. The number of pumping sites, as estimated with ^{3}H-ouabain binding, is not affected by serum. Thus the number of specific labeled sites was 3.47 ± 0.39 (24) $\times 10^5$ sites per cell at 0.5% serum and 2.74 ± 0.31 (25) $\times 10^5$ at 10% serum. Therefore, serum should act on the functional capacity of each pump.

Table I. Effect of Serum on Different Na-Transporting Mechanisms in MDCK Cells

Mechanism	Sodium efflux (μmole h^{-1} cm^{-2})		
	0.5% CS	10% CS	p
All	0.61 ± 0.04 (20)	1.22 ± 0.06 (15)	< 0.001
Ouabain insensitive[a]	0.41 ± 0.03 (15)	0.47 ± 0.03 (10)	> 0.05
Ouabain insensitive[b], sodium independent	0.23 ± 0.01 (20)	0.39 ± 0.26 (15)	< 0.01

[a]Ouabain 5×10^{-4} M, in the incubation period.
[b]Ouabain 5×10^{-4} M, in the incubation period, washout in Na^+-free saline.

The calculation of the active outflux of Na^+ using tracer kinetics depends on two parameters; the rate constant k (in min^{-1}) and the size of the sodium pool S (in $nEq\ cm^{-2}$) (Cereijido *et al.*, 1980). Table II shows that both parameters are increased by serum.

Taub *et al.* (1979) have developed a hormonally defined serum-free medium (HDM) for the growth of MDCK cells. We have confirmed that this medium supports the growth and multiplication of MDCK cells. Furthermore, we have observed by means of Rb^+ influx studies that the activity of the pump in cells cultured in HDM is not significantly different from the activity in cells cultured with 10% calf serum (Table III).

7. VIRUSES AND IONIC FLUXES

Permeability changes produced by viruses in animal cells have been known for more than two decades (Klemperer, 1960; Hatanaka *et al.*, 1969; Isselbacher, 1972; Kalkar *et al.*, 1973; Negreanu *et al.*, 1974; Okada *et al.*, 1975; Pasternak and Micklem, 1973; Imprain *et al.*, 1980). It was even proposed that viruses can redirect the machinery of the cells in order to make viral components (Carrasco, 1978), through these permeability changes. The cellular machinery may either be directly affected by these permeability changes or indirectly affected by the resulting modifications of the ion content. The study of the effect of viruses on ion transport is of particular interest in MDCK cells, because the budding of influenza (WSN strain) and vesicular stomatitis viruses (VSV) from these cells has the remarkable property of being polarized: influenza buds through the apical and VSV through the basolateral surface, respectively (Rodríguez–Boulán and Sabatini, 1978; Rodríguez–Boulán and Pendergast, 1980). Therefore, the questions arise as to whether (1) influenza and VSV have an influence on ion permeability in MDCK cells, and (2) these viruses have specific effects on the ion-

Table II. Effect of Serum on the Rate Constant of Na Efflux and on the Na Transport Pool in MDCK Cells[a]

Serum (%)	k (min^{-1})	S (nmole cm^{-2})
0.5	0.21 ± 0.01 (85)	32.8 ± 1.3 (85)
10	0.25 ± 0.01 (111)	57.9 ± 1.2 (111)

[a]The Na pool was calculated on the basis of the ^{22}Na washout as described by Cereijido *et al.* (1980). The rate constant for the active flux was obtained by subtraction of the rate constants for the total and ouabain-inhibited outflux (p < 0.001) for both k and S.

*Table III. Influx of Potassium in
Monolayers of MDCK Cells Incubated at
Low or High Serum and in Hormonally
Defined, Serum-Free Medium[a]*

Media	K influx (nmole h^{-1} μg prot^{-1})
HDM	648.4 ± 54.8 (5)
DMEM + 0.5% CS	476.4 ± 31.1 (5)
DMEM + 10% CS	640.3 ± 65.6 (6)

[a]Monolayers were incubated for 6 days in the hormonally defined, serum-free medium (HDM) developed by Taub *et al.*
(1979) or in DMEM with 0.5 or 10% C.S.

translocating mechanisms located at either the apical or basolateral regions of the
cellular membrane.

We measured unidirectional fluxes of Na$^+$ and K$^+$ during two critical
stages: virus penetration (first hour postinfection) and budding (fifth hour)
(López-Vancell *et al.*, unpublished results).

At 1 hr postinfection VSV (which buds from the basolateral membrane)
produces a significant increase in the rate of K pumping (a function located on
the basolateral membrane) without affecting either the passive K fluxes or the Na
fluxes (Figs. 2, 4). Later (fifth hour) the picture changes: infection with VSV
decreases all K fluxes and Na fluxes (Figs. 3,4). In contrast, at the early period
postinfection (first hour) influenza (which buds from the apical membrane) increases the passive permeability to Na (an apical membrane function) without
affecting the rate of pumping or the fluxes of K$^+$. Even at the peak of the
budding period (fifth hour), influenza affects neither the passive permeability to
K$^+$ nor that of Na$^+$.

Therefore, viruses do affect membrane permeability, yet the changes observed may not be ascribed solely to a perturbation of the specific translocating
mechanisms for Na$^+$ and K$^+$ which operate in the same membrane that viruses
use to penetrate and leave the MDCK cells.

8. THE MECHANISM OF RESPONSE TO ADH

Epithelial cells, like those of the frog urinary bladder and the mammalian
collecting tubule, respond to antidiuretic hormone (ADH or arginine vasopressin) by increasing the transepithelial movement of Na$^+$. There is now
evidence that MDCK cells have at least some components of the system that is
responsible for the cellular response to ADH, including (1) biochemical and

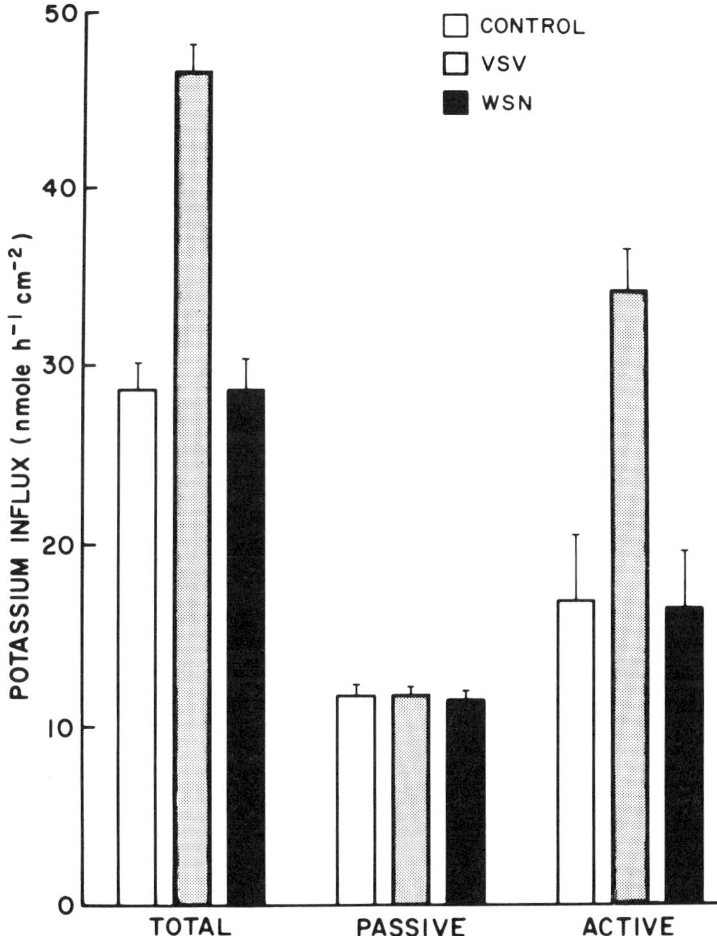

Figure 2. Potassium influx as calculated with the uptake of ^{86}Rb in confluent monolayers cultured in Petri dishes 3 days after plating. Columns represent values under control, VSV, and WSN infected conditions, 1 hr before flux measurements. Passive influx is the influx obtained in monolayers treated with 5×10^{-4} M ouabain added 20 min before starting the measurement of the uptake. Active influx is the fraction of the total influx inhibited by ouabain.

morphological studies suggesting that MDCK cells are derived from the distal and collecting tubule of the kidney (Rindler *et al*, 1977, 1979a; Valentich, 1981); (2) the observation that the MDCK cells also express membrane antigens corresponding to the thick portion of the Henle loop and distal tubule (Herzlinger *et al.*, 1982); (3) the observed stimulatory affect of ADH on the activity of adenylate cyclase and on cAMP levels in MDCK cells (Rindler *et al.*, 1977, 1979a);

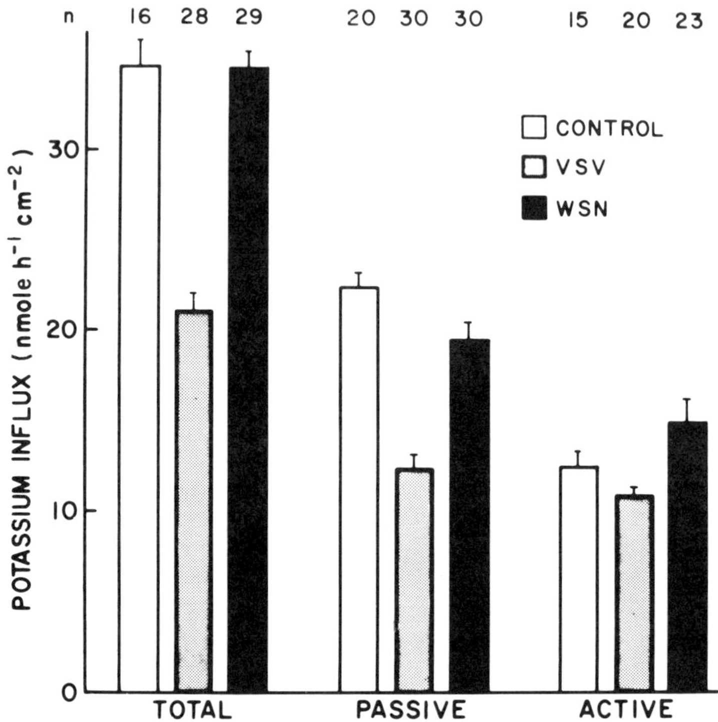

Figure 3. Potassium influx calculated as described in Fig. 2, 5 hr after infection with VSV or WSN.

(4) the observation of a stimulatory effect of ADH on the synthesis of prostaglandins (Hassid, 1981) as was previously observed in the collecting tubule of the rabbit (Holt and Lechene, 1981); and (5) the stimulatory effect of agents that increase intracellular cAMP levels (e.g., theophylline and 1-isobutyl-3-methylxanthine) on the blistering activity of MDCK monolayers (Lever, 1979a,b; Valentich *et al.*, 1979).

This led us to measure unidirectional fluxes of Na^+ in monolayers of MDCK cells treated with either cAMP or ADH (Bolívar *et al.*, unpublished results). Cyclic AMP (1 mM) produces a significant ($p < 0.05$) increase of the rate constant for Na outflux from 0.56 ± 0.03 (16) to 0.70 ± 0.05 (15) min^{-1}. ADH also increases the rate of Na^+ outflux. However, relative to cAMP, the ADH effect is smaller and delayed. In monolayers incubated with low Na, a more dramatic effect of ADH (50 mU/ml) on the rate constant is observed (Table IV). The inhibitory effect of high Na^+ concentrations on the response to ADH may be explained by the permeability-decreasing effect of high sodium on epithelial cells, as discussed previously.

The response of MDCK cells to ADH, a 40% increase in the rate constant for Na^+ outflux, is not as marked as the response exhibited by the frog urinary

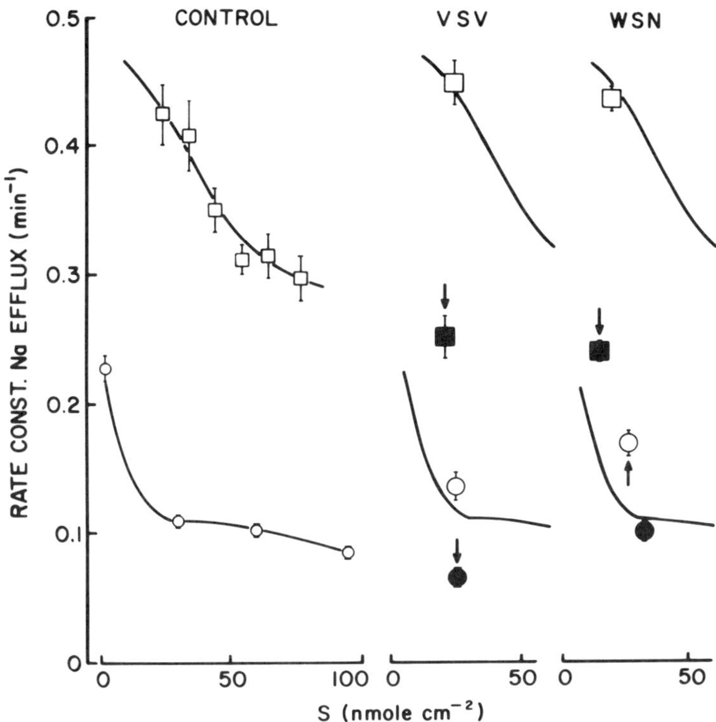

Figure 4. Rate constant for the outflux of sodium as calculated with the washout of ^{22}Na in monolayers under control, VSV, and WSN infected conditions. Rate constants are represented as a function of the Na pool in the cells (S) as it was noticed that in the control condition (left panel) they vary with this parameter. The lines fitted by eye to the experimental points under control conditions were represented again in the middle and right panels to permit comparisons with values obtained in WSN and VSV infected monolayers. In these panels open and filled symbols correspond to monolayers infected 1 or 5 hr previously, respectively. Square symbols represent rate constants for the total outflux, and circles correspond to those of passive outflux. Arrows were added as reminders that the signaled point represents a decrease or an increase from control values.

Table IV. Effect of ADH on the Rate Constant (k) for Na Outflux on Monolayers of MDCK Cells

	Time with ADH (min)	k (hr⁻¹)	Difference with control
Control[a]		11.0 ± 0.3 (19)	
ADH[b]	15	13.2 ± 0.4 (10)	$p < 0.001$
ADH[b]	150	15.5 ± 1.0 (10)	$p < 0.005$

[a]Incubating medium (mM): NaCl 8.43, KCl 3.5, NaHCO$_3$ 1.57, CaCl$_2$ 1.8, hepes 10, glucose 5, sucrose 260, T=37°C, pH=7.4.
[b]50 mU/ml.

bladder, but is larger than the ADH response observed in the distal tubule of the rat (Constanzo and Windhager, 1980). However, the possibility exists that some MDCK cells possess a much more dramatic response to ADH than observed in our experiments. Some MDCK cell cultures seem to be composed of different subpopulations of cells (see Valentich, 1981). Only some of these subpopulations may thus respond to the hormone. Cloning procedures should prove important in clarifying this point.

REFERENCES

Abaza, N. A., Leighton, J., and Schultz, S. G., 1974, Effects of ouabain on the function and structure of a cell line (MDCK) derived from canine kidney, *In Vitro* **10**:172–183.

Carrasco, L., 1978, Membrane leakiness after viral infection and a new approach to the development of antiviral agents, *Nature* **272**:694–699.

Cereijido, M., Herrera, F. C., Flanigan, W. J., and Curran, P. F., 1964, The influence of Na concentration on Na transport across frog skin, *J. Gen. Physiol.* **47**:879–893.

Cereijido, M., Robbins, E. S., Dolan, W. J., Rotunno, C. A., and Sabatini, D. D., 1978a, Polarized monolayers formed by epithelial cells on a permeable and translucent support, *J. Cell Biol.* **77**:853–880.

Cereijido, M., Rotunno, C. A., Robbins, E. S., and Sabatini, D. D., 1978b, Polarized epithelial membranes produced in vitro, in: *Membrane Transport Processes*, Volume 1 (J. F. Hoffman, ed.), Raven Press, New York, pp. 433–461.

Cereijido, M., Ehrenfeld, J., Meza, I., and Martínez–Palomo, A., 1980, Structural and functional membrane polarity in cultured monolayers of MDCK cells, *J. Membr. Biol.* **52**:147–159.

Cereijido, M., Ehrenfeld, J., Fernández–Castelo, S., and Meza, I., 1981, Fluxes, junctions and blisters in cultured monolayers of epithelioid cells (MDCK), *Ann. N.Y. Acad. Sci.* **372**:422–441.

Constanzo, L. F., and Windhager, E. E., 1980, Effects of PTH, ADH, and cyclic AMP on distance tubular Ca and Na reabsorption, *Am. J. Physiol.* **239**:F478–F485.

Cunningham, D. D., and Pardee, A. B., 1969, Transport changes rapidly initiated by serum addition to "contact-inhibited" 3T3 cells, *Proc. Natl. Acad. Sci.* **64**:1049–1056.

Field, H., 1978, Some speculations on the coupling between sodium and chloride transport processes in mammalian teleost intestine, in: *Membrane Transport Processes*, Volume 1 (J. F. Hoffman, ed.), Raven Press, New York, pp. 277–292.

Frizzel, R. A., Field, M., and Schultz, S. G., 1979, Sodium coupled chloride transport by epithelial tissues, *Am. J. Physiol.* **236**:F1–F8.

Hassid, A., 1981, Transport-active renal tubular epithelial cells (MDCK and LLC-PK$_1$) in culture. Prostaglandin biosynthesis and its regulation by peptidehormones and ionophore, *Prostaglandins* **21**:985–1001.

Hatanaka, M., Huebner, R. J., and Gilden, R. V., 1969, Alterations in the characteristics of sugar uptake by mouse cells infected by murine sarcoma viruses, *J. Natl. Cancer Inst.* **43**:1091–1096.

Herzlinger, D. A., Easton, T. G., and Ojakian, G. K., 1982, The MDCK epithelial cell line expresses a cell surface antigen of the kidney distal tubule, *J. Cell Biol.* **93**:269–277.

Hodgkin, A., and Keynes, R., 1955, The potassium permeability of a giant nerve fibre, *J. Physiol.* **128**:71–88.

Holt, W. F., and Lechene, C., 1981, ADH-PGE$_2$ interactions in cortical collecting tubule. I. Depression of sodium transport, *Am. J. Physiol.* **241**:F452–F460.

Imprain, C. C., Foster, K. A., Micklem, K. J., and Pasternak, C. A., 1980, Nature of virally mediated changes in membrane permeability to small molecules, *Biochem. J.* **187**:847–860.

Isselbacher, K. J., 1972, Increased uptake of aminoacids and 2-deoxy-D-glucose by virus. Transformed cells in culture, *Proc. Natl. Acad. Sci. USA* **69**:585–589.

Kalkar, H. M., Ullrey, D., Kijomoto, S., and Hakomori, S., 1973, Carbohydrate catabolism and the enhancement of uptake of galactose in hamster cells transformed by polyoma virus, *Proc. Natl. Acad. Sci. USA* **70**: 839–843.

Klemperer, H. G., 1960, An effect of phoritzin on influenza virus elution and on neuroaminidase activity, *Virology* **12**:495–498.

Lamb, J. F., Odgen, P., and Simmons, N. L., 1981, Autoradiographic localisation of [^{3}H] ouabain bound to cultured epithelial cell monolayers of MDCK cells, *Biochim. Biophys. Acta* **644**:333–340.

Leighton, J., Brada, Z., Estes, L. W., and Justh, G., 1969, Secretory activity and oncogenicity of a cell line (MDCK) derived from canine kidney, *Science* **163**:472–473.

Lever, J. E., 1979a, Inducers of mammalian cell differentiation stimulate dome formation in a differentiated kidney epithelial cell line (MDCK), *Proc. Natl. Acad. Sci. USA* **76**:1323–1327.

Lever, J. E., 1979b, Cyclic AMP and inducers of mammalian cell differentiation stimulate dome formation in mammary and renal epithelial cell cultures, in: *Hormones and Cell Culture,* Cold Spring Harbor Conferences on Cell Proliferation, Volume 6 (G. Sato and R. Ross, eds.), Cold Spring Harbor Press, New York, pp. 727–738.

Lever, J. E., 1979c, Regulation of dome formation in differentiated epithelial cell cultures, *J. Supram. Struct.* **12**:259–272.

Lever, J. E., 1981, Regulation of dome formation in kidney epithelial cell cultures, *Ann. N.Y. Acad. Sci.* **372**:371–383.

Madin, S. H., and Darby, N. B., 1958, CCL-34, as catalogued in: *American Type Culture Collection of Strains,* 1975, Volume 2 (H. O. Hatt, ed.), Library of Congress, Rockville, Maryland, p. 30.

McRoberts, J. A., Erlinger, S., Rindler, M. J., and Saier, Jr., M. H., 1982, Furosemide-sensitive salt transport in the Madin–Darby canine kidney cell line, *J. Biol. Chem.* **257**:2260–2266.

Misfeldt, D. S., Hamamoto, S. T., and Pitelka, D. R., 1976, Transepithelial transport in cell culture, *Proc. Natl. Acad. Sci. USA* **73**:1212–1216.

Misfeldt, D. S., Tunner, C., and Franbach, D., 1984, The biophysics of domes formed by the renal cell line MDCK, *Fed. Proc.* **43**:2217–2220.

Negreanu, Y., Reinhertz, Z., and Kohn, A., 1974, Effects of adsorption of u.d.-inactivated parainfluenza (Sendai) virus on the incorporation of aminoacids in animal host cells, *J. Gen. Virol.* **22**:265–270.

Okada, Y., Koseki, I., Kim, J., Maeda, Y., Hashimoto, T., Kanno, Y., and Matsui, Y., 1975, Modification of cell membranes with viral envelopes during fusion of cells with HDJ (Sendai virus). I. Interaction between cell membranes and virus in the early stage, *Exp. Cell Res.* **93**:368–378.

Pasternak, C. A., and Micklem, K. J., 1973, Permeability changes during cell fusion, *J. Membr. Biol.* **14**:293–303.

Rabito, C. A., Tchao, R., Valentich, J., and Leighton, J., 1978, Distribution and characteristics of the occluding junctions in a monolayer of a cell line (MDCK) derived from canine kidney, *J. Membr. Biol.* **43**:351–365.

Rabito, C. A., Tchao, R., Valentich, J., and Leighton, J., 1980, Effect of cell-substratum interaction on hemicyst formation by MDCK cells, *In Vitro* **16**:461–468.

Rindler, M. J., and Saier, M. H., Jr., 1981, Evidence for Na$^+$, H$^+$ antiport in cultured dog kidney cells (MDCK), *J. Biol. Chem.* **256**:10820–10825.

Rindler, M. J., Chuman, L. M., and Saier, M. H., Jr., 1977, Hormone responsiveness of an established but differentiated kidney epithelial cell line (MDCK), *Fed. Proc.* **36**:911.

Rindler, M. J., Chuman, L. M., Shaffer, L., and Saier, M. H., Jr., 1979a, Retention of differentiated properties in an established dog kidney epithelial cell line (MDCK), *J. Cell Biol.* **81:**635–648.

Rindler, M. J., Taub, M., and Saier, M. H., Jr., 1979b, Uptake of ^{22}Na$^+$ by cultured dog kidney cells (MDCK), *J. Biol. Chem.* **254:**11431–11439.

Rindler, M. J., McRoberts, J. A., and Saier, M. H., Jr., 1982, (Na$^+$–K$^+$) cotransport in the Madin–Darby canine kidney cell line. Kinetic characterization of the interaction between Na$^+$ and K$^+$, *J. Biol. Chem.* **257:**2254–2259.

Rodríquez–Boulán, E., and Pendergast, M., 1980, Polarized distribution of viral envelope proteins in the plasma membrane of infected epithelial cells, *Cell* **20:**45–54.

Rodríguez–Boulán, E., and Sabatini, D. D., 1978, Asymmetric budding of viruses in epithelial monolayers: a model system for study of epithelial polarity, *Proc. Natl. Acad. Sci. USA* **75:**5071–5075.

Rozengurt, E., and Heppel, L. A., 1975, Serum rapidly stimulates ouabain-sensitive ^{86}Rb$^+$ influx in quiescent 3T3 cells, *Proc. Natl. Acad. Sci. USA* **72:**4492–4495.

Simmons, N. L., 1981a, Ion transport in ''tight'' epithelial monolayers of MDCK cells, *J. Membr. Biol.* **59:**105–114.

Simmons, N. L., 1981b, Stimulation of Cl$^-$ secretion by exogenous ATP in cultured MDCK epithelial monolayers, *Biochim. Biophys. Acta* **646:**231–242.

Simmons, N. L., Brown, C. D. A., and Rugg, E. L., 1984, The action of adrenaline upon MDCK cells, *Fed. Proc.,* in press.

Smith, J. B., and Rozengurt, E., 1978, Lithium transport by fibroblastic mouse cells: characterization and stimulation by serum and growth factors in quiescent cultures, *J. Cell. Physiol.* **97:**441–450.

Taub, M., and Saier, M. H., Jr., 1979, Regulation of ^{22}Na$^+$ transport by calcium in an established kidney epithelial cell line, *J. Biol. Chem.* **254:**11440–11444.

Taub, M., Chuman, L., Saier, M. H., Jr., and Sato, G. H., 1979, The growth of kidney epithelial cell line (MDCK) in hormone supplemented, serum-free medium, *Proc. Natl. Acad. Sci. USA* **76:**3338–3342.

Ussing, H. H., 1954, Active transport of inorganic ions, *Symp. Soc. Exp. Biol.* **8:**407.

Valentich, J. D., 1981, Morphological similarities between the dog kidney cell line MDCK and the mammalian cortical collecting tubule, *Ann. N.Y. Acad. Sci.* **372:**384–405.

Valentich, J. D., Tchao, R., and Leighton, J., 1979, Hemicyst formation stimulated by cyclic AMP in dog kidney cell line MDCK, *J. Cell. Physiol.* **110:**291–304.

Villereal, M. L., 1981, Sodium fluxes in human fibroblasts: Kinetics of serum-dependent and serum independent pathways, *J. Cell. Physiol.* **108:**251–259.

4

Application of the Microbiological Approach to the Study of Passive Monovalent Salt Transport in a Kidney Epithelial Cell Line, MDCK

MILTON H. SAIER, JR.

1. PERSPECTIVES IN KIDNEY PHYSIOLOGY

The mammalian kidney is a complex organ consisting of several tissues and a large number of distinct cell types. A multiplicity of epithelial cell types line fluid-filled tubules, differing morphologically and functionally depending on the segment of the tubule in which they arise. They are connected by junctional complexes which make for tissue integrity and allow intercellular communication (Fig. 1). They are polar, possessing plasma membranes of differing protein compositions on the mucosal and serosal sides of the cell. Asymmetry is essential to epithelial cell function since different transport systems are associated with the two membranes. Underlying the epithelial cell layer is a basement membrane, a complex network of proteins and carbohydrate-rich macromolecules, which confer a relatively static shape to the tubule. Underlying the basement membrane are fibroblasts which together with the epithelial cells participate in basement membrane biogenesis, synthesize collagen, and comprise much of the tissue bulk. Muscle, nerve, and endothelial cells participate in organ mobility, communication, and nutrition, respectively. The complexity of the kidney is awe inspiring. An understanding of its function will require the concerted efforts of

MILTON H. SAIER, JR.   •   Department of Biology, University of California, San Diego, La Jolla, California 92093.

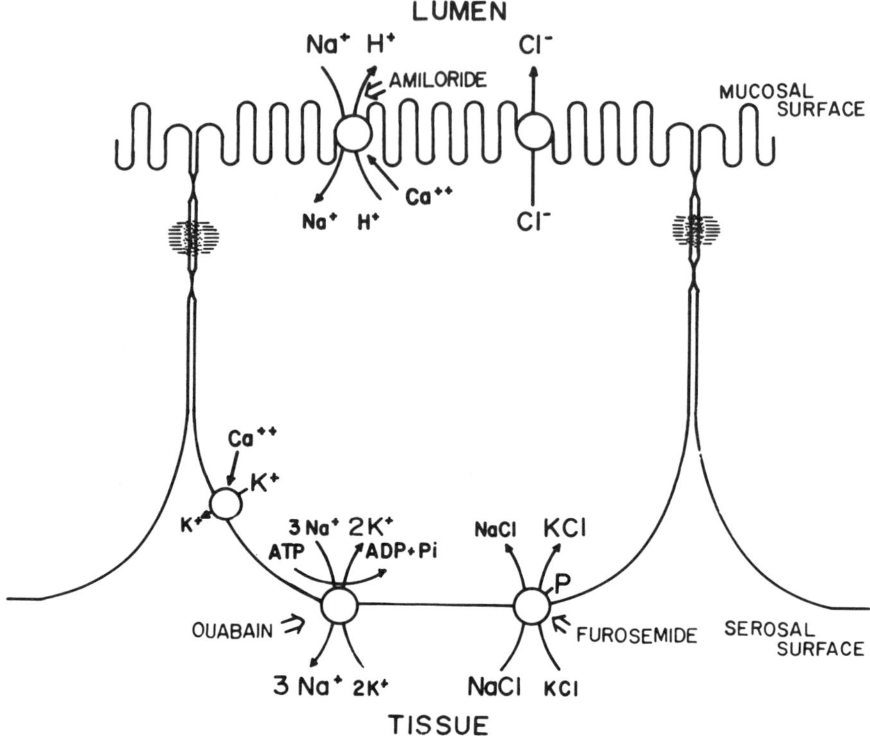

Figure 1. Schematic depiction of elements that comprise a kidney tubule. Shown are epithelial elements, the five major monovalent salt transport systems, and their asymmetrical cellular locations as they are believed to exist in the MDCK cells. The location of the Ca^{2+}-activated K^+ channel (Saier and Boyden, 1984) is not known.

numerous investigators applying a variety of experimental approaches to the same problem.

2. DEFINITION OF EXPERIMENTAL CELL SYSTEMS FOR EXAMINATION OF KIDNEY EPITHELIAL FUNCTION

With the ultimate goal of understanding kidney function at the molecular level, the kidney physiologist is faced with four fundamental biological questions (Table I). First, what are the mechanisms regulating growth of each of the cell types in the kidney during embryological development and subsequently in the young and adult organisms? The chemical agents and physical forces acting on the cell to stimulate or restrict its growth must be identified, the mechanisms of

Table I. Major Questions, Problems, and Experimental Approaches to the Study of Kidney Structure and Function

A. Fundamental problems in kidney physiology
   1. Growth regulation
   2. Tubule biogenesis
   3. Cell polarity
   4. Differentiated functions
B. Major experimental roadblocks facing the kidney physiologist
   1. Small tubular size
   2. Cell heterogeneity
   3. Cell mortality
   4. Poor cell viability
C. Experimental approaches to the study of kidney physiology
   1. Physiological manipulation
   2. Genetic dissection
   3. Biochemical analysis
   4. Biophysical observation

their recognition must be determined, and the nature of the intracellular relay systems conveying "knowledge" of their presence to the target organelles and enzyme systems must be elucidated. Second, what cellular characteristics are responsible for tubule biogenesis? The molecular basis for all cell–cell and cell–substrate interactions must be defined throughout development in order to understand tissue morphogenesis, and the recognition processes that are retained in the adult tissue must be identified. Third, what developmental events give rise to cell morphogenesis? Epithelial cells are universally asymmetric with distinct luminal and basolateral plasma membrane compositions. How does this polarity arise, and what forces and cellular characteristics are responsible for its maintenance? Finally, each cell within a tissue exhibits differentiated characteristics that confer upon the tissue as a whole its functionality. Mechanisms of salt and fluid transport and their regulation are of primary importance to the kidney physiologist.

Although the fundamental questions facing the kidney physiologist are fairly well defined, progress in answering these questions has been slow. There are several technical problems which are in part responsible for this state of affairs (Table I). First, mammalian kidney tubules are small, rendering physiological manipulations difficult. Second, cell heterogeneity makes assignment of specific differentiated functions to a particular cell type almost impossible. Third, cellular mortality prevents effective use of genetic techniques for the dissection of kidney functions. And finally, the poor viability of some cell types, probably attributable to incomplete knowledge of their nutritional needs and growth regulatory characteristics, has hampered tissue culture approaches.

Any biological system that is to provide a maximal yield of information from minimal effort must exhibit four desirable traits (Table I): First, it must be readily amenable to physiological manipulation. For a complex system such as a kidney epithelium, the capacity for cycling cells and their progeny through *in vitro* and *in vivo* environments would provide maximal potential for information retrieval following experimental manipulation. Further, the growth environment should be well defined and controllable. Second, genetic dissection must be feasible. The isolation and characterization of mutants provide a powerful tool facilitating the analysis of structure–function relationships, and techniques for the maintenance of permanent stocks of genetically altered cell lines must therefore be available. In order to allow genetic manipulation the ideal experimental system exhibits the immortality characteristics of a microorganism, unless, of course, mortality itself is the subject of investigation. Capacity for *in vitro* and *in vivo* cycling (mentioned previously) should allow assignment of the relevance of a genetically modified cellular trait to intact organ function. Third, detailed biochemical analyses must be possible, necessitating the preparation of relatively large amounts of tissue, pure cell cultures, cell structural components, and catalytic elements. Finally, the selected experimental system must be readily subject to biophysical observation. Techniques such as electron, light, and fluorescent microscopy must be feasible in order to establish correct structure–function relationships. A concerted approach employing several or all of these techniques far surpasses in potential any approach that employs only one. Totality is greater than the sum of its parts.

3. THE MICROBIOLOGICAL APPROACH APPLIED TO KIDNEY FUNCTION

Over the past 8 yr, our laboratory has taken the microbiological approach to the study of kidney epithelial cell function for the reasons discussed in Section 2. This approach has been highly profitable as the yield of information has been considerable (Table II). We have posed all four of the major questions facing the kidney physiologist using an established kidney epithelial cell line of dog origin, the MDCK line, and have made substantial progress toward answering these questions (Table II). First, we have carried out extensive studies to define the growth-regulatory characteristics of the MDCK cells. Serum-free growth media for this and another kidney epithelial line of presumed proximal tubular origin have been determined (Taub *et al.*, 1979; Taub and Saier, 1978) and shown to be applicable to primary cultures of kidney epithelial cells (Taub *et al.*, 1981). Transformants, altered in their *in vivo* and *in vitro* growth characteristics, have been isolated and studied (Saier, 1981; Taub *et al.*, 1981; Saier *et al.*, 1982; Boerner and Saier, 1982a,b), and possible mechanisms of second-messenger

*Table II. Application of the Microbiological Approach to Kidney Physiology
Employing MDCK Cells*

1. Growth regulation
   a. Hormonally defined medium developed (Taub *et al.*, 1979; Chuman *et al.*, 1982; Taub and Saier, 1978)
   b. Genetically altered transformants isolated *in vivo* and *in vitro* (Saier, 1981; Taub *et al.*, 1981; Saier *et al.*, 1982; Boerner and Saier, 1982a,b)
   c. *In vivo–in vitro* cycling possible (Stiles *et al.*, 1976a,b; Stiles *et al.*, 1977)
   d. Essential hormonal agents and probable second messengers defined (Rindler *et al.*, 1979a)
2. Tubule biogenesis
   a. Tubulelike structures form spontaneously *in vivo* (Stiles *et al.*, 1976a; Rindler *et al.*, 1979a)
   b. *In vitro* tubular systems deemed feasible
3. Cell polarity
   a. Structural asymmetry established by freeze–fracture (U *et al.*, 1979, 1980)
   b. Functional asymmetry established in tissue culture (Rindler *et al.*, 1979a) and *in vivo* (Stiles *et al.*, 1976a)
   c. Agents involved in polarity maintenance investigated (U *et al.*, 1980)
4. Differentiated functions
   a. Na^+/H^+ antiporter defined (Rindler *et al.*, 1979b; Taub and Saier, 1979; Rindler and Saier, 1981; Taub, 1978; Taub and Saier, 1981; Erlinger and Saier, 1982)
   b. NaCl/KCl symporter characterized (Erlinger and Saier, 1982; Rindler *et al.*, 1982; McRoberts *et al.*, 1982; McRoberts and Saier, 1982)
   c. Neutral amino acid transport systems characterized (Boerner and Saier, 1982a,b)
   d. Hormone receptors defined (Taub *et al.*, 1979; Rindler *et al.*, 1979a)

control of growth have been proposed (for reviews see Saier *et al.*, 1982; Boerner and Saier, 1982a; and McRoberts *et al.*, 1981).

Second, the MDCK cells have been shown to spontaneously form fluid-filled sacs *in vivo* under appropriate conditions in the athymic nude mouse (Stiles *et al.*, 1976a,b; Stiles *et al.*, 1977). Structural features of these sacs resemble tubules found in normal kidney tissue. Thus, adjacent epithelial cells were joined by junctional complexes consisting of both tight and gap junctions. A membrane studded with microvilli faced the lumen, but an apparently normal basement membrane lined the basal region. The fact that tubulelike structures form spontaneously *in vivo* leads to the possibility that analogous structures will form *in vitro* provided that proper conditions (possibly including other cell types) are found. Such a system would provide the first *in vitro* system for studying the biogenesis and maintenance of kidney tubules.

Third, the technique of freeze–fracture has been used to confirm structural and functional studies of others (Stiles *et al.*, 1976a; Rindler *et al.*, 1979a; McRoberts *et al.*, 1981) demonstrating the asymmetry of an MDCK cell monolayer after growth in the tissue culture environment (U *et al.*, 1979, 1980). Work has been initiated to determine the cellular structural elements involved in the formation and maintenance of polarity (U. *et al.*, 1980).

Finally, a variety of kidney-specific proteins have been demonstrated in the MDCK cells. These proteins include membrane-associated ectoenzymes (Rindler *et al.*, 1979a), several hormone systems (Rindler *et al.*, 197a), amino acid transport systems (Boerner and Saier, 1982a&b), and the salt transport systems that are responsible for salt and fluid absorption in the kidney (Rindler *et al.*, 1979b; Taub and Saier, 1979; Rindler and Saier, 1981; Taub, 1978; Taub and Saier, 1981; Erlinger and Saier, 1982; Rindler *et al.*, 1982; McRoberts *et al.*, 1982; McRoberts and Saier, 1982). Because we have concentrated much of our effort on the elucidation of salt transport mechanisms, these systems will be discussed in considerable detail in Sections 4 and 5.

4. THE AMILORIDE-SENSITIVE Na^+/H^+ ANTIPORTER

Leighton *et al.* first demonstrated that confluent MDCK cells form hemispherical blisters or domes (see McRoberts *et al.*, 1981, for a comprehensive review). This behavior was attributed to the vectorial flux of salt and fluid from the mucosal (top) surface of the monolayer to the serosal (bottom) surface. The

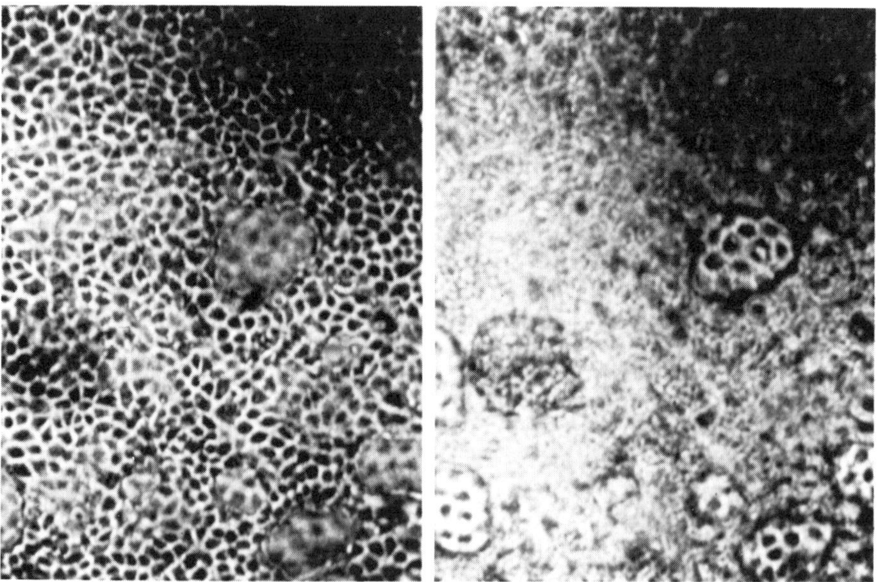

Figure 2. Hemispherical blister formation by MDCK cells grown in tissue culture to confluency. The medium used was DME/F12 supplemented with 10% fetal bovine serum. The hemicysts were photographed in two different focal planes (photography by Michael J. Rindler).

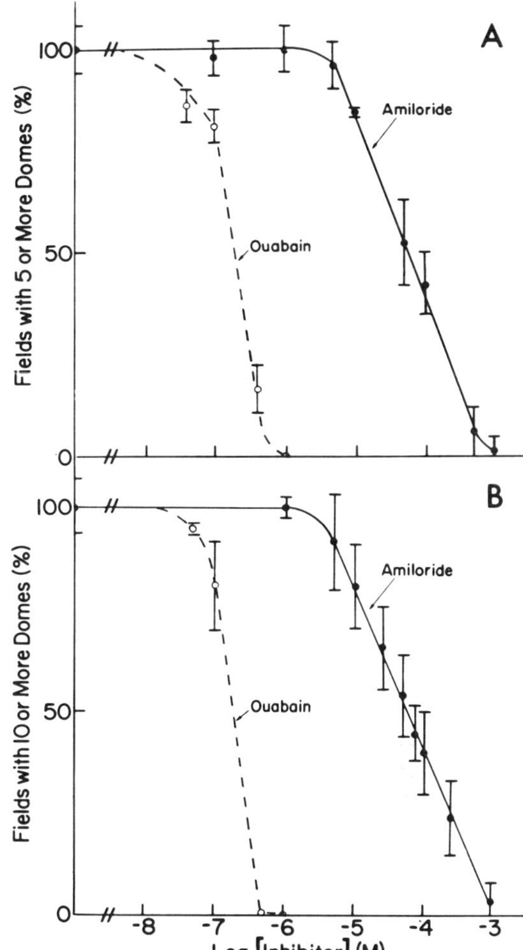

Figure 3. Inhibition of hemicyst formation by inhibitors of salt transport in MDCK cells. Ouabain, an inhibitor of the Na^+/K^+ ATPase, and amiloride, an inhibitor of the Na^+/H^+ antiporter, blocked dome formation at the concentrations that inhibit the respective transport systems. Furosemide, an inhibitor of the NaCl/KCl symport system, did not block dome formation. Growth conditions were as described in Fig. 2. (From M. J. Rindler *et al.*, 1979b, with permission.)

hydrostatic pressure that built up between monolayer and plastic dislodged groups of cells giving rise to the blisters shown in Fig. 2.

Inhibitor studies (Fig. 3) revealed that ouabain and amiloride specifically blocked blister formation, whereas loop diuretics such as furosemide were without effect or even stimulated blister formation. These observations suggested that the two transport systems responsible for vectorial Na^+ transport (mucosal to serosal) were the amiloride-sensitive Na^+/H^+ antiporter (mucosal surface) and the Na^+/K^+ ATPase (serosal surface).

The latter system had been subjected to detailed analyses in numerous laboratories and had been shown to exhibit a serosal location in MDCK cells

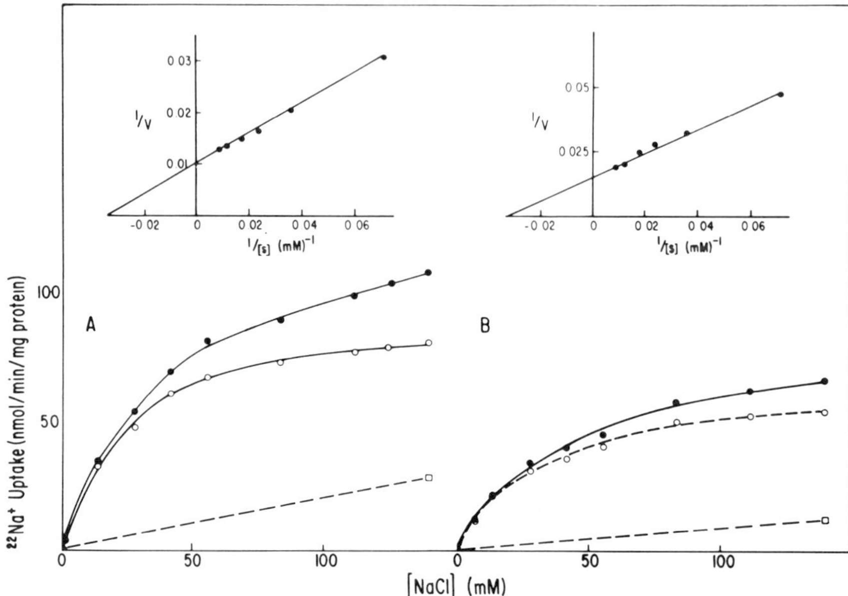

Figure 4. Na$^+$ uptake via the Na$^+$/H$^+$ antiporter as a function of sodium concentration. Cells were preincubated under either (A) steady-state or (B) sodium-depleted conditions. Assays were conducted in buffer containing ^{22}Na$^+$ (0.5–1uCi/ml) and various concentrations of NaCl (with sucrose maintaining constant osmolarity) for 1 min at 23°C. It proved necessary to subtract a linear component ([]- - -[]) from the data in order to achieve a good fit for the Lineweaver–Burk treatment of the data (insets). All values are accurate to within 10% standard error of the mean. (From Rindler *et al.*, 1979b, with permission.)

(McRoberts *et al.*, 1981). Thus, we initially concentrated our attention on the amiloride-sensitive system.

Extensive kinetic studies revealed that under normal physiological conditions the system catalyzed two isotope exchange processes: Na$^+$/Na$^+$ exchange (Fig. 4A) and Na$^+$/H$^+$ exchange (Fig. 4B). Li$^+$ could substitute for Na$^+$, and in fact, Li$^+$ bound to the antiporter with higher apparent affinity than did Na$^+$ (Rindler *et al.*, 1979b). Abnormalities in this system may be related to the manic–depressive condition in humans.

The maximal velocity of Na$^+$/Na$^+$ exchange was about twice that of Na$^+$/H$^+$ exchange under the conditions employed in Fig. 4, and it was therefore concluded that the system exhibited the phenomenon of accelerative exchange transport (Saier, 1979), a process that has been shown to be characteristic of a variety of chemiosmotically coupled transport systems. It has been suggested that accelerative transport is a diagnostic feature of carrier-mediated transport mechanisms (Saier, 1979). These processes are to be contrasted with channel-type transport mechanisms, which may be incapable of catalyzing accelerative exchange.

Further studies revealed that the amiloride-sensitive transport activity was subject to nonspecific inhibition by extracellular Ca^{2+}, but fairly specific activation by intracellular Ca^{2+} (Fig. 5 and Taub and Saier, 1979). It is not yet known whether calmodulin mediates the Ca^{2+} activation of Na^+/H^+ exchange.

In preliminary studies it was found that amiloride was toxic to MDCK cells. Mutant strains could be isolated after mutagenesis which were partially resistant to amiloride (Taub, 1978). Studies with [^{14}C]amiloride revealed that these variants accumulated less amiloride than did the parental strain (Fig. 6). Further studies revealed that the activity of the Na^+/H^+ exchange protein was reduced in the mutants (Taub and Saier, 1981). It is possible that amiloride, which is cationic, is transported via the exchanger. Other transport activities were also altered by the mutations which conferred amiloride resistance (Taub and Saier, 1981). The primary defect in these mutants has yet to be identified.

In order to develop a biochemical assay for the amiloride-sensitive Na^+ transport protein, reconstitution of the system in artificial liposomes was attempted (F. Grenier and M. H. Saier, Jr., unpublished results). As shown in Fig. 7A, a pH-dependent amiloride-sensitive $^{22}Na^+$ transport activity was demonstrable in whole cells. Similar activity could be demonstrated in plasma membrane vesicles prepared by nitrogen cavitation (Fig. 7B). Passage of these vesicles through a French pressure cell did not diminish the initial rate of $^{22}Na^+$ uptake (Fig. 7C) suggesting that the system was functionally symmetrical, transporting Na^+ with equal efficiency in both the inward and outward directions.

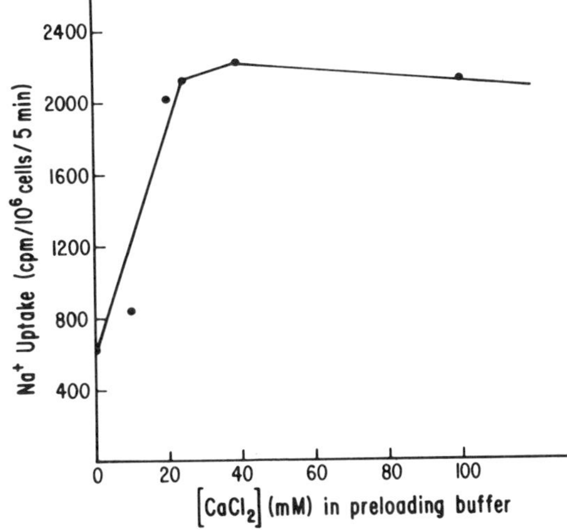

Figure 5. The effect of preloading MDCK cells with Ca^{2+} on the rate of $^{22}Na^+$ uptake. MDCK cells were incubated 20 min in 10mM Tris.HCl buffer, pH 7.3, containing 6% sucrose and $CaCl_2$ (0–10 mM). After the preincubation period the cells were washed three times with 10 mM Tris buffer, pH 7.3. $^{22}Na^+$ uptake is plotted as a function of the calcium concentration used for preloading the MDCK cells. (From Taub and Saier, 1979, with permission.)

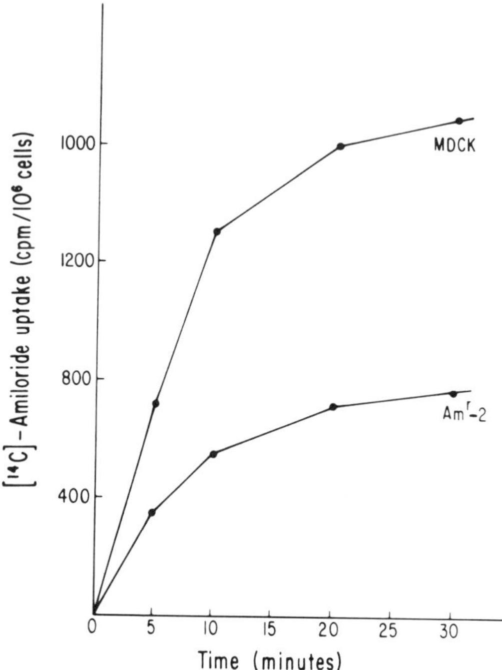

Figure 6. Time course of amiloride uptake in MDCK and mutant AM^r-2 cells. The uptake of 3.5×10^{-5} M [^{14}C]amiloride was studied over a 25-min period in Tris buffer. Uptake determinations were made in duplicate every 5 min during the uptake period. (From Taub, 1978, with permission.)

The transporter was solubilized with the neutral detergent, octyl glucoside, and reconstituted in phospholipid liposomes employing the octyl glucoside dilution procedure of Racker (Racker *et al.*, 1979). As shown in Fig. 7D, the pH dependency and amiloride sensitivity of this preparation were retained. It can therefore be concluded that a single protein, or a tightly associated protein complex that does not readily dissociate in the presence of detergent, is responsible for Na^+/H^+ exchange in kidney epithelial cells. Because other chemiosmotically driven cotransport systems are known to be attributable to a single polypeptide chain (Newman *et al.*, 1981), it seems reasonable that amiloride-sensitive Na^+/H^+ exchange is catalyzed by a single, intregral, transmembrane protein. Possibly, sensitivity to Ca^{2+} is conferred by a second protein which binds to the transporter and allosterically controls its activity. A schematic representation of the system is depicted in Fig. 8.

5. THE LOOP-DIURETIC-SENSITIVE NaCl/KCl SYMPORTER

During study of conditions influencing Na^+/H^+ exchange in the MDCK cells, a K^+-stimulated, Cl^--dependent component of the $^{22}Na^+$ flux was dem-

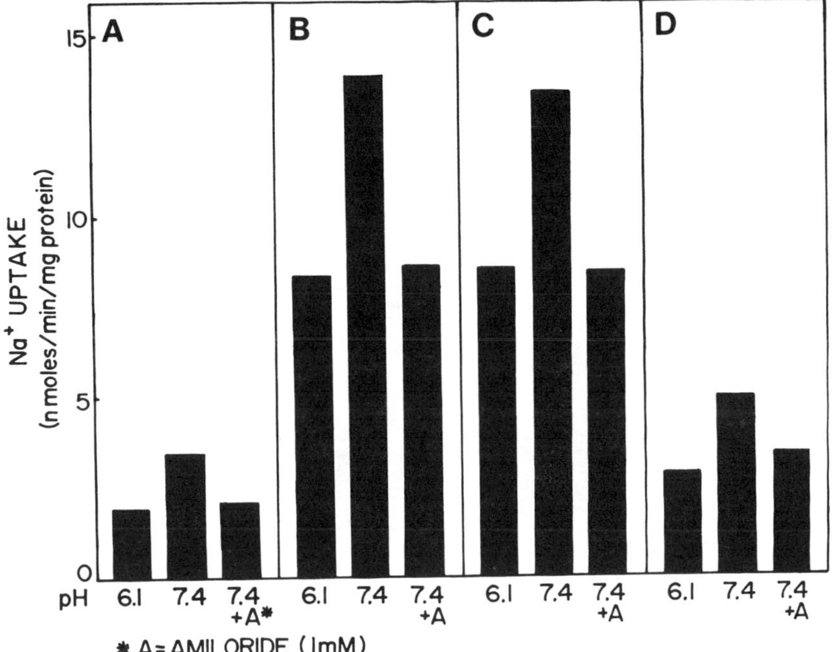

Figure 7. Uptake of Na⁺ via the Na⁺/H⁺ antiporter into (A) MDCK cells, (B) plasma membrane vesicles prepared from kidney tissue by nitrogen cavitation, (C) French pressed vesicles prepared from B, and (D) reconstituted liposomes containing octyl glucoside solubilized proteins from kidney tissue. (F. C. Grenier and M. H. Saier, Jr., unpublished results).

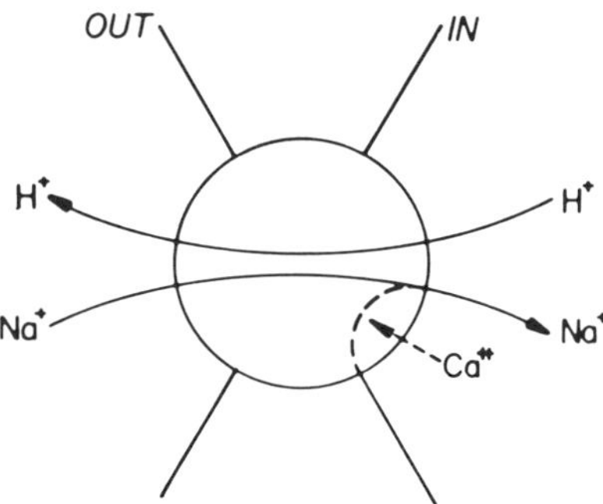

Figure 8. Schematic model of the amiloride-sensitive Na⁺/H⁺ antiporter present in the plasma membrane of MDCK cells.

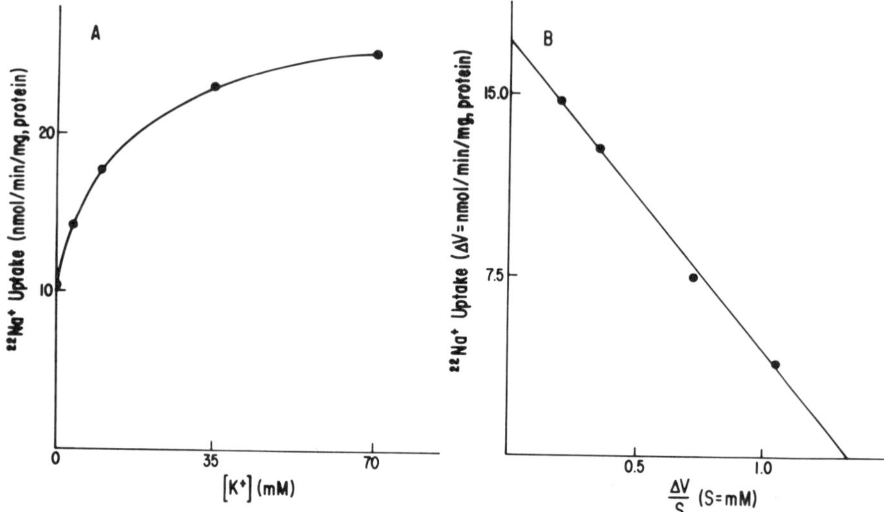

Figure 9. The effect of potassium on ^{22}Na$^+$ uptake. MDCK cells were preincubated under conditions that resulted in depletion of internal Na$^+$. The assay buffer contained 14 mM NaCl and various concentrations of KCl (with choline chloride added to maintain isoosmolarity), and uptake was conducted for 3 min at 23°C. (A) The rate of ^{22}Na$^+$ uptake as a function of potassium concentration. The increase in the ^{22}Na$^+$ uptake rate caused by addition of K$^+$ is plotted on the ordinate in (B), with V divided by the K$^+$ concentration plotted on the abscissa. The value of the ^{22}Na$^+$ uptake rate in the absence of K$^+$ was 21 nmol/mg protein per 3 min. (From Rindler *et al.*, 1979b, with permission.)

onstrated (Fig. 9). The activity was inhibitable by loop diuretics such as furosemide and bumetanide but was insensitive to amiloride inhibition. A separate transport system was implicated. The Na$^+$ uptake rate catalyzed by this new system exhibited a hyperbolic dependency on the K$^+$ concentration (Fig. 9A) but a sigmoidal dependency on the Cl$^-$ concentration (Fig. 10A). Whereas the K$^+$ dependency linearized when V was plotted versus V/[K$^+$] (Fig. 9B), the Cl$^-$ dependency linearized only if V was plotted versus V/[Cl$^-$]^2 (Fig. 10B). This result provided the first evidence that one K$^+$, one Na$^+$, and two Cl$^-$ ions were transported together by a tightly coupled, obligatorily concerted mechanism. This stoichiometry was confirmed by several independent techniques, and a complete kinetic description of the system was presented (McRoberts *et al.*, 1982, 1983). Recently, accelerative exchange transport (Saier, 1979) catalyzed by this system has been demonstrated (J. A. McRoberts, unpublished results) suggesting a carrier-type mechanism. The results indicate that the loop-diuretic-sensitive salt symport system in MDCK cells may be among the most complex transport systems known in nature. Its wide distribution in numerous animal

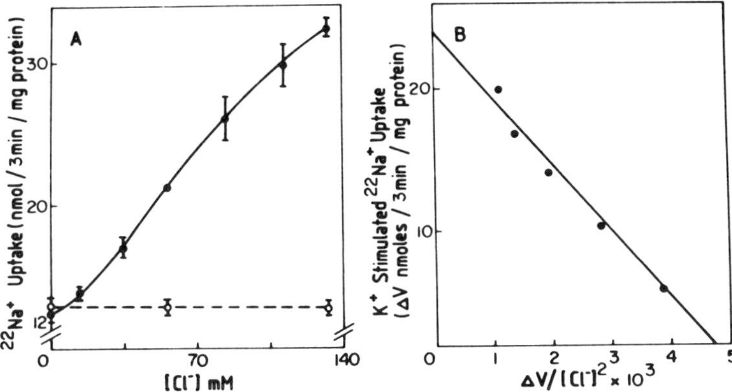

Figure 10. Dependence of K$^+$-stimulated ^{22}Na$^+$ uptake on Cl$^-$ concentration. (A) ^{22}Na$^+$ uptakes were performed in Tris·SO$_4$ buffer containing 7 mM Na gluconate in the presence (●) or absence (○) of 133 mM K$^+$ and varying concentrations of Cl$^-$. The Cl$^-$ concentration was varied by replacing KCl with K gluconate (○) or N-methylglucamine chloride with sucrose (○). Uptakes were carried out in triplicate and terminated after 3 min using a MgCl$_2$ wash procedure. (B) The difference between ^{22}Na$^+$ uptakes in the presence and absence of K$^+$, ΔV, is plotted as a function of $\Delta V/[Cl]^2$. The average value for uptake in the absence of K$^+$ at different Cl$^-$ concentrations was used to obtain the difference. (From McRoberts *et al.*, 1982, with permission.)

species and cell types indicate that it may be of critical importance to the maintenance of body salt and fluid (Saier and Boyden, 1984). It has been implicated in essential hypertension.

By the use of metabolic inhibitors, the furosemide-sensitive symport system was shown to be dependent on a source of metabolic energy such as ATP (Table III). Because the system appears to catalyze salt transport with comparable efficiency in both the inward and outward directions, it is suggested that ATP acts via a protein kinase-catalyzed reaction to activate the system. Such a postulate is substantiated by the observation that the system in avian erythrocytes is activated by cyclic AMP, presumably by a kinase-mediated mechanism (Saier and Boyden, 1984). A cyclic AMP dependency of the system in MDCK cells has not been demonstrable in spite of considerable efforts in this direction (S. Erlinger and J. A. McRoberts, unpublished experiments).

Genetic dissection of the loop-diuretic-sensitive symporter is feasible. Tissue culture cells cannot grow in media containing less than 0.25 mM K$^+$, but mutagenesis gives rise to variants that can grow in low-K$^+$ media. Growth characteristics of two such mutants, isolated from the MDCK cells, are shown in Fig. 11 (McRoberts *et al.*, 1983). These mutants can grow in low-K$^+$ media, but they sometimes exhibit low maximal growth rates as observed for strain LK–A3 in Fig. 11. Other mutants isolated grow optimally at rates comparable to the parental cell line as for mutant LK–Cl in Fig. 11. Both strains were shown to be

Table III.   The Effect of Metabolic Inhibitors on Loop-Diuretic-Sensitive NaCl/KCl Symport Activity in MDCK Cells[a]

Agent	Concentration	Residual transport activity (%)	Cellular ATP level (%)
None	—	100	100
Iodoacetate	1 mM	95	77
Oligomycin	0.2 μg/ml	107	95
Antimycin	5 μg/ml	108	73
Iodoacetate + oligomycin	1 mM 0.2 μg/ml	8	< 3
Iodoacetate + antimycin	1 mM 5 μg/ml	6	< 2

[a]Values reported were calculated from $^{86}Rb^+$ influx measurements although comparable values were obtained for $^{22}Na^+$ influx. The results tabulated were calculated from data obtained under uniform conditions as reported in Rindler *et al.*, 1982.

devoid of the loop-diuretic-sensitive NaCl/KCl symport activity. Evidently the system functions primarily as an efflux system in media containing low K^+. This fact apparently explains the ability of mutants lacking the system to grow in low-K^+ media (McRoberts *et al.*, 1983; Saier and Boyden, 1984).

The complexity of the system renders mechanistic interpretation of cur-

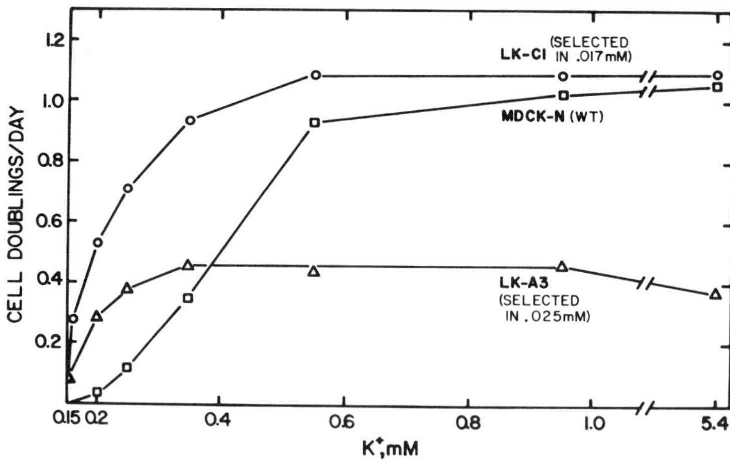

Figure 11. Growth of MDCK cells and two mutants selected on the basis of enhanced growth rates in medium containing low potassium (< 0.2 mM). Growth rate is plotted as a function of the K^+ concentration in the external medium. The two mutants shown are LK–C1 and LK–A3, both of which lack demonstrable NaCl/KCl symport activity. (James A. McRoberts and Milton H. Saier, Jr., unpublished results.)

rently known characteristics difficult. Clearly the system catalyzes unidirectional NaCl/KCl symport and probably accelerative exchange transport when all ions are present on both sides of the membrane. These results suggest a carrier-mediated process as shown in Fig. 12. The reversible process is activated by cytoplasmic ATP, most likely by a protein kinase-catalyzed reaction. Plasma membrane vesicles from kidney tissue or from MDCK cells, prepared as described in Fig. 7, are devoid of bumetanide-inhibitable salt symport activity under a variety of conditions, even though the amiloride-sensitive Na^+/H^+ antiporter (Rindler and Saier, 1981) and the A amino acid/Na^+ symporter (Boerner and Saier, 1982a,b) are active (F. Grenier, P. Boerner, and M. H. Saier, unpublished results). These observations may reflect the need for protein phosphorylation in order to demonstrate activity, or they may suggest that unrecognized factors or conditions are essential to the activity of the system. The complexity of the salt symporter renders it intriguing, and detailed analyses may reveal novel mechanistic features, particularly with respect to coupling and regulation of transport. Even the nature of the translocation process can only be guessed at at this time (Saier and Boyden, 1984). One can anticipate that without a multipronged approach to the mechanism of loop-diuretic-sensitive salt transport, there is little hope of understanding the multifaceted details of the process.

6. CONCLUSIONS

There is a need for a multipronged approach to the study of transport processes in animal cells. Only when information is available using the resources of the cell physiologist, the biochemist, the geneticist, and the biophysicist will our yield of information be maximal. The microbiological approach, applied to established cell lines in tissue culture, provides the multiplicity of approaches needed to obtain and focus the information about salt transport which will reveal the molecular details of translocation, coupling, and regulation.

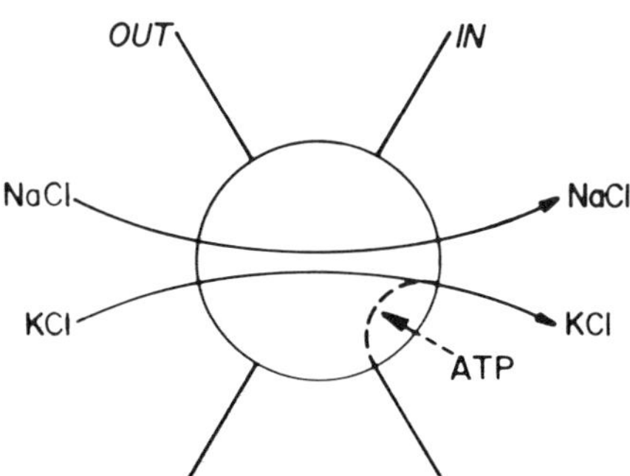

Figure 12. Schematic model for the loop-diuretic-sensitive NaCl/KCl symporter found in the plasma membrane of the MDCK cells.

ACKNOWLEDGMENTS.   It is a pleasure to acknowledge the contributions of the investigators who have carried out the studies summarized in this review. Together they have made a unique contribution toward developing and applying microbiological techniques to the study of animal cell physiology. These individuals and their areas of contribution are as follows: C. Stiles and W. Desmond, growth regulation *in vivo;* C. Stiles, L. Chuman, and C. Taylor, growth regulation *in vitro;* L. Shaffer, electron microscopy; M. Rindler, *in vitro* differentiated properties and hormonal responses; M. Rindler, M. Taub, and F. C. Grenier, Na^+/H^+ antiport; M. Rindler and J. McRoberts, NaCl/KCl symport; B. Ü and M. Ellisman, cell polarity; M. Taub and L. Chuman, growth in serum-free medium; B. Ü, M. Rindler, M. Taub, J. McRoberts, S. Erlinger, and P. Boerner, isolation and characterization of transformants; P. Boerner, amino acid transport.

REFERENCES

Boerner, P., and Saier, M. H., Jr., 1982a, Nutrient transport and growth regulation in kidney epithelial cells (MDCK) cultured in a defined medium, *Cold Spring Harbor Conferences on Cell Proliferation* **9:**555–565.

Boerner, P., and Saier, M. H., Jr., 1982b, Growth regulation and amino acid transport in epithelial cells: I. Influence of culture conditions and transformation on A, ASC, and L transport activities, *J. Cell Physiol.* **12:**240–246.

Chuman, L., Fine, L. G., Cohen, A. H., and Saier, M. H., Jr., 1982, Continuous growth of proximal tubular kidney epithelial cells in hormone-supplemented serum-free medium, *J. Cell Biol.* **94:**506–510.

Erlinger, S., and Saier, M. H., Jr., 1982, Decrease in protein content and cell volume of cultured dog kidney epithelial cells during growth. Importance for transport measurements, *In Vitro* **18:**196–202.

McRoberts, J. A., Taub, M., and Saier, M. H., Jr., 1981, The Madin–Darby canine kidney (MDCK) cell line, in: *Functionally Differentiated Cell Lines* (G. Sato, ed.), Alan R. Liss, New York, pp. 117–139.

McRoberts, J. A., Erlinger, S., Rindler, M. J., and Saier, M. H., Jr., 1982, Furosemide-sensitive salt transport in the Madin-Darby canine kidney cell line: Evidence for the cotransport of Na^+, K^+, and Cl^-, *J. Biol. Chem.* **257:** 2260–2266.

McRoberts, J. A., Tran, C. T., and Saier, M. H., Jr., 1983, Characterization of low potassium-resistant mutants of the Madin–Darby canine kidney cell line with defects in NaCl/KCl symport, *J. Biol. Chem.* **258:**12320–12326.

Newman, M. J., Foster, D. L., Wilson, T. H., and Kaback, H. R., 1981, Purification and reconstitution of functional lactose carrier from *Escherichia coli, J. Biol. Chem.* **256:**11804–11808.

Racker, E., Violand, B., O'Neal, S., Alfonzo, M., and Telford, J., 1979, Reconstitution, a way of biochemical research; some new approaches to membrane-bound enzymes, *Arch. Biochem. Biophys.* **198:**470–477.

Rindler, M. J., and Saier, M. H., Jr., 1981, Evidence for Na^+/H^+ antiport in cultured dog kidney cells (MDCK), *J. Biol. Chem.* **256:**10820–10825.

Rindler, M. J., Chuman, L. M., Shaeffer, L., and Saier, M. H., Jr., 1979a, Retention of differentiated properties in an established dog kidney epithelial cell line (MDCK), *J. Cell Biol.* **81:**635–648.

Rindler, M. J., Taub, M., and Saier, M. H., Jr., 1979b, Uptake of ^{22}Na$^+$ by cultured dog kidney cells (MDCK), *J. Biol. Chem.* **254: 254**:11431–11439.

Rindler, M. J., McRoberts, J. A., and Saier, M. H., Jr., 1982, (Na$^+$,K$^+$)-cotransport in the Madin–Darby canine kidney cell line: Kinetic characterization of the interaction between Na$^+$ and K$^+$, *J. Biol. Chem.* **257**:2254–2259.

Saier, M. H., Jr., 1979, Sugar transport mediated by the bacterial phosphoenolpyruvate-dependent phosphotransferase system, in: *Microbiology*, American Society for Microbiology, Washington, D.C., pp. 72–75.

Saier, M. H., Jr., 1981, Growth and differentiated properties of a kidney epithelial cell line (MDCK), *Am. J. Physiol.* **240**:C106–C109.

Saier, M. H., Jr., and Boyden, D., 1984, Mechanism, regulation and physiological significance of the loop diuretic-sensitive NaCl/KCl symport system in animal cells, *Molecular and Cellular Biochemistry* **59**:11–32.

Saier, M. H., Jr. Erlinger, S., and Boerner, P., 1982, Studies on growth regulation and the mechanism of transformation of the kidney epithelial cell line, MDCK: Importance of transport function to growth, in: *Membranes in Growth and Development* (J. F. Hoffman, ed.), in press.

Stiles, C. D., Desmond, W., Chuman, L. M., Sato, G., and Saier, M. H., Jr., 1976a, Growth control of heterologous tissue culture cells in the congenitally athymic nude mouse, *Cancer Research* **36**:1353–1360.

Stiles, C. D., Desmond, W., Chuman, L. M., Sato, G., and Saier, M. H., Jr., 1976b, Relationship of cell growth behavior *in vitro* to tumorigenicity in athymic nude mice, *Cancer Research* **36**:3300–3305.

Stiles, C. D., Roberts, P. E., Saier, M. H., Jr., and Sato, G., 1977, Growth regulation and suppression of metastasis in the congenitally athymic nude mouse, in: *Modern Trends in Human Leukemia II* (R. Neth, R. C. Gallo, K. Mannweiler, and W. C. Maloney, eds.), Lehmanns Verlag, Munchen, pp. 185–194.

Taub, M., 1978, Isolation of amiloride-resistant clones from dog kidney epithelial cells, *Somatic Cell Genetics* **4**:609–616.

Taub, M., and Saier, M. H., Jr., 1978, An established but differentiated kidney epithelial cell line (MDCK), in: *Methods in Enzymology*, Volume LVIII, Academic Press, New York, pp. 552–560.

Taub, M., and Saier, M. H., Jr., 1979, Regulation of ^{22}Na$^+$ transport by calcium in an established kidney epithelial cell line, *J. Biol. Chem.* **254**:11440–11444.

Taub, M., and Saier, M. H., Jr., 1981, Amiloride-resistant Madin–Darby canine kidney (MDCK) cells exhibit decreased cation transport, *J. Cell Physiol.* **106**:191–199.

Taub, M., Chuman, L., Saier, M. H., Jr., and Sato, G., 1979, Growth of Madin–Darby canine kidney epithelial cell (MDCK) line in hormone supplemented, serum-free medium, *Proc. Natl. Acad. Sci. USA* **76**:3338–3342.

Taub, M., U., B., Chuman, L., Rindler, M. J., Saier, M. H., Jr., and G. Sato, 1981, Alterations in growth requirements of kidney epithelial cells in defined medium associated with malignant transformation, *J. Supra. Struc. Cellular Biochem.* **15**:63–72.

U, H. S., Saier, M. H., Jr., and Ellisman, M. H., 1979, Tight junction formation is closely linked to the polar redistribution of intramembranous particles in aggregating MDCK epithelia, *Exp. Cell Res.* **122**:384–390.

U, H. S., Saier, M. H., Jr., and Ellisman, M. H., 1980, Tight junction formation in the establishment of intramembranous particle polarity in aggregating MDCK cells. Effect of drug treatment, *Exper. Cell Res.* **128**:223–235.

III

TRANSPORT OF NEUTRAL SOLUTES BY LLC–PK$_1$ CELLS

The LLC–PK$_1$ line exhibits in culture some of the differentiated transport properties of proximal tubule cells. The LLC–PK$_1$ cells exhibit transepithelial sugar typical of the proximal tubule. The Na$^+$-dependent sugar transport system, which is involved in transepithelial sugar transport, has been characterized by means of labeled flux studies. The mechanism and bioenergetics of sugar transport in LLC–PK$_1$ monolayers have also been examined in an Ussing flux chamber. Changes in transepithelial electrical resistance and potential have been measured in such a chamber. Transepithelial amino acid transport is also a characteristic of the proximal tubule *in vivo*. The LLC–PK$_1$ cells have been shown to possess several amino acid transport systems which are localized at the basolateral membrane. However, apical transport systems for neutral amino acids have not been observed. This may be a developmental problem. The development of polarity of the Na$^+$-dependent sugar transport system and other apical membrane proteins in LLC–PK$_1$ cells has been the topic of extensive study. Inducers of differentiation and phosphodiesterase inhibitors accelerate the appearance of these membrane proteins in LLC–PK$_1$ cells.

5

Sugar Transport in the Renal Epithelial Cell Culture

JAMES M. MULLIN and ARNOST KLEINZELLER

1. INTRODUCTION

1.1. Inherent Advantages

In 1975, MacKnight, Civan, and Leaf pointed out that any decisive conclusions on possible compartmentation of luminally and antiluminally derived Na^+ within the toad bladder epithelial cell were obfuscated by the different cell types encountered in this tissue. Whether the issue is compartmentation within a cell or the existence of multiple transport systems in a cell (e.g., the antiluminal sugar transporters of proximal tubule cells), the endemic problem lies in the assumptions that must be made when examining *cellular* properties in a *tissue* preparation. The kidney is particularly troublesome, for here one has not only epithelial and nonepithelial cell types, but a progression of morphologically and/or functionally different transporting epithelia along the nephron, and even within the proximal tubule itself (see, e.g., Woodhall *et al.*, 1978).

Examination of the various preparations in use for studying renal sugar transport [clearances, cortical slices, isolated (and/or teased) tubules, perfused tubules (*in vivo* and *in vitro*), and vesicles] reveals inherent problems. The

JAMES M. MULLIN • Department of Human Genetics, Yale University School of Medicine, New Haven, Connecticut 06510   ARNOST KLEINZELLER • Department of Physiology, School of Medicine, University of Pennsylvania, Philadelphia, Pennsylvania 19104.   Work supported in part by National Institutes of Health Grant 12619 and the Whitehall Foundation (both to A. K.) and National Science Foundation Postdoctoral Fellowship SPI-8166058 (to J. M. M.).

clearance technique and its more sophisticated version the multiple-indicator dilution method treat the whole kidney as one black box and hence cannot elucidate events at different portions of the nephron. Cortical slices or suspensions of isolated proximal tubules suffer from some cellular heterogeneity as seen by Woodhall *et al.* (1978), even when the tubules have been purified by fractional centrifugation. Moreover, aspects of the siddedness of transport processes are difficult to approach in these preparations without additional experimental handles. The perfused tubule (*in vivo* and *in vitro*), although still probably suffering from cellular heterogeneity, does offer distinct advantages to the studies of the siddedness of transport processes. However, the amount of biological material is rather limited for studies requiring the knowledge of intracellular events. Finally, vesicles of microvillus and basal–lateral membranes are only as homogeneous as the cell preparation from which they are derived.

In our view, the renal epithelial cell culture offers such unique advantages to the transport physiologist as (1) homogeneity of cell *type;* (2) the ability to grow as much cellular material as required; (3) the ability to manipulate the biological material generally and to study the siddedness of transport processses (by clamping monolayers of cells in Ussing-type chambers); (4) the option of returning to the identical biological material with which a study was performed by thawing cells maintained in liquid nitrogen; (5) the opportunity to examine transport parameters in both cycling and noncycling (i.e. subconfluent and confluent cultures) cells rather than working only with the growth-arrested cells of the kidney; and, finally, (6) taking advantage of a growing epithelial population, the opportunity to enact and select for mutants in given transport parameters. Limitations to these advantages will now be considered.

1.2. Inherent Disadvantages

Any model system is only as useful as the discipline exercised by the investigator in recognizing its innate limitations. The emphasis in sugar transport in epithelial cell cultures has until recently been weighted more toward the continuous cell line rather than the primary culture. It is thus noteworthy that the *cell lines* most widely used in epithelial transport, MDCK, LLC–PK$_1$, and MDBK, have an unknown origin with respect to a particular site along the nephron. All originated as explants which grew out from a cortical mince (Hull *et al.*, 1976; Madin and Darby, 1958). Moreover, it is interesting to note that in the primary cultures from which these continuous cell lines emerged, specific steps were not taken to (1) prevent fibroblast overgrowth of the culture, or (2) use a physical, chemical, or biological agent to enact a partial transformation which would give rise to a continuous cell line. [The MDBK cell line was actually first described as fibroblastic (Madin and Darby, 1958).] Rather a somewhat ill-defined "spontaneous event" seems to have provided for both occurrences. Such is the hazy early history of our most utilized epithelial cell lines.

The issue of "physiological relevance" thus lurks off to the side of this new field of study. Of basic concern here is the fact that genes expressed *in vivo* may not be expressed *in vitro,* and new phenotypes may actually emerge as a result of culture conditions. In the thesis project that first demonstrated the existence of an active sugar transport system in a cell line, the LLC–PK$_1$ cell line was chosen for study simply because of its normal karyotype (Hull *et al.,* 1976) unlike the MDCK and MDBK cell lines, which have lost several chromosomes (J. Mullin, 1980). It has recently come to our attention that the LLC–PK$_1$ cell line is considered by some to be aneuploid. However, karyotypes of a clone derived from this cell line at Yale University are diploid and contain the marker chromosomes peculiar to *Sus scrofa* (S. Katz and J. Mullin, unpublished results, see Fig. 1). It was reasoned that an *active* sugar transport system was not necessary for survival *of the cell,* and hence it would be wise to work with cells that had retained the full complement of at least chromosomes, if not genes. However, although this property of the renal proximal tubule has been retained in this cell line, it is noteworthy that the hormone response pattern to adenyl cyclase is not typical of the proximal tubule (Goldring *et al.,* 1978), nor is there any gluconeogenic activity in these cells (Mullin *et al.,* 1982a,b), although growth of the cells on floating collagen gels may change these results. In regard to both homogeneity and expression of differentiated properties in cell cultures, we would add that even if one's culture is homogeneous according to cell type, there will be differences in the level of differentiation from one cell to another (see Section 3.4, the discussion of alkaline phosphatase in LLC–PK$_1$ cells) in a confluent culture and differences in the position of cells in the cell cycle in subconfluent cultures (thus sometimes making synchronized cultures desirable).

The proper conclusion to this introduction is thus a caveat. We advance the LLC–PK$_1$ cell line as being useful as a model for a given function of the kidney. It is seemingly unwise, however, to advance a cell line as a model of a given organ, or conversely to ignore the merits of a cell line with an interesting marker property, because it lacks certain other properties of the desired organ.

2. AVENUES OF APPROACH

2.1. Primary Cultures

The outgrowth of epithelial cells from explants of the kidney, or segments of the kidney, has its own intrinsic advantages and drawbacks. In 1978, Foster and Willis demonstrated ouabain- and phlorizin-sensitive sugar transport in primary cultures of rabbit renal cortex. Horster and Wilson (1981) focused on

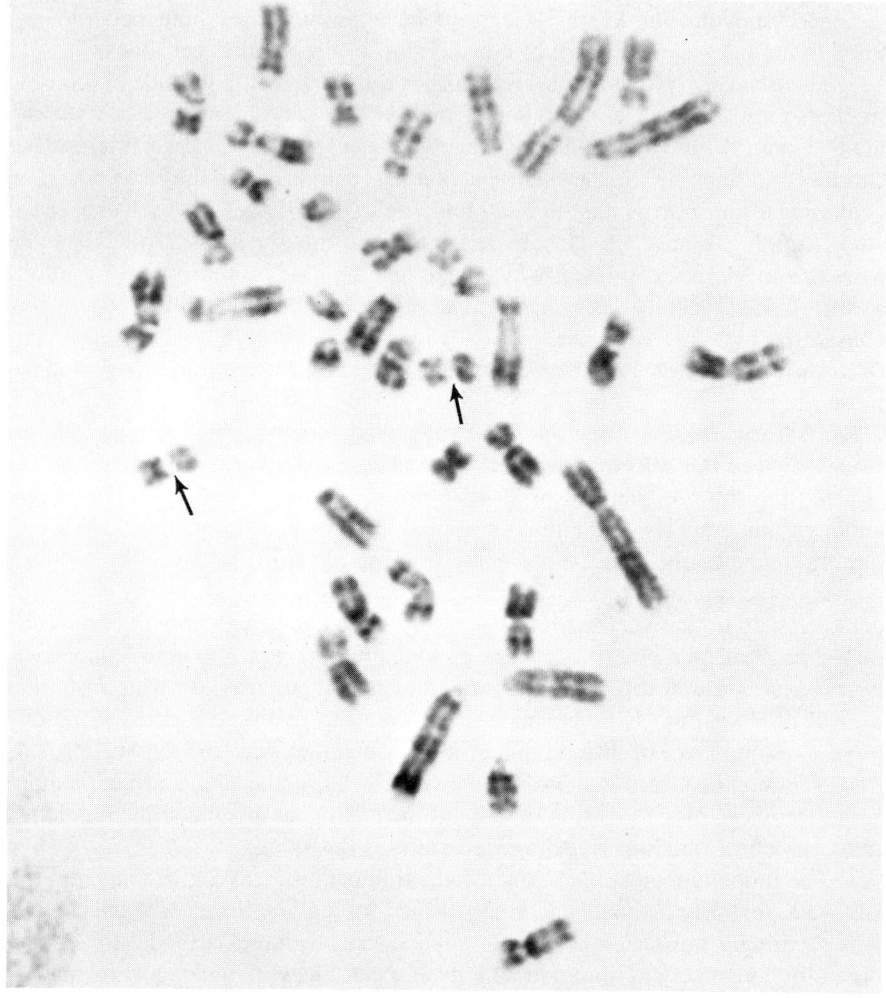

Figure 1. Metaphase spread of LLC–PK$_1$ clone Sui. Note that the diploid number of chromosomes ($2n = 38$) is in evidence, as well as the marker chromosomes (see arrows) for *Sus scrofa.*

microdissection of nephron segments to perform primary cultures of biological material of whose origin one is reasonably certain (unlike the MDCK and LLC–PK$_1$ cell lines). By development of serum-free media, the disadvantage of fibroblast overgrowth can be eliminated, and moreover, growth of epithelia from desired nephron segments can be promoted (Chuman *et al.*, 1982). By combining these approaches of microdissection of defined nephron segments with defined serum-free media, Chung *et al.* (1982) have produced an ''alternative'' to the LLC–PK$_1$ cell line, in the form of a primary culture from rabbit proximal tubule which not only actively transports alpha-methyl-D-glucoside by a Na$^+$- and energy-dependent, phlorizin-sensitive process, but also demonstrates para-thyroid-hormone-sensitive cAMP synthesis.

There are, however, three inherent disadvantages: primary cultures usually cease growth after only a few divisions, hence the biological material is imper-manent (and cannot be cloned, genetically manipulated, or returned to at a later date); second, the primary culture may be heterogeneous to cell type even if prepared from a defined nephron segment (Woodhall *et al.*, 1978); and third, a phenotypic "instability" often appears in primary cultures, in that the properties of interest sometimes disappear. These drawbacks seemingly nullify the reasons we advanced in the Introduction for using renal cell cultures to investigate epithelial transport. Evidence for such views is reflected by the following observations.

Becker and Willis (1979) reported that in the course of 72 hr, a primary culture of ground squirrel kidney cortex loses a feature characteristic of the parent tissue (high Na,K-ATPase activity) and gains a feature common to many undifferentiated cells (furosemide-sensitive K:K exchange).

Chung *et al.* (1982) demonstrated persistence for 28 days of highly charac-teristic features of the proximal tubule in their rabbit kidney primary cultures, but after 9 days, the amount of cellular *material* was constant, demonstrating the impermanence of the *culture*. However, Jefferson *et al.* (1980) have utilized the defined culture media of the primary cell culturist for a different inarguable end, namely to obtain *new* established (continually growing) cell lines, in their case with antigenic characteristics of the distal tubule. We feel that the use of primary cultures in sugar transport studies is legitimate as long as they further an end not attainable by more conventional renal preparations such as perfused tubules, and these guidelines have been discussed by Chung *et al.* (1982).

2.2. Established Cell Lines

An established cell line, by virtue of the fact that it is "immortal," is a mixed blessing. It confers all the advantages previously discussed for epithelial cell cultures. On the other hand, it implies by some criteria that the culture is already spontaneously, though partially, transformed (Stiles *et al.*, 1976; Mullin *et al.*, 1980a). The question of the established cell line's physiological relevance has already been discussed, and admittedly there exist many properties charac-teristic of the tissue of origin in the neonatal or fetal state (Mullin, 1980). Moreover, unpublished results of J. Mullin and A. Kleinzeller have shown that clonal variants can be selected from the LLC–PK$_1$ cell line, implying that even at high passage levels, this cell line may not be a completely homogeneous popula-tion. (The clones "Sui" and "Porky" differ from LLC–PK$_1$ in morphology and growth rate but have retained the capacity for active sugar transport. "Sui" is now the subject of ongoing genetic studies relating to active sugar transport.) It is thus crucial to work within a defined "window" of passage levels of the cell line, returning regularly to one's frozen cell stocks to avoid accruing spontaneous changes in one's biological material. It is advisable to clone before beginning a specific study, to achieve a more homogeneous cell population with which to work.

The study of sugar transport in renal epithelial cell lines has focused largely on the LLC–PK$_1$ cell line as it is the only established cell line that now possesses a Na$^+$-dependent, active sugar transport system. Rindler *et al.* (1979) ruled out such a system in the MDCK cell line, and the MDBK, RK$_{13}$, and LLC–RK$_1$ also do not possess such a system (unpublished data). The useful and easy method used in our laboratory for screening new renal cell lines or primary cultures for active sugar transport is to examine the uptake of 0.1 mM ^{3}H-3-O-methyl-D-glucose and 0.1 mM ^{14}C-alpha-methyl-D-glucoside in a double-label experiment. A ratio of the C-1 analog/C-3 analog which is significantly greater than 1 serves as a suitable initial test for an active sugar transport system as found in mammalian proximal tubule (Kleinzeller and Mullin, unpublished results).

An active sugar transport system in the intestine and kidney is quite useful for the organism, but our ability to grow the LLC–PK$_1$ cells on sugars that are not substrates for the Na$^+$-dependent sugar transporter demonstrates that this system is not particularly vital *for the cell*. Hence, the fact that there has not been any recent outpouring of new *cell lines* with this property may not be totally baffling.

3. SUGAR TRANSPORT IN THE LLC–PK$_1$ CELL LINE

3.1. *General Characteristics of LLC–PK$_1$ Cells*

As already mentioned, the LLC–PK$_1$ cell line, derived from the cortical mince of a male pig, is near diploid (Hull *et al.*, 1976; Katz and Mullin, unpublished results), a noteworthy fact in our decision to originally study it and an unusual trait in an established cell line. Like the MDCK and MDBK cells, LLC–PK$_1$ cells manifest dome formation in confluent monolayers, although domes can also sometimes be seen in isolated islands (J. Mullin, unpublished results). The cells have a generation time of 24–30 hr at 37°C (depending on culture conditions). Although the cells have survived in suspension culture, significant growth is not noted. The cells produce plasminogen activator at confluence, but evidence for renin or prostaglandin production is negative (Hull *et al.*, 1976). It is noteworthy that the plasminogen activator is released by the cells at the area of the cell-to-cell junctions (Paul *et al.*, 1979). Inability of LLC–

PK$_1$ cells to produce prostaglandins may be linked to a defect in their cyclooxygenase activity (Lifschitz, 1982). The electrical resistance across confluent LLC–PK$_1$ monolayers has been reported to be 400 ohms $\times$ cm^2, with a potential difference less than 2 mV, lumen negative (Handler *et al.*, 1980). Higher potential differences (two- to threefold, see Fig. 2) can be measured if the cells are refed with medium containing 30% fetal bovine serum, rather than the normal 10% level. Calf serum does not produce this effect (Adler, Mullin, and Kleinzeller, unpublished results).

The polarity of these cells is demonstrated by transmission electron microscopy and the sidedness of K$^+$-sensitive ^{3}H-ouabain binding (7.8 pmol/mg protein) in the study of Mills *et al.* (1979). Cell H$_2$O was here reported unaffected by ouabain, although sharp drops in cellular K$^+$ and reciprocal increases in cellular Na$^+$ occurred. Functioning of the Na$^+$ pump in the Na$^+$ transport mode was later found to be a requirement for ouabain binding to the cells (Mills *et al.*, 1981).

Concerning properties of these cells relevant to specific nephron segments, the following results exist. Histochemically, high levels of gamma-glutamyl transpeptidase and moderate levels of leucine aminopeptidase show similarity to the proximal tubule (Perantoni and Berman, 1979). Unpublished results of Mullin and Kleinzeller have shown moderate levels of alkaline phosphatase and beta-galactosidase by histochemical and spectrophotometric methods (and these levels are up to tenfold lower when the cultures are subconfluent). Cyclic AMP content of the cells is elevated by serum calcitonin and vasopressin, but not by parathyroid hormone or epinephrine, a pattern of response similar to the thick ascending limb (Goldring *et al.*, 1978). .Rapid desensitization of adenylate cyclase activity was observed upon treatment of cells with high vasopressin concentrations (Roy *et al.*, 1981). Insulin and serum have been reported to increase the vasopressin receptors in a subline of LLC–PK$_1$ cells capable of

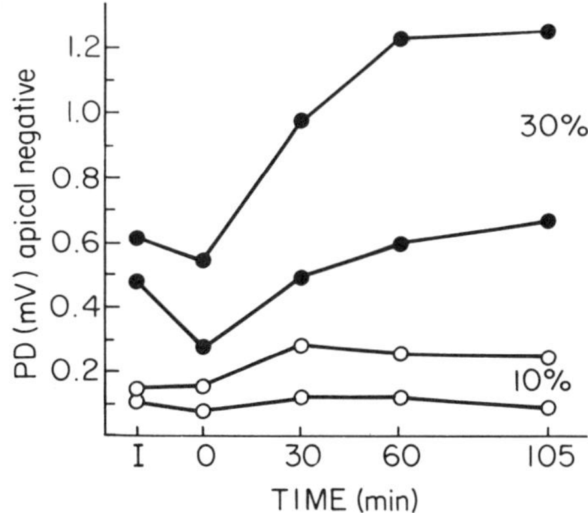

Figure 2. Influence of fetal calf serum on transepithelial potential difference in LLC–PK$_1$ monolayers. (I) Initial reading; 0–105 min: time after refeeding cells with 30% fetal bovine serum (FBS). O, Grown in 10% FBS; ●, grown in 30% FBS. The two traces for each experimental condition represent two separate monolayers.

growth in a serum-free, defined medium (Roy *et al.*, 1980). Evidence has not been obtained for the ability of these cells to synthesize glucose from lactate or glycerol, a prominent marker property of the proximal tubule. Rather, amino acids and urea are synthesized from these metabolic intermediates, processes that would aid in removal of nitrogen from the cell (Mullin *et al.*, 1982a).

A special note in regard to the categorizing of differentiated properties of this or any cell line is the line of approach typified by the mammary epithelial studies of Emerman and Pitelka (1977). Here it was observed that certain morphological and functional differentiated properties are not manifested when cells are cultured conventionally on solid plastic or glass. Only on collagen gels, where cells have their antiluminal surfaces exposed and can undergo shape changes, does a fuller, more characteristic differentiated state emerge. This variable is worth keeping in mind in evaluating studies with the $LLC-PK_1$ cell line which would relate to one or another nephron segment. One should also note in transport *assays* with the cells whether they are sitting on solid or permeable subsurfaces, and hence whether the substrate is readily available to an antiluminal transport system, and also if fluid under, as well as in, the cells will be sampled when the monolayer is assayed after uptake of the substrate (e.g., refer to the amino acid transport study of Sepulveda and Pearson, 1982, versus that of Rabito and Karish, 1982).

Under a miscellaneous heading the following four findings are noteworthy. First, a serum-free growth medium for $LLC-PK_1$ cells has been formulated requiring seven factors: transferrin, insulin, selenium, hydrocortisone, triiodothyronine, vasopressin, and cholesterol (Chuman *et al.*, 1982). Second, $LLC-PK_1$ cells are the only cell line known to make a glucose conjugate of the powerful tumor promoter TPA (O'Brien *et al.*, 1982), a curious phenomenon in light of this cell line's equally unique *active* sugar transport system. Third, Cook *et al.* (1982) have determined that confluent $LLC-PK_1$ cultures are not electrically coupled, show no spreading of fluorescein dye injected into single cells, and do not manifest morphological evidence of gap junctions. Fourth, in holding $LLC-PK_1$ cultures for a week past confluence, we observed the formation of "wormlike" structures across the monolayer. Transmission electron microscopy shows that the structures may be tubular formations with microvilli facing inside the "tubule." The important point here is that the normally inaccessible antiluminal membrane may be facing *upward* in these structures, thus making the antiluminal membrane quite accessible, as, e.g., in patch clamp studies. The corollary of this observation would be that at this stage of the culture we may not any longer be dealing with a true monolayer with uniform polarity. It should be noted that no such "wormlike" structures have been seen with MDCK cells even 6 wk after confluence was reached.

Finally, before going on to sing the praises of $LLC-PK_1$ cell's sugar transport activity, a brief look "under the rug" is in order. Several laboratories in addition to our own have reported the existence of a degeneration of LLC–PK$_1$ monolayers in plaque-like formations (personal communication by J. Cook, F. Sepulveda, T. O'Brien, and E. Adler). The degeneration may be linked to the budding of c-type virus particles from the cells (Mullin *et al.*, 1981). In any case, the phenomenon has yet to be explained or treated but deserves recognition since its occurrence is both unpredictable and insidious in studies with these cells.

3.2. Na$^+$-Dependent, Luminal Sugar Transport

At present, studies of this transport system of the LLC–PK$_1$ cells take one of three approaches: studies of intracellular uptake with cells cultured on collagen-coated filters; studies of transepithelial fluxes with those filters clamped in Ussing-type chambers; studies of uptake into luminally enriched membrane vesicles of the cells.

Realization of the existence of this system came from the first approach. Confluent monolayers of LLC–PK$_1$ cells were determined to have an intracellular volume of 0.979×10^{-4} ml H$_2$O/μg DNA (approximately 1.3 pl H$_2$O/cell) by bathing cells on collagen-coated Nucleopore filters in the presence of radiolabeled 3-O-methyl-D-glucose according to Kletzien *et al.* (1975). From this determined volume, incubations of confluent monolayers on Nucleopore filters in the presence of 1 mM radiolabeled (nonmetabolizable) alpha-methyl-D-glucoside (AMG) were observed to result in intracellular AMG concentrations of as high as 10 mM after 30 min at 25°C (Mullin *et al.*, 1979). The uptake process thus resulted in accumulation of the substrate against its concentration gradient. Later studies then indicated the uptake process to be Na$^+$ and energy dependent and sensitive to inhibition by the Na,K-ATPase inhibitor, ouabain, and the luminal sugar transport inhibitor phlorizin. The specificity of the uptake of AMG was also similar to that observed in mammalian proximal tubule (Mullin *et al.*, 1980a; Rabito and Ausiello, 1980). Uptake of 3-O-methyl-D-glucose was found not to proceed against a concentration gradient and to be Na$^+$ independent and phlorizin insensitive although some affinity for the Na$^+$-dependent sugar transport system was noted (Mullin *et al.*, 1980a). These findings are consistent with the observation of Amsler and Cook (1982a) that if 3-O-methyl-glucose uptake is examined in the presence of cytochalasin B (blocking antiluminal efflux), a weak concentration gradient of this glucose analog is seen. Vanadate, a known inhibitor of renal Na,K ATPase (Grantham and Glynn, 1979), did not exert an inhibitory effect on AMG uptake by these cells, even though it was found to be a most potent compound in causing collapse of the three-dimensional domes of the LLC–PK$_1$ monolayers. Vanadate was also without effect on Rb$^+$ uptake by these cells, a process that is ouabain sensitive, presumably owing to rapid reduction to the vanadyl species within the cell (Mullin *et al.*, 1980b).

An extensive kinetic analysis of AMG uptake by the LLC–PK$_1$ cells revealed the following features: constant rate of uptake for 120 min (1 mM AMG, 37°C); a K_M of 0.75 mM and a V_{max} of 3.03 μmol/hr/mg DNA; an effect of external Na^+ on the K_m but not the V_{max} of the transport system; a transstimulation of uptake by preloading cells with AMG; and an inhibition of uptake by the sulfhydryl-group modifying agents N-ethylmaleimide and p-hydroxymercurybenzoate (Rabito and Ausiello, 1980). Rabito (1981) localized the active sugar transport process to the luminal surface by bathing either the luminal or antiluminal surfaces of the cells with AMG and observing only insignificant uptake (0.003 μmol/hr/mg DNA) from the antiluminal side. Phlorizin binding was also localized luminally. A K_i for phlorizin inhibition of AMG uptake of 1.1 μM was observed here. High-affinity (Na^+-dependent) phlorizin binding was reported to be 0.30 μmol/gm DNA. Amsler and Cook (1982a) reported a K_i of 0.2 μM for phlorizin inhibition of AMG uptake. Sanders and Misfeldt (1981) reported a 2:1, Na:glucose stoichiometry, which differend from the 1:1 Na^+-dependent phlorizin-binding stoichiometry.

The transepithelial transport studies of Misfeldt and Sanders (1981) demonstrated that the transepithelial potential differences (2.8 mV) and the short-circuit current (13 μA/cm^2) were dependent on the presence of Na^+ *and sugar* in the luminal bathing solution. Phlorizin (luminal) caused dissipation of both electrical parameters. Glucose analogs which were able to stimulate short-circuit current paralleled the specificity studies reported previously for the active sugar transport system. An important contribution of glycolytically produced ATP for transepithelial sugar transport has also been suggested by this group (Sanders *et al.*, 1981). The close coupling of D-glucose and Rb^+ uptake at the luminal and antiluminal cell surfaces, respectively, has been investigated as well (Sanders and Misfeldt, 1982).

As dealt with by Moran *et al.* (1982), plasma membrane vesicle preparations allow circumventing the complexities that surround changing the extracellular Na^+ levels when dealing with whole cells. Their vesicular uptake studies with AMG show a characteristic "overshoot" phenomenon with an initial 100-mM NaCl gradient and a specificity similar to that seen with whole cells. Kinetic analyses indicated the existence of only one Na^+-dependent transporter with a K_m of 2 mM, a V_{max} of 3 nmole/min/mg protein, and a Na:sugar stoichiometry of 2:1. Lever (1982) also performed such studies with the following noteworthy differences: a very lengthy "overshoot" period of AMG uptake (initial 50mM NaSCN gradient), and a K_m of 10 mM with a V_{max} of 5 nmole/min/mg protein. Lever's study also reports a pH optimum of 6.5 for AMG uptake into the vesicles and very poor substitution for Na^+, by K^+, Li^+, tris$^+$, Rb^+, or choline.

3.3. Na^+-Independent, Antiluminal Sugar Transport

As yet, little information exists concerning sugar uptake at the antiluminal face of LLC–PK$_1$ cells. In AMG efflux studies Weiss and Cook (1982) have shown the reversibility of the Na^+-dependent sugar transport system but also compiled the following information on a passive carrier-mediated system: (1) 3-O-methyl-D-glucose, but not AMG, is a substrate; and (2) the system is inhibitable by cytocholasin B but not phlorizin. Efflux of AMG from preloaded cells was insensitive to cytocholasin B and did not saturate at intracellular concentrations as high as 10 mM. Mullin *et al.* (1982b) adopted the reverse tactic of examining AMG *uptake* into confluent LLC–PK$_1$ cells when it was presented only to the antiluminal face. Here a shared glucose–AMG transport system was not found, as glucose exerted no inhibition of antiluminal AMG uptake. Both studies suggest that a carrier-mediated antiluminal "exit" pathway for AMG does not exist in these cells. These latter Ussing-chamber studies also raised the caveat that it is questionable whether a low-molecular-weight substrate such as AMG, when presented only to the antiluminal face of the monolayer, does not pass between the cells of this leaky "epithelium" and enter at the unintended luminal cell face. Even with this factor in mind, unpublished results of Kleinzeller and Mullin indicate that an antiluminal glucose transport system does exist in these cells (although its specificity excludes AMG). Figure 3 gives the respective unidirectional fluxes for AMG and glucose into the LLC–PK$_1$ cells. Glucose transported into the cells was also rapidly phosphorylated with subsequent conversion of glucose carbons to (intracellular) lactate, alanine, and glutamate. These facts (namely absence of antiluminal AMG transport but presence of antiluminal glucose transport and metabolism) help to explain why AMG is strongly accumulated within renal proximal tubule cells, whereas glucose is not.

3.4. The Development of Na$^+$-Dependent Sugar Transport in LLC–PK$_1$ Cells—Expression of Polarity in Cultured Epithelia

The LLC–PK$_1$ cell line is unique in its ability to actively transport sugar. With recognition of the limitations we have discussed, it represents an addition to the techniques available to the (renal) transport physiologist. It is potentially equally useful to the cell biologist interested in the development of epithelial

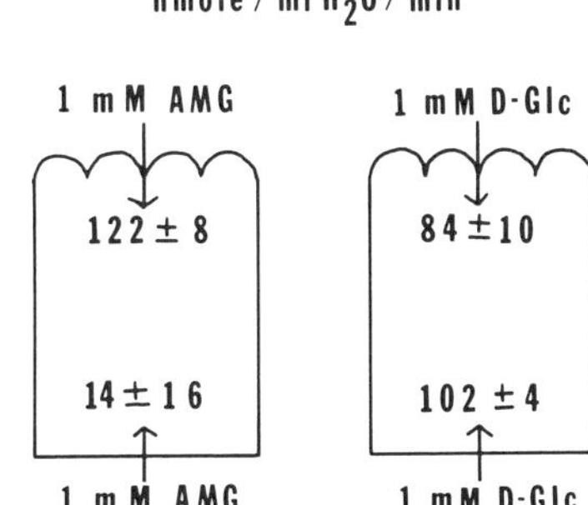

Figure 3. Unidirectional fluxes of D-glucose and alpha-methyl-D-glucoside into intracellular water of confluent LLC–PK$_1$ monolayers at 25°C.

polarity. To date, most studies of epithelial polarity have focused on primary cultures or the MDCK cell line. The LLC–PK$_1$ cell line, however, has a singular advantage here, because in this cell line both morphologic polarity *and Na$^+$-dependent active sugar transport* have correlated with the confluence of the culture (and perhaps the quiescence of the cells) (Mullin *et al.*, 1980a). This study simply showed that LLC–PK$_1$ cells at low density (isolated islands where *all* cells incorporated ^{3}H-thymidine into DNA) did not possess the ability to accumulate AMG against its concentration gradient, nor was AMG uptake by these cycling cells strongly inhibitable by phlorizin. Weiss and Cook (1982) have shown that this increase in concentrative capacity is due to an augmentation of influx and not a decrease in AMG efflux, as there was no evidence for carrier-mediated AMG efflux at any stage. Amsler and Cook (1982b) observed a correlation between the capacity for active AMG uptake and the number of Na$^+$-dependent phlorizin binding sites. Lever (1982) has shown that membrane vesicles from subconfluent LLC–PK$_1$ cultures had barely detectable Na$^+$-dependent AMG uptake. The proper conclusion seems that the appearance of ability to actively transport sugar in LLC–PK$_1$ cultures at confluence is linked to an "event" at the luminal surface of the cells, specifically the "appearance" of the Na$^+$-dependent transport system. Cook *et al.* (1982) have ruled out a more generalized explanation, namely an increase in the inwardly directed Na$^+$ gradient, by demonstrating at confluence a decrease in the (Na$^+$-dependent) uptake of the nonmetabolizable amino acid amino isobutyric acid.

In more extensive studies of this phenomenon, Amsler and Cook (1982a) observed that HMBA, an inducer of differentiation in Friend erythroleukemic cells, and phosphodiesterase inhibitors accelerated the development of concentrative sugar transport in LLC–PK$_1$ cultures. The studies with phosphodiesterase inhibitors moreover *slowed* the growth rate of the cells such that *subconfluent* cultures were demonstrating active sugar transport.

As we move these observations into the wider arena of epithelial polarity, a major question emerges. All previous findings dealt with the population as a whole. As Amsler and Cook (1982a) point out, an increase in concentrative ability of the *population* may represent the recruitment of more and more cells into the differentiated state or the increased differentiation of single cells. Results of Mullin *et al.* (1983) suggest the former hypothesis, since (1) spectrophotometric assays of the luminal membrane markers alkaline phosphatase and beta-galactosidase in confluent LLC–PK$_1$ monolayers have shown eightfold and threefold increases, respectively, in the activity of these two enzymes with onset of confluence (see also Yoneyama and Lever, 1983); and (2) increases in alkaline phosphatase activity of the culture at confluence were shown histochemically to occur on a cell-by-cell basis, not a gradual uniform increase in *all* cells of the culture as a whole. Furthermore, histochemistry (Mullin *et al.*, 1983) provided evidence against *terminal* differentiation of LLC–PK$_1$ cells. Sterilely "wounding'' a confluent monolayer, we observed after 36 hr that the cycling cells bordering on and filling in the ''wound'' no longer demonstrated alkaline phosphatase activity histochemically, whereas the other cells of the monolayer, even those directly behind this band of dividing cells, remained heavily stained.

The finding that phosphodiesterase inhibitors can yield development of active sugar transport in *subconfluent* cultures is not totally surprising as both our laboratories and others have observed on occasion the appearance of domes in isolated islands of cells. Clearly there is much left to be understood. The simple equating of confluence and polarity does not fully explain the data, but neither does the statement of Rodriguez–Boulan *et al.* (1981) that attachment of the cell to the substratum or to another cell allows for *full* expression of polarity. For example, Herzlinger and Ojakian (1982) show uniform distribution of a glycoprotein over the surface of subconfluent MDCK cells, whereas monoclonal antibodies to this glycoprotein label only the antiluminal membrane of the cells of the confluent cultures.

The *criteria* for epithelial polarity, either active sugar transport in pig kidney epithelia (Mullin *et al.*, 1980a; Amsler and Cook, 1982a), directional budding of viruses in MDCK cells (Rodriquez–Boulan and Sabatini, 1978), or the morphologic polarity of thyroid epithelia (Nitsch and Wollman, 1980), should be recognized as such—criteria. There is a *connection* between cell attachment and epithelial polarity (as evidenced by virus budding in MDCK cells), just as there is a *connection* between cell quiescence and epithelial polarity (as evidenced by active sugar transport and alkaline phosphatase activity in LLC–PK$_1$ cells). Future research may show that neither position is wholly without truth.

ACKNOWLEDGMENTS.   We gratefully acknowledge the help of Ms. Lois Porten in the preparation of the manuscript. We sincerely thank Drs. John Cook, Kurt Amsler, Douglas Jefferson, Walter Scott, Thomas O'Brien, and Elizabeth Adler for making available unpublished data. We would also like to thank Sandy Katz for his assistance in karyotyping cells.

REFERENCES

Amsler, K., and Cook, J. S., 1982a, Development of Na$^+$-dependent hexose transport in a cultured line of porcine kidney cells, *Amer. J. Physiol.* **242**:C94–C101.

Amsler, K., and Cook, J. S., 1982b, Development of Na$^+$-dependent phlorizin-binding sites in a clone of LLC–PK$_1$ cells, *J. Cell Biol.* **95**:30a.

Becker, J. H., and Willis, J. S., 1979, Properties of Na–K pump in primary cultures of kidney cells, *J. Cell Physiol.* **99**:427–440.

Chuman, L., Fine, L. G., Cohen, A. H., and Saier, M. H., 1982, Continuous growth of proximal tubular kidney epithelial cells in hormone-supplemented serum-free medium, *J. Cell Physiol.* **94**:506–510.

Chung, S. D., Alavi, N., Livingston, D., Hiller, S., and Taub, M., 1982, Characterization of primary rabbit kidney cultures that express proximal tubule functions in a hormonally defined medium, *J. Cell Biol.* **95**:118–126.

Cook, J. S., Amsler, K., Weiss, E. R., and Schaffer, C., 1982, Development of Na^+-dependent hexose transport in vitro, in: *Membranes in Growth and Development*, Liss, New York, pp. 551–567.

Emerman, J. T., and Pitelka, D. R., 1977, Maintenance and induction of morphological differentiation in dissociated mammary epithelium on floating collagen membranes, *In Vitro*, **13**:316–328.

Foster, R. F., and Willis, J. S., 1978, Retention of differentiated membrane transport activity in primary cultures of mammalian renal cortex, *In Vitro*, **14**:335.

Goldring, S. R., Dayer, J.-M., Ausiello, D. A., and Krane, S. M., 1978, A cell strain cultured from porcine kidney increases cyclic AMP content upon exposure to calcitonin or vasopressin, *Biochem. Biophys. Res. Communs.* **83**:434–440.

Grantham, J. J., and Glynn, I. M., 1979, Renal Na,K-ATPase: Determinants of inhibition by vanadium, *Amer. J. Physiol.* **236**:F530–F536.

Handler, J. S., Perkins, F. M., and Johnson, J. B., 1980, Studies of renal cell function using cell culture techniques, *Amer. J. Physiol.* **238**:F1–F9.

Herzlinger, D. A., and Ojakian, G. K., 1982, Development and maintenance of epithelial polarity in MDCK cells, *J. Cell Physiol.* **95**:105a.

Horster, M. F., and Wilson, B. D., 1981, Nephron epithelia in culture: Growth of loop of Henle cells in hormonally defined serum free medium, *Cell Biol. Intl. Reports* **5**:765.

Hull, R. N., Cherry, W. R., and Weaver, G. W., 1976, The origin and characteristics of a pig kidney cell strain, LLC–PK$_1$, *In Vitro* **12**:670–20.

Jefferson, D. M., Cobb, M. H., Gennaro, J. F., and Scott, W. N., 1980, Transporting renal epithelial culture in hormonally defined serum-free medium, *Science* **210**:912–914.

Kletzien, R. F., Parza, M. W., Becker, J. E., and Potter, V. R. 1975, A method using 3-O-methyl-D-glucose and phloretin for the determination of intracellular water space of cells in monolayer culture, *Anal. Biochem.* **68**:537–544.

Lever, J. E., 1982, Expression of a differentiated transport function in apical membrane vesicles isolated from an established kidney epithelial cell line, *J. Biol. Chem.* **257**, 8680–8686.

Lifschitz, M. D., 1982, LLC–PK$_1$ cells derived from pig kidneys have a defect in cyclooxygenase, *J. Biol. Chem.* **257**:12611–12615.

MacKnight, A. D. C., Civan, M. M., and Leaf, A., 1975, The Na^+-transport pool in toad urinary bladder epithelial cells, *J. Memb. Biol.* **20**:365–385.

Madin, S. H., and N. B. Darby, 1958, Established kidney cell lines of normal adult bovine and ovine origin, *Proc. Soc. Exptl. Biol. Med.*, **98**:574–576.

Mills, J. W., MacKnight, A. D. C., Dayer, J.-M., and Ausiello, D. A., 1979, Localization of [^{3}H]ouabain-sensitive Na^+ pump sites in cultured pig kidney cells, *Amer. J. Physiol.* **236**:C157–C162.

Mills, J. W., MacKnight, A. D. C., Farrell, J. A., Dayer, J. M., and Ausiello, D. A., 1981, Interaction of ouabain with the Na^+ pump in intact epithelial cells, *J. Cell Biol.* **88**:637–643.

Misfeldt, D. S., and Sanders, M. J., 1981, Transepithelial transport in cell culture: D-glucose transport by a pig kidney cell line (LLC–PK$_1$), *J. Memb. Biol.* **59**:13–18.

Moran, A., Handler, J. S., and Turner, R. J., 1982, Na^+-dependent hexose transport in vesicles from cultured renal epithelial cell line, *Amer. J. Physiol.* **243**:C293–C298.

Mullin, J. M., 1980, Transport-related properties and development of polarity in an established epithelial cell line of renal origin, Ph.D. thesis, University of Pennsylvania.

Mullin, J. M., Diamond, L., and Kleinzeller, A., 1979, Uptake of alpha-methyl-D-glucoside and 3-O-methy-D-glucose by an established pig renal epithelial cell line, *Fed. Proc.* **38**:1058.

Mullin, J. M., Weibel, J., Diamond, L., and Kleinzeller, A., 1980a, Sugar transport in the LLC–PK₁ renal epithelial cell line: Similarity to mammalian kidney and the influence of cell density, *J. Cell Physiol.* **104**:375–389.

Mullin, J. M., Diamond, L., and Kleinzeller, A., 1980b, Effects of ouabain and ortho-vanadate on transport-related properties of the LLC–PK₁ renal epithelial cell line, *J. Cell Physiol.* **105**:1–6.

Mullin, J. M., Weibel, J., Diamond, L., and Kleinzeller, A., 1981, Transport related properties and the development of polarity of an established epithelial cell line of renal origin, *Adv. Physiol. Sci.* (IUPS Symposium, Budapest, Hungary) **11**:175–179.

Mullin, J. M., Cha, C.-J. M., and Kleinzeller, A., 1982a, Metabolism of L-lactate by LLC–PK₁ renal epithelia, *Amer. J. Physiol.* **242**:C41–C45.

Mullin, J. M., Fluk, L., and Kleinzeller, A., 1982b, Sidedness of lactate and glucose transport and metabolism in LLC–PK₁ cells, *The Physiologist* **25**(4):294a.

Mullin, J. M., Fluk, L., and Tchao, R., 1983, Mitotic cells in domes of confluent epithelial (LLC–PK₁) cultures, *J. Cell Biol.* **97**:312a.

Nitsch, L., and Wollman, S. H., 1980, Ultrastructure of intermediate stages in polarity reversal of thyroid epithelium in follicles in suspension culture, *J. Cell Biol.* **86**:875–880.

O'Brien, T. G., Saladik, D., Sina, J. F., and Mullin, J. M., 1982, Formation of a glucuronide conjugate of TPA by LLC–PK₁ renal epithelial cells in culture, *Carcinogenesis* **3**:1165–1169.

Paul, D. C., Bobbitt, J. L., Williams, D. C., and Hull, R. N., 1979, Immunocytochemical localization of plasminogen activator on procine kidney cell strain: LLC–PK₁ (LP₁₀₀), *J. Histochem. Cytochem.* **27**:1035–1040.

Perantoni, A., and J. G. Berman, 1979, Properties of Wilm's tumor line (TuWi) and pig kidney line (LLC–PK₁) typical of normal kidney tubular epithelium, *In Vitro* **15**:446–454.

Rabito, C. A., 1981, Localization of the Na⁺-sugar cotransport system in a kidney epithelial cell line, *Biochim Biophys. Acta* **649**:286–296.

Rabito, C. A., and Ausiello, D. A., 1980, Na⁺-dependent sugar transport in a cultured epithelial cell line from pig kidney. *J. Membrane Biol.* **54**:31–38.

Rabito, C. A., and Karish, M. V., 1982, Polarized amino acid transport by an epithelial cell line of renal origin (LLC–PK₁), *J. Biol. Chem.* **257**:6802–6808.

Rindler, M. J., Chuman, L. M., Shaffer, L., and Saier, M. H., 1979, Retention of differentiated properties in an established dog kidney epithelial cell line (MDCK), *J. Cell Biol.* **81**:635–648.

Rodriquez–Boulan, E. J., and Sabatini, D. D., 1978, Asymmetric budding of viruses in epithelial monolayers: A model system for study of epithelial polarity, *Proc. Natl. Acad. Sci. USA* **75**:5071–5075.

Rodriguez–Boulan, E. J., Baskiet, K. T., and Sabatini, D. D., 1981, Asymmetric budding of enveloped viruses from isolated epithelial cells attached to a collagen substrate, *J. Cell Biol.* **91**:121a.

Roy, C., Preston, A. S. and Handler, J. S., 1980, Insulin and serum increase the number of receptors for vasopressin in a kidney derived line of cells grown in a defined medium, *Proc. Natl. Acad. Sci. USA* **77**:5979–5983.

Roy, C., Hall, D., Karish, M., and Ausiello, D. A., 1981, Relationship of (8-lysine) vasopressin receptor transition to receptor funciional properties in a pig kidney cell line (LLC–PK₁), *J. Biol. Chem.* **256**:3423–3427.

Sanders, M. J., and Misfeldt, D. S., 1981, Transepithelial transport in cell culture: Different stoichiometries of Na:phlorizin binding and Na:D-glucose cotransport, *J. Cell Biol.* **91**:415a.

Sanders, M. J., and Misfeldt, D. S., 1982, Ouabain-sensitive ⁸⁶Rb(K) influx is linked to transepithelial Na transport in pig kidney cell line, *Biochim Biophys. Acta* **685**:383–385.

Sanders, M. J., Misfeldt, D. S., and Simon, L. M., 1981, Transepithelial transport in cell culture: Bioenergetics of Na,D-glucose co-transport, *Fed. Proc.* **40**:373.

Sepulveda, F. V., and Pearson, J. D., 1982, Characterization of neutral amino acid uptake by cultured epithelial cells from pig kidney, *J. Cell Physiol.* **112**:182–188.

Stiles, C. D., Desmond, W., Chuman, L. M., Sato, G., and Saier, M. H., 1976, Growth control of heterologous tissue culture cells in the congenitally nude mouse, *Cancer Res.* **36**:1353–1360.

Weiss, E. R., and Cook, J. S., 1982, Pathways for efflux of hexoses during differentiation of LLC–PK$_1$ cells in culture, *J. Cell Biol.* **95**:31a.

Woodhall, R. B., Tisher, C. C., Simonton, C. A., and Robinson, R. R., 1978, The relationship between para-aminohippurate secretion and cellular morphology in rabbit proximal tubules, *J. Clin. Invest.* **61**:1320–1329.

Yoneyama, Y., and Lever, J. E., 1983, Modulation of gene expression in cell culture; cell confluence and inducers of cell differentiation trigger the development of several microvillar hydrolases in an established kidney epithetial cell line (LLC–PK$_1$). *J. Cell Biol.* **97**:55a.

6

Amino Acid Transport in Cultured Kidney Tubule Cells

FRANCISCO V. SEPÚLVEDA and JEREMY D. PEARSON

1. THE LLC–PK$_1$ CELL LINE

Continuous cultures of epithelial cells that retain the ability to develop differentiated functions *in vitro* provide useful model systems in which to study the regulation of such functions. The LLC–PK$_1$ line is one such cell line, which was derived from pig kidney by Hull and his co-workers in 1958 (Hull *et al.*, 1976). It has been shown to have a near diploid chromosome number and does not produce tumors in immunosuppressed mice. When grown on a solid substratum, LLC–PK$_1$ cells form monolayers that exhibit epithelial morphology: numerous microvilli are seen on the apical plasma membrane facing the culture medium, cells are connected by tight junctions near the apical surface, and beneath these the basolateral membrane foldings delineate a complex intercellular space (see, e.g., Fig. 1 in Misfeldt and Sanders, 1981). Unlike other epithelial cells, LLC–PK$_1$ cells do not form permeable intercellular junctions as judged by their inability to transfer uridine nucleotides transcellularly (Sepúlveda and Pearson, 1984a). This correlates with the absence of gap junctions from electron microscope images of freeze-fractured monolayers (unpublished results from our laboratory). In addition to their general epithelial morphology, LLC–PK$_1$ cells form domelike structures upon reaching confluency (Hull *et al.*, 1976). These fluid-

FRANCISCO V. SEPÚLVEDA and JEREMY D. PEARSON • ARC Institute of Animal Physiology, Babraham, Cambridge CB2 4AT, England. *Present address for J. D. P.:* MRC Clinical Research Centre, Harrow HA1 3UJ, England.

filled cavities are believed to arise from active transport of water across the cells into the space between the monolayer and the substratum (Abaza *et al.*, 1974).

The morphology of LLC–PK$_1$ monolayers and their capacity to transport water correlate with the presence of Na^+/K^+-pumping sites, as revealed by ouabain binding, solely on the basolateral aspect of the plasma membrane, facing the substratum (Mills *et al.*, 1979). LLC–PK$_1$ cells also possess a Na^+-dependent, phloridzin-sensitive hexose transport system that is located in the part of the plasma membrane facing the culture medium (Mullin *et al.*, 1980; Rabito and Ausiello, 1980; Misfeldt and Sanders, 1981; Rabito, 1981). This is analogous to the localization of Na^+-dependent hexose transport in the brush border membranes of proximal tubule cells *in vivo* (Aronson and Sacktor, 1975; Kinne *et al.*, 1975; Turner and Silverman, 1977). These characteristics suggest that LLC–PK$_1$ cells express *in vitro* some functions of the cells in the proximal part of kidney tubules and that they may be used as convenient models for the *in vitro* study of proximal tubule function. One of the most important properties of the proximal nephron is the complete removal of amino acids from the glomerular filtrate (see Ullrich, 1979; Silbernagl *et al.*, 1975; and Silbernagl, 1979, for reviews on amino acid reabsorption). In the following sections the way in which LLC–PK$_1$ cells handle neutral and basic amino acids is examined and compared to amino acid transport in epithelial and other cells.

2. AMINO ACID TRANSPORT SYSTEMS IN MAMMALIAN CELLS

The studies of Oxender and Christensen (1963) established that in Ehrlich ascites tumor cells the uptake of neutral amino acids is mediated by two transport systems having overlapping specificities. They called these two agencies A and L, because of their preference for alanine and leucine, respectively; these denominations are, however, arbitrary because of the broad specificity of the two agencies. System A preferentially takes up neutral amino acids with short polar, or linear, side chains such as L-alanine and L-threonine, whereas system L prefers substrates such as L-leucine, L-isoleucine, and nonpolar amino acids. The two systems differ markedly in their capacity to produce intracellular accumulation, sensitivity to metabolic inhibitors, pH dependence, and requirements for alkali metal ions. Later work by Christensen and his colleagues (Christensen *et al.*, 1967) disclosed that an additional Na^+-dependent transport system is present in Ehrlich cells. This novel carrier was designated the ASC system; it mediates the uptake of alanine, serine, cysteine, and other small neutral amino acids but, unlike system A, does not tolerate *N*-methylated amino acids. Intracellular accumulation of amino acids against their concentration gradients occurs through Na^+-dependent systems: the cation and the amino acid are cotransported and the electrochemical potential difference for Na^+ drives the uphill transport of amino acids (Reid *et al.*, 1974; Philo and Eddy, 1978).

Table I summarizes the properties of the three main transport systems for neutral amino acids. Amino acid transport systems are subject to regulation. In most cases so far studied it has been found that this is exerted primarily through effects on system A; a comprehensive review on this subject has already been published (Guidotti *et al.*, 1978). Table I also summarizes the characteristics of a cationic transport system (y^+) that has been characterized recently in fibroblasts and in HTC hepatoma cells (White and Christensen, 1982; White *et al.*, 1982). This transport system has two interesting features: it mediates intracellular accumulation in the presence or absence of Na^+, and in the presence of Na^+ is susceptible to inhibition by neutral amino acids such as homoserine.

3. AMINO ACID TRANSPORT BY EPITHELIA

The transepithelial transport of amino acids in the proximal tubule and in the intestinal epithelium involves the permeation of two plasma membranes in series: the brush border and the basolateral membrane, respectively. A model was proposed by Crane (1962) to describe the active absorption of glucose in the small intestine; this hypothesis involves the coupling of the fluxes of Na^+ and the sugar at the brush border membrane so that the sugar accumulates intracellularly at the expense of the Na^+ gradient. Exit of sugar molecules through

*Table I. Neutral and Cationic Amino Acid Transport Systems
of Mammalian Cells[a]*

	Preferred substrates	Specific substrate	Na^+ dependence	Li^+ dependence	Exchange capacity
System A	Ala, Pro, Ser, Gly, Met, AIB[c]	MeAIB[b]	+	+	−
System ASC	Ala, Ser, Cys, Thr	Cys[d]	+	−	+
System L	Leu, Ile, Phe	BCH[e]	−	−	+
System y^+	Arg, Lys, Orn	Homoarginine GPA[f]	−	−	+

[a]For references to the original work delineating these transport systems see Christensen (1975, 1979).
[b]MeAIB is 2-(methylamino)-isobutyric acid.
[c]AIB is α-amino-isobutyric acid.
[d]Cysteine has been shown to be transported exclusively by the ASC system in isolated hepatocytes (Kilberg, 1982). This is not the case in other cells such as the hepatoma cell line (HTC) (Handlogten *et al.*, 1981), cultured human fibroblasts (Franchi–Gazzola *et al.*, 1982), or LLC–PK_1 cells (see below).
[e]BCH [2-aminobicyclo-(2,2,1)-heptane-2-carboxylic acid] has recently been shown to be taken up by at least two Na^+-independent transport systems in rat hepatocytes in primary culture (Handlogten *et al.*, 1982).
[f]GPA is 4-amino-1-guanylpiperidine-4-carboxylic acid.

the basolateral membrane follows the concentration gradient and does not require the input of energy. Later work (reviewed by Schultz and Curran, 1970) extended Crane's model to explain the intestinal absorption of amino acids. The available evidence concerning amino acid transport by proximal tubule epithelium favors an analogous mechanism to account for the removal of amino acids from the glomerular ultrafiltrate. Epithelial cells in the proximal tubule accumulate amino acids against a concentration gradient, and active accumulation is inhibited by removal of Na^+ from the bathing medium or by the presence of metabolic inhibitors (Rosenberg *et al.*, 1961; Fox *et al.*, 1964; Hillman *et al.*, 1968a,b; Barfuss and Schafer, 1979).

Isolated proximal tubule brush border membrane vesicles also possess an active Na^+ dependent amino acid transport mechanism. Transient accumulation of amino acids in the vesicles depends on both an inwardly directed gradient of Na^+ concentration and an inside-negative electrical potential difference (Evers *et al.*, 1976; Fass *et al.*, 1977; Hammerman and Sacktor, 1977). Studies with isolated basolateral membrane vesicles show, in contrast, that most amino acid transport across that membrane is Na^+ independent; the small, but significant, Na^+-dependent component has been attributed to contamination by brush border membranes, which can be as high as 25% (Evers *et al.*, 1976; Slack *et al.*, 1977).

The number and specificities of the transport systems for neutral amino acids in brush border and basolateral membranes of the proximal tubules have not been elucidated. The two main approaches so far used to address this question involve work with isolated kidney brush border membrane vesicles (Murer and Kinne, 1980) and the use of indirect electrophysiological analysis (Hoshi *et al.*, 1976; Frömter, 1982). Work with both techniques suggests that multiple transport systems occur in the luminal membrane (Mircheff *et al.*, 1982; Samaržija and Frömter, 1982). Neutral L-α-amino acid transport is apparently mediated mainly by a single broad-specificity, Na^+-dependent transport system that shows higher affinities for hydrophobic amino acids. A similar situation seems to prevail for the transport of neutral L-α-amino acids across the brush border membrane of the small intestine (Sigrist-Nelson *et al.*, 1975; Sepúlveda and Smith, 1978; Paterson *et al.*, 1979, 1980; Stevens *et al.*, 1982). It therefore appears that the transport systems for neutral amino acids in epithelial brush border membranes do not fall into the classification discussed in Section 2 and that they are restricted to cells specializing in transepithelial transport (for review see Smith *et al.*, 1983).

Less information concerning the multiplicity of transport systems at the basolateral membrane is available. The most complete study has been carried out by Mircheff *et al.* (1980) employing intestinal basolateral membrane vesicles: the specificity pattern for amino acid transport differs from that in brush border membranes, and the authors suggested that the A, ASC, and L systems are present. The possibility that similar transport systems occur in renal basolateral membranes has been proposed by Foulkes and Gieske (1973) and Schafer and Barfuss (1980). This might be expected from the striking morphological and functional similarities between epithelial cells of the small intestine and the proximal tubule. Cationic amino acids seem to be reabsorbed in the proximal tubule through different pathways from those used by neutral amino acids (Silbernagl, 1979). Recent studies with brush border membrane vesicles isolated from proximal tubule cells demonstrate that part of the transport of lysine is Na^+ dependent (Mircheff *et al.*, 1982). In addition to this difference from transport system y^+, lysine transport by renal brush border membrane vesicles is totally inhibited by alanine, suggesting that the cationic amino acid transport agency (or agencies) present in the apical membrane of the renal tubule is not the same as the y^+ transport system of other cells not involved in transepithelial transport of amino acids.

4. AMINO ACID TRANSPORT SYSTEMS IN LLC–PK$_1$ CELLS

It has been shown that LLC–PK$_1$ cells can take up amino acids by transport systems that resemble the ASC, A, and L agencies (Sepúlveda and Pearson, 1982; Rabito and Karish, 1982). The general approach followed in detecting the presence of different transport systems is illustrated here with reference to cysteine transport by LLC–PK$_1$ cells. Figure 1 shows the uptake of 500 μM cysteine by subconfluent cultures of LLC–PK$_1$ cells in either Na^+-containing or in Na^+-free Hanks' medium. In the presence of Na^+, cysteine uptake is linear for about 60 sec. In the absence of Na^+, uptake is markedly depressed and remains linear for 2–5 min. The Na^+ dependence of cysteine uptake is intermediate between that of alanine uptake, which is almost totally Na^+ dependent, and that of leucine, which shows no dependence on the presence of Na^+ (Sepúlveda and Pearson, 1982). The concentration dependence of cysteine uptake is shown in Figure 2. If the Na^+ dependent and Na^+ independent uptakes are assumed to occur via discrete systems, Na^+-dependent transport takes place with an apparent K_m of 256 μM, whereas Na^+-independent transport has an apparent K_m of 2.35 mM. It should be noted that these constants provide only a general description of the concentration dependence; they do not necessarily relate to the affinities of the transport systems involved, which have yet to be fully characterized (see below, Section 4.).

The effects of other amino acids on cysteine uptake are shown in Table II. The most effective inhibitors of cysteine uptake in the absence of Na^+ are the hydrophobic amino acids, which are also preferred substrates of transport system L, whereas glutamic acid and lysine are poor inhibitors. The best inhibitors in the presence of Na^+ are the ASC substrates serine, alanine, and ANB. Thus the

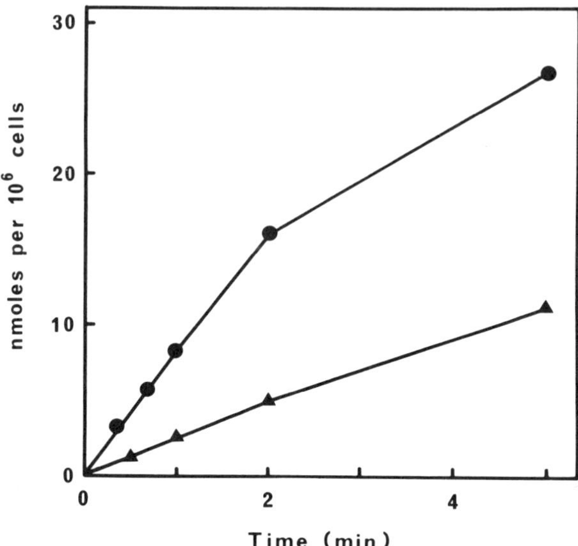

Figure 1. The time course of cysteine uptake by subconfluent monolayers of LLC–PK$_1$ cells, in a medium containing Na$^+$ (circles) or one in which all Na$^+$ had been replaced by choline (triangles). The amino acid concentration was 500 μM. Uptake measurements performed as described by Sepúlveda and Pearson (1982).

results in Table II suggest that a major proportion of cysteine uptake in the presence of Na$^+$ is mediated by the ASC system whereas the L system is probably responsible for uptake in the absence of Na$^+$.

The additional participation of the A system in cysteine uptake in the presence of Na$^+$, and the virtually exclusive mediation of cysteine uptake by system L in the absence of Na$^+$, are shown by the experiments illustrated in Figure 3. About 35% of cysteine uptake in the presence of Na$^+$ is inhibited by MeAIB (a specific substrate for system A; see Table I). When Na$^+$ in the bathing medium is replaced by choline, MeAIB is no longer an effective inhibitor whereas BCH (a substrate for system L) completely inhibits the uptake of cysteine. The interaction between cysteine and MeAIB is reciprocal: Rabito and Karish (1982) have shown that cysteine inhibits MeAIB uptake by LLC–PK$_1$ cells in the presence of Na$^+$. The interpretation of this inhibition is, however, complicated by the fact that, surprisingly, only 45% of MeAIB uptake in the cited reference is apparently dependent on the presence of Na$^+$.

A distinguishing feature of the ASC system (see Table I) is its ability to mediate countertransport, uptake being accelerated when suitable substrates for exchange are present inside the cell. Thus alanine uptake by LLC–PK$_1$ cells is greatly stimulated by preincubation of the cells with alanine (Sepúlveda and

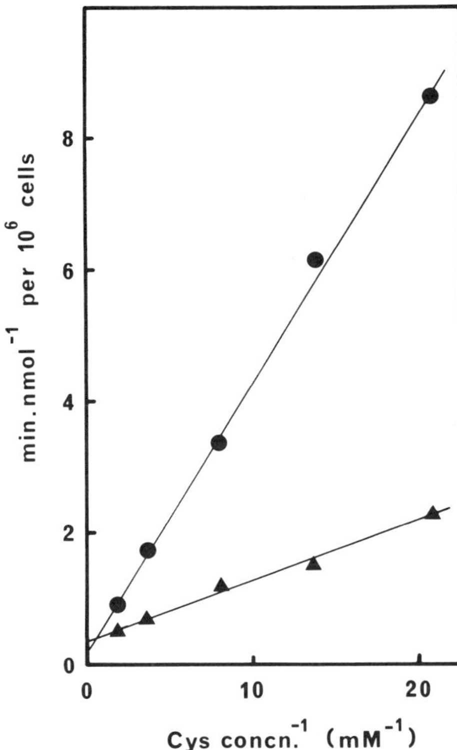

Figure 2. Double reciprocal plot showing the concentration dependence of cysteine uptake by subconfluent monolayers of LLC–PK_1 cells. Circles represent uptake measurements made in Na^+-free, choline-substituted medium. Triangles represent Na^+-dependent uptake. Incubation time was 30 sec, and the experimental details are similar to those described by Sepúlveda and Pearson (1982). Apparent Km values derived from the lines illustrated are 2.35 mM (circles) and 256 μM (triangles).

Pearson, 1982), but preloading with leucine has no effect. An interesting feature of the interaction between alanine and leucine is that the latter behaves as an inhibitor when added with alanine to the outside of the cells; a similar finding was reported originally for the ASC system in Ehrlich cells by Thomas and Christensen (1971). The enhancement of alanine uptake by intracellular alanine is fully dependent on the presence of extracellular Na^+.

We have used similar experimental techniques to look for the presence of cationic amino acid transport systems. Arginine and lysine are each taken up rapidly by LLC–PK_1 cells. The concentration dependence of lysine uptake can be resolved into non-saturable and saturable components, the K_m of the saturable portion being about 200 μM. Uptake rates of cationic amino acids are similar in the presence and absence of Na^+ in the bathing medium. Preloading the cells with cationic amino acids stimulates the subsequently measured uptake of lysine, and neutral amino acids have no stimulatory effect. Table III shows that lysine uptake by LLC–PK_1 cells either in the presence or absence of Na^+ is very little affected by amino acids other than arginine and ornithine. It is noteworthy that

Table II.   Inhibition of Cysteine Uptake[a]

| | Percent inhibition of cysteine uptake | |
Inhibitor	$+Na^+$	$-Na^+$
Leucine	69.5	93.3
Isoleucine	59.5	94.1
Phenylalanine	42.7	91.7
Methionine	78.2	93.9
Alanine	71.0	32.0
Serine	76.4	29.4
Cysteine	96.8	97.1
ANB	87.2	77.6
Glycine	34.4	0
Glutamic acid	—	0
Lysine	5.8	9.5

[a]Uptake of cysteine was measured at 250 μM either in the presence or in the absence of Na^+. Concentrations of inhibitors were 10 mM except for ANB (DL-α-amino-*n*-butyric acid; 20 mM).

neither homoserine, which in the presence of Na^+ strongly inhibits transport through system y^+ (White *et al.*, 1982), nor alanine, which blocks lysine transport by proximal tubule apical membrane vesicles (Mircheff *et al.*, 1982), have any effect on lysine transport by LLC–PK$_1$ cells.

In summary, LLC–PK$_1$ cells possess characteristic transport agencies for the uptake of both neutral and cationic amino acids, but these systems do not

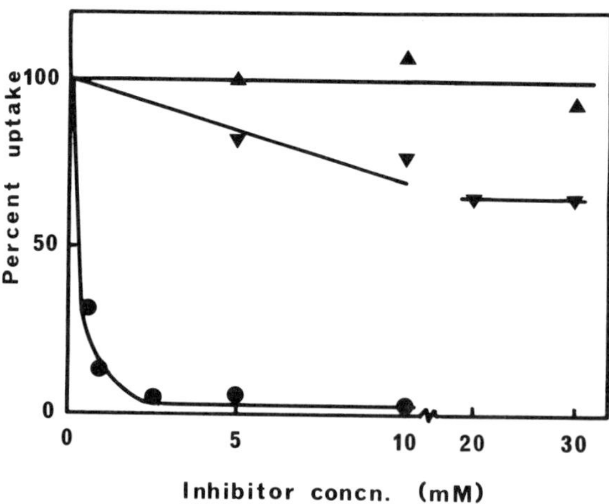

Figure 3. Effect of increasing concentrations of MeAIB (triangles) or BCH (circles) on cysteine uptake by subconfluent monolayers of LLC–PK$_1$ cells. Uptake from 250 μM cysteine was measured in the presence or absence of Na^+ during 30-sec incubations. Uptake in the absence of Na^+ (circles and upright triangles) or the Na^+-dependent component of cysteine uptake (inverted triangles) is shown. Values are expressed as percentage of the control uptake measured in the absence of inhibitors.

Table III.   Inhibition of 25 μM Lysine Uptake

Inhibitor		Percent inhibition of lysine uptake	
		$+Na^+$	$-Na^+$
Leucine	(5 mM)	6.0	2.1
BCH	(10 mM)	−24.1	5.3
Alanine	(10 mM)	1.0	−19.2
Serine	(10 mM)	−12.9	−5.3
ANB	(10 mM)	−18.0	−6.8
Glutamate	(10 mM)	2.7	−19.9
Glutamine	(5 mM)	0.0	3.6
Arginine	(2 mM)	96.6	92.2
Ornithine	(2 mM)	91.7	87.2
Homoserine	(5 mM)	−0.4	−5.0

Uptake of lysine was measured from a 25 μM solution either in the presence or absence of Na^+.

seem to be the same as those that function to reabsorb amino acids in the proximal tubule *in vivo*.

5. PRESENCE IN LLC–PK$_1$ CELLS OF AN ENZYME POSTULATED TO MEDIATE AMINO ACID TRANSPORT

γ-Glutamyltransferase (EC-2.3.2.2.) is an enzyme of broad specificity which catalyzes the transfer of the γ-glutamyl group of glutathione and a variety of γ-glutamyl peptides to a wide range of amino acids and peptide acceptors (see Meister and Tate, 1976; and Curthoys and Hughey, 1979, for reviews). The enzyme is a particulate glycoprotein that is present at very high activities in the proximal nephron and it has been shown to be located exclusively in the brush border membrane of proximal tubule cells (Marathe *et al.*, 1979; Tsao and Curthoys, 1980). High levels of γ-glutamyltransferase have been observed associated with brush border membranes of other epithelia involved in active transport of amino acids (Curthoys and Shapiro, 1975; Cornell and Meister, 1976; Garvey *et al.*, 1976; Sepúlveda and Burton, 1982). The preferential association of γ-glutamyltransferase with cellular barriers involved in amino acid transport has led Meister (1973) to postulate that this enzyme, together with the associated intracellular γ-glutamyl cycle, might mediate amino acid transport.

γ-Glutamyltransferase activity was detected histochemically in LLC–PK$_1$ cells by Perantoni and Berman (1979). We have shown (Sepúlveda *et al.*, 1982) that γ-glutamyltransferase is present in confluent LLC–PK$_1$ cultures at activities comparable to those in pig kidney cortex homogenates. The properties of the LLC–PK$_1$ cell enzyme resemble closely those of the enzyme purified from mammalian kidney. It is interesting that during LLC–PK$_1$ cell growth *in vitro* there is an increase in the activity of γ-glutamyltransferase to reach highest values in confluent monolayers (Sepúlveda *et al.*, 1982). The development of enzyme activity is paralleled by the increase of α-methylglucoside uptake by the cells (Figure 4). The development of this Na$^+$-dependent hexose transport system during confluent monolayer formation in LLC–PK$_1$ cells has previously been observed by Mullin *et al.* (1980)and Amsler and Cook (1982). The former authors proposed that the development of the hexose transport system *in vitro*, as the cells progress from the subconfluent state, represents a process of differentiation analogous to that occurring in proximal tubule epithelium *in vivo*. The increase in γ-glutamyltransferase levels mimics the ontogeny of the appearance of enzyme in the kidney (Tate and Meister, 1975), which suggests that

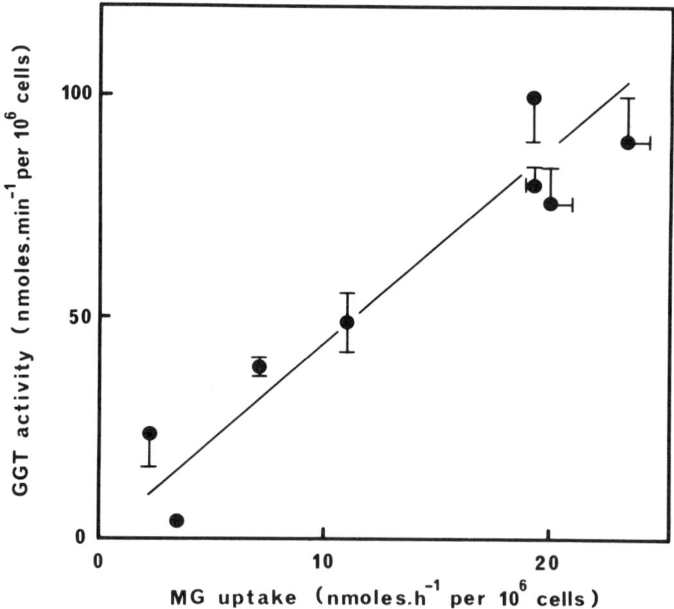

Figure 4. Correlation between α-methyl-D-glucoside (MG) uptake and γ-glutamyltransferase (GGT) activity in LLC–PK$_1$ cells during growth (modified from Sepúlveda *et al.*, 1982). Cells were seeded at low density in plastic dishes and sampled every 2 days to assess sugar transport and enzyme activity.

γ-glutamyltransferase is another expression of a differentiated phenotype corresponding to proximal tubule cells that has been retained by the LLC–PK$_1$ line.

The question of the possible involvement of γ-glutamyltransferase and the γ-glutamyl cycle in amino acid transport is still in debate (Kalra *et al.*, 1981, Curthoys and Hughey, 1979; Silbernagl, 1982). We have concluded that the lack of correlation between amino acid uptake and the activity of the enzyme suggested that the latter could not be a major mediator of neutral amino acid transport in LLC–PK$_1$ cells (Sepúlveda *et al.*, 1982). More recently Rabito *et al.* (1984) used a histochemical technique to show that γ-glutamyltransferase is associated exclusively with the apical membrane of LLC–PK$_1$ monolayers. This taken together with the location of amino acid transport systems (see section 6) makes the possible function of the enzyme in the amino acid transport process in LLC–PK$_1$ cells unlikely. Whether this also implies that the enzyme does not participate in the reabsorption of neutral amino acids in the nephron requires further study.

6. SITE OF TRANSPORT OF AMINO ACIDS IN LLC–PK$_1$ CELLS

The uptakes of alanine and of leucine by LLC–PK$_1$ cells fall as the cells grow from a low culture density, attaining steady levels at confluence (Sepúlveda *et al.*, 1982). A similar observation has been made for the uptake of α-aminoisobutyric acid by LLC–PK$_1$ cells (Lever, 1982). One possible explanation for these findings is that down regulation of the transport systems takes place as confluency is reached. Analogous changes in the activity of the A, ASC, and, to a minor extent, the L systems in MDCK cells have also been reported (Boerner and Saier, 1982). This canine renal cell line also has an epithelial morphology, but it exhibits biochemical characteristics of the distal rather than the proximal tubule (see Chapter 3). Boerner and Saier interpreted their results in terms of the coordinated regulation of amino acid transport and cell growth: they support the view that nutrient transport may play a key role in regulating cell proliferation, and that an increased nutrient uptake may be a triggering event for the establishment of malignant cell growth (Pardee, 1964; Holley, 1972).

Although increased amino acid uptake has been observed in transformed cells (Isselbacher, 1972; Borghetti *et al.*, 1980), a causal relationship between amino acid transport regulation and growth control has not been demonstrated. An alternative explanation for the cell-density-related change in amino acid transport observed in LLC–PK$_1$ cells and in MDCK cells is based on the knowledge that as growth takes place toward confluency, the cells become closely apposed and attached via tight junctions. Subsequent development of membrane asymmetry could result in part of the transport systems becoming hidden from the bathing medium and thus undetectable using normal methods of measurement. We have employed an autoradiographic technique (King *et al.*, 1981) to locate the site of alanine transport in LLC–PK$_1$ cells grown on a solid substratum. On macroscopic examination, the silver grains are apparently uniformly distributed over subconfluent cultures, but they are confined to the edges of the monolayer when the cultures are fully confluent (Fig. 5). Microscopic examination of the autoradiographs reveals that most of the uptake in subconfluent cultures is performed by cells that have sides not in contact with neighboring cells (Figure 6a), and similarly, in confluent cultures, silver grains are found predominantly over cells located at the periphery of the monolayer. As expected, no uptake of alanine is observed in the absence of Na$^+$ (Figures 5b, 6b). These results, when considered together with those of Rabito and Karish (1982), who by growing LLC–PK$_1$ cells on a permeable support demonstrated that most of the neutral amino acid uptake in confluent monolayers takes place through the basolateral membrane, suggest that the fall in uptake observed with growth is a consequence of cellular polarization rather than, or in addition to, regulation related to growth. This is confirmed by the fact that in monolayers grown on a permeable substratum, the distribution of transported amino acid assessed by autoradiography seems uniform (Sepúlveda and Pearson, 1984b). It could be argued that the location of the amino acid transport system in cells grown on a

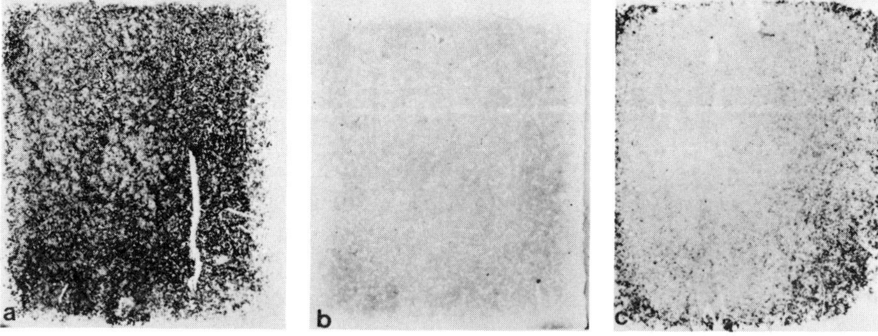

Figure 5. Photograph of autoradiographs of LLC–PK$_1$ monolayers grown on glass cover slips. Size of cover slips is 22 × 26 mm. Monolayers a and b were grown to a density of 350,000 cells/cover slip, monolayer c to confluency at a density of 750,000 cells/cover slip. The cells were incubated for 1 min in the presence of 250 µM alanine and 100 µCi/ml [³H] alanine in Na$^+$-containing Hanks' medium (a and c) or Na$^+$-free, choline-substituted medium (b). The incubation was stopped by washing in ice-cold choline–Hanks' buffer and immediate fixation in a 4% glutaraldehyde-containing buffer (King *et al.*, 1981). The washed and dried monolayers were coated with Ilford K2 photographic emulsion, exposed for 18 days, developed, and stained with eosin. Dark areas show regions of high density of silver grains. Uptake of alanine measured in parallel cover slips gave 2.17 ± 0.25, 0.05 ± 0.06, and 0.91 ± 0.17 nmoles min^{-1}/10^6 cells for conditions a, b, and c, respectively (means ± S.E.M. of three replicates).

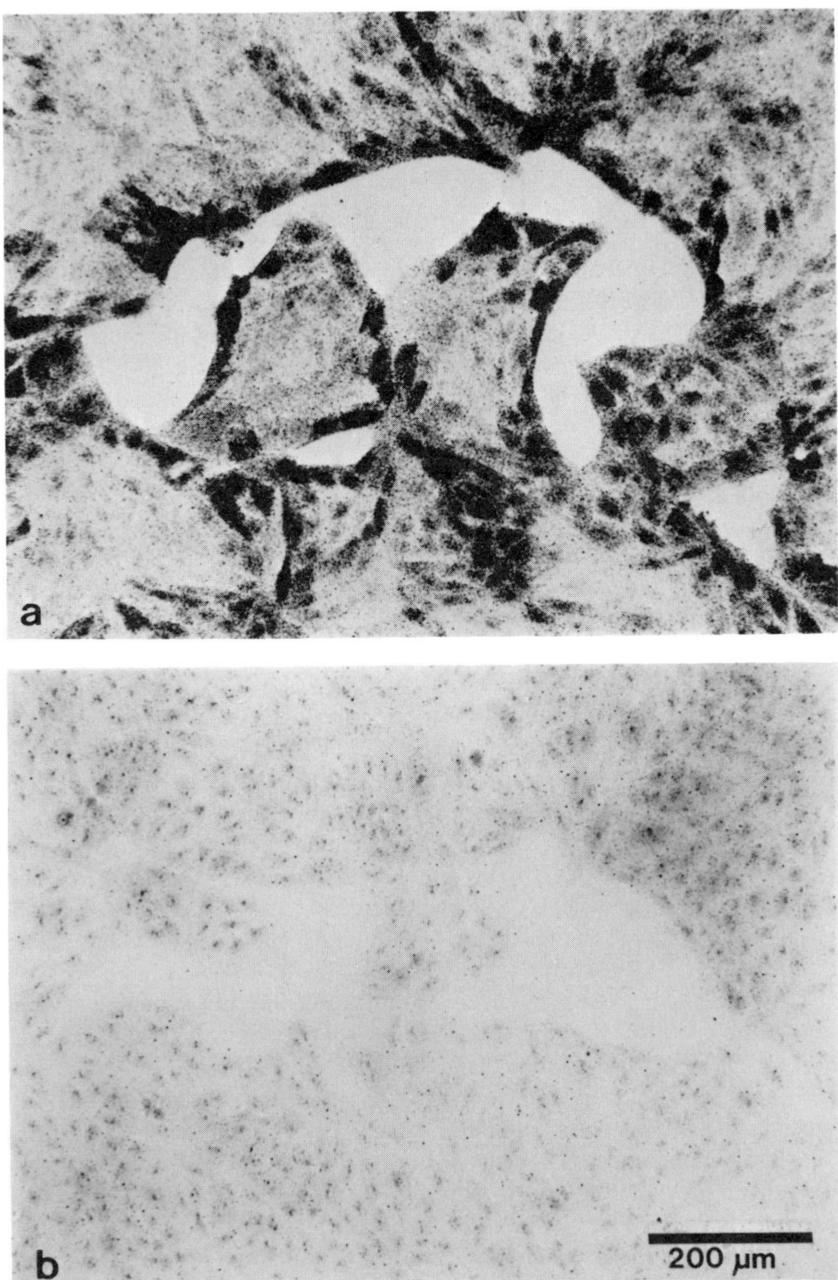

Figure 6. Photomicrographs of monolayers of LLC–PK$_1$ cells grown on glass cover slips, taken from the specimens shown in Fig. 5a and b. A high density of grains is associated, in the presence of Na$^+$, with cells at the border between the monolayer and empty noncolonized spaces. Only a few silver grains uniformly distributed on the monolayers are seen in the absence of Na$^+$. This could represent background radioactivity or a weak uptake not detected in conventional flux measurements.

permeable substratum might be different from that in cells grown on impermeable surfaces. We have demonstrated, however, (Sepúlveda and Pearson, 1984b) that substantial increases in alanine transport occur when cell monolayers are suspended or when the permeability of the tight junctions is increased by brief pretreatment in solution lacking Ca^{2+} ions. There is a parallel increase in ouabain-sensitive K^+ uptake that suggests a common (basolateral) location for the transport systems for alanine and K^+. The results presented above do not invalidate the idea that down regulation of transport takes place during growth. In fact, Amsler *et al.* (1983) have demonstrated growth regulation of transport system A in LLC–PK$_1$ cells grown on a permeable substratum, where the problems associated with accessibility to the site of transport should be minimized.

7. CONCLUSIONS

The availability of cultured cell lines of kidney tubule epithelial origin provides a simple *in vitro* system in which to study basic problems of epithelial physiology and biochemistry. The results discussed previously show that LLC–PK$_1$ cells, a line showing many features of proximal tubule cells, take up neutral amino acids via the well-characterized, A, ASC, and L transport systems and cationic amino acids via a specific transport system that shares some (but not all) of the characteristics of the y^+ system of other cells. In LLC–PK$_1$ cells, these transport systems seem to be localized at the basolateral aspect of the cell facing the substratum. It is reasonable to speculate that this location is the same as that found in the proximal tubule *in vivo*. The fact that no neutral amino acid transport systems with characteristics similar to those of proximal nephron or intestinal brush border membranes have yet been detected in LLC–PK$_1$ cells suggests that the expression of apical and basolateral membrane components is under independent control.

Little is known about the generation or maintenance of membrane asymmetry in epithelial cells. While it has been shown that tight junctions provide a barrier for lateral diffusion in the plane of the membrane (Dragsten *et al.*, 1981), experiments on viral budding from MDCK cells suggests that interaction of epithelial cells with a substratum is a sufficient signal to induce the polarization of membrane components (Rodriguez-Boulan *et al.*, 1983). Some information exists on the biosynthesis of enzymes preferentially associated with the brush border membrane of kidney and intestine (see Kenny and Maroux, 1982, for a review), but very little is understood concerning how these proteins are directed to their specific membrane location.

The observation that both Na^+-dependent hexose transport and γ-glutamyltransferase activity can be induced to develop in LLC–PK$_1$ cells under the appropriate growth conditions suggests that it may also be possible to establish culture conditions, perhaps requiring specific hormones, under which apical transport systems for neutral amino acids are expressed. This aim should be aided by the recent availability of a serum-free, hormone-supplemented medium for the culture of $LLC-PK_1$ cells (Chuman *et al.*, 1982). If this approach is successful, the $LLC-PK_1$ cell line will be of further use as a model system in which to characterize the transport systems responsible for neutral amino acid extraction in the nephron and their relationship to γ-glutamyltransferase activity.

REFERENCES

Abaza, N. A., Leighton, J., and Schultz, S. G., 1974, Effects of ouabain on the structure and function of a cell line (MDCK) derived from dog kidney, *In Vitro* **10**:172–183.

Amsler, K., and Cook, J. S., 1982, Development of a Na^+-dependent hexose transport in a cultured line of porcine kidney cells, *Am. J. Physiol.* **242**:C94–C101.

Amsler, K., Shafer, C., and Cook, J. S., 1983, Growth-dependent AIB and meAIB uptake by LLC-PK_1 cells: Effect of differentiation inducers and of TPA, *J. Cell Physiol.* **114**:184–190.

Aronson, P. S., and Sacktor, B., 1975, The Na^+-gradient dependent transport of D-glucose in renal brush border membranes, *J. Biol. Chem.* **250**:6032–6039.

Barfuss, D. W., and Schafer, J. A., 1979, Active amino acid absorption by proximal convoluted and proximal straight tubules, *Am. J. Physiol.* **236**:F149–F162.

Boerner, P., and Saier, M. H., 1982, Growth regulation and amino acid transport in epithelial cells: Influence of culture conditions and transformation on A, ASC and L transport activities, *J. Cell. Physiol.* **113**:240–246.

Borghetti, A. F., Piedimonte, G., Tramacere, M., Severini, A., Ghiringhelli, P., and Guidotti, G. G., 1980, Cell density and amino acid transport in 3T3, SV3T3 and SV3T3 revertant cells, *J. Cell. Physiol.* **105**:39–49.

Christensen, H. N., 1975, Recognition sites for material transport and information transfer, *Curr. Top. Membr. Transp.* **6**:227–258.

Christensen, H. N., 1979, Exploiting amino acid structure to learn about membrane transport, *Adv. Enzymol.* **49**:41–101.

Christensen, H. N., Liang, M., and Archer, E. G., 1967, A distinct Na^+-requiring transport system for alanine, serine, cysteine and similar amino acids, *J. Biol. Chem.* **242**:5237–5246.

Chuman, L., Fine, L. G., Cohen, A. H., and Saier, M. H., 1982, Continuous growth of proximal tubular kidney epithelial cells in hormone-supplemented serum-free media, *J. Cell. Biol.* **94**:506–510.

Cornell, J. S., and Meister, A., 1976, Glutathione and γ-glutamyl cycle enzymes in crypt and villus tip cells of rat jejunal mucosa, *Proc. Natl. Acad. Sci. USA* **73**:420–422.

Crane, R. K., 1962, Hypothesis of mechanism of intestinal active transport of sugars, *Fed. Proc.* **21**:891–895.

Curthoys, N. P., and Hughey, R. P., 1979, Characterisation and physiological function of rat renal γ-glutamyltranspeptidase, *Enzyme* **24**:383–403.

Curthoys, N. P., and Shapiro, R., 1975, γ-Glutamyltranspeptidase in intestinal brush border membranes, *FEBS Lett.* **58**:230–233.

Dragsten, P. R., Blumenthal, R., and Handler, J. S., 1981, Membrane asymmetry in epithelia: Is the tight junction a barrier to diffusion in the plamsa membrane? *Nature* **294**:718–722.

Evers, J., Murer, H., and Kinne, R., 1976, Phenylalanine uptake in isolated renal brush border vesicles, *Biochim. Biophys. Acta* **426**:598–615.

Fass, S. J., Hammerman, M. R., and Sacktor, B., 1977, Transport of amino acids in renal brush border membrane vesicles. Uptake of the neutral amino acid L-alanine, *J. Biol. Chem.* **252**:583–590.

Foulkes, E. C., and Gieske, T., 1973, Specificity and metal sensitivity of renal amino acid transport, *Biochim. Biophys. Acta* **318**:439–445.

Fox, M., Thier, S., Rosenberg, L., and Segal, S., 1964, Ionic requirements for amino acid transport in the rat kidney cortex slice. I. Influence of extracellular ions, *Biochim. Biophys. Acta* **79**:167–176.

Franchi–Gazzola, R., Gazzola, G. C., Dall'Asta, V., and Guidotti, G. G., 1982, The transport of alanine, serine and cysteine in cultured human fibroblasts, *J. Biol. Chem.* **257**:9582–9587.

Frömter, E., 1982, Electrophysiological analysis of rat renal sugar and amino acid transport. I. Basic phenomena, *Pflügers Arch.* **393**:179–189.

Garvey, T. Q., Hyman, P. E., and Isselbacher, K. J., 1976, γ-Glutamyltranspeptidase of rat intestine: localisation and possible role in amino acid transport, *Gastroenterology* **71**:778–785.

Guidotti, G. G., Borghetti, A. F., and Gazzola, G. C., 1978, The regulation of amino acid transport in animal cells, *Biochim. Biophys. Acta* **515**:329–366.

Hammerman, M. R., and Sacktor, B., 1977, Transport of amino acids in renal brush border membrane vesicles, uptake of L-proline, *J. Biol. Chem.* **252**:591–595.

Handlogten, M. E., Garcia–Canero, R., Lancaster, K. T., and Christensen, H. N., 1981, Surprising differences in substrate selectivity and other properties of systems A and ASC between rat hepatocytes and the hepatoma cell line HTC, *J. Biol. Chem.* **256**:7905–7909.

Handlogten, M. E., Weissbach, L., and Kilberg, M. S., 1982, Heterogeneity of Na$^+$-independent 2-aminobicyclo-(2,2,1)-heptane-2-carboxylic acid and L-leucine transport in isolated rat hepatocytes in primary culture, *Biochem. Biophys. Res. Comm.* **104**:307–313.

Hillman, R. E., Albrecht, I., and Rosenberg, L. E., 1968, Transport of amino acids by isolated rabbit renal tubules, *Biochim. Biophys. Acta* **150**:528–530.

Hillman, R. E., Albrecht, I., and Rosenberg, L. E., 1968b, Transport of amino acids and analysis of multiple glycine systems in isolated mammalian renal tubules, *J. Biol. Chem.* **243**:5566–5571.

Holley, R. W., 1972, A unifying hypothesis concerning the nature of malignant growth, *Proc. Natl. Acad. Sci. USA* **69**:2840–2841.

Hoshi, T., Sudo, K., and Suzuki, Y., 1976, Characteristics of changes in the intracellular potential associated with transport of neutral dibasic and acidic amino acids in *Triturus* proximal tubules, *Biochim. Biophys. Acta* **448**:492–504.

Hull, R. N., Cherry, W. R., and Weaver, G. W., 1976, The origin and characteristics of pig kidney cell strain, LLC–PK$_1$, *In Vitro* **12**:670–677.

Isselbacher, K. J., 1972, Increased uptake of amino acids and 2-deoxy-D-glucose by virus-transformed cells in culture, *Proc. Natl. Acad. Sci. USA* **69**:585–589.

Kalra, V. K., Sikka, S. C., and Sethi, G. S., 1981, Transport of amino acids in γ-glutamyltranspeptidase-implanted human erythrocytes, *J. Biol. Chem.* **256**:5567–5571.

Kenny, A. J., and Maroux, S., 1982, Topology of microvillar membrane hydrolases of kidney and intestine, *Physiol. Rev.* **62**:91–128.

Kilberg, M. S., 1982, Amino acid transport in isolated rat hepatocytes, *J. Membr. Biol.* **69**:1–12.

King, I. S., Sepúlveda, F. V., and Smith, M. W., 1981, Cellular distribution of neutral and basic amino acid transport systems in rabbit ileal mucosa, *J. Physiol. (London)* **319**:355–368.

Kinne, R., Murer, H., Kinne–Saffran, E., Thees, M., and Sachs, G., 1975, Sugar transport by renal plasma membrane vesicles. Characterisation of the systems in the brush-border microvilli and basal-lateral plasma membranes, *J. Membr. Biol.* **21**:375–395.

Lever, J. E., 1982, Expression of a differentiated transport function in apical membrane vesicles isolated from an established kidney epithelial cell line. Sodium electrochemical potential-mediated active sugar transport, *J. Biol. Chem.* **257**:8680–8686.

Marathe, G. V., Nash, B., Haschemeyer, R. H., and Tate, S. S., 1979, Ultrastructural localisation of γ-glutamyltranspeptidase in rat kidney and duodenum, *FEBS Lett.* **107:**436–440.

Meister, A., 1973, On the enzymology of amino acid transport, *Science* **180:**33–39.

Meister, A., and Tate, S. S., 1976, Glutathione and related γ-glutamyl compounds: Biosynthesis and utilisation, *Ann. Rev. Biochem.* **45:**559–604.

Mills, J. W., Macknight, A. D. C., Dayer, J.-M., and Ausiello, D. A., 1979, Localisation of [^{3}H]ouabain-sensitive Na$^+$ pump sites in cultured pig kidney cells, *Am. J. Physiol.* **236:**C157–C162.

Mircheff, A. K., Van Os, C. H., and Wright, E. M., 1980, Pathways for alanine transport in intestinal basal lateral membrane vesicles, *J. Membr. Biol.* **52:**83–92.

Mircheff, A. K., Kippen, I., Hirayama, B., and Wright, E. M., 1982, Delineation of sodium-stimulated amino acid transport pathways in rabbit kidney brush border vesicles, *J. Membr. Biol.* **64:**113–122.

Misfeldt, D. S., and Sanders, M. J., 1981, Transepithelial transport in cell culture: D-glucose transport by a pig kidney cell line (LLC–PK$_1$), *J. Membr. Biol.* **59:**13–18.

Mullin, J. M., Weibel, J., Diamond, L., and Kleinzeller, A., 1980, Sugar transport in the LLC–LK$_1$ renal epithelial cell line; similarity to mammalian kidney and the influence of cell density, *J. Cell Physiol.* **104:**375–389.

Murer, H., and Kinne, R., 1980, The use of isolated membrane vesicles to study epithelial transport processes, *J. Membr. Biol.* **55:**81–95.

Oxender, D. L., and Christensen, H. N., 1963, Distinct mediating systems for the transport of neutral amino acids by the Ehrlich cells, *J. Biol. Chem.* **238:**3686–3699.

Pardee, A. B., 1964, Cell division and hypothesis of cancer, *Natl. Cancer Inst. Monogr.* **14:**7–14.

Paterson, J. Y. F., Sepúlveda, F. V., and Smith, M. W., 1979, Two-carrier influx of neutral amino acids into rabbit ileal mucosa, *J. Physiol. (London)* **292:**339–350.

Paterson, J. Y. F., Sepúlveda, F. V., and Smith, M. W., 1980, A sodium-independent low affinity transport system for neutral amino acids in rabbit ileal mucosa, *J. Physiol. (London)* **298:**333–346.

Perantoni, A., and Berman, J. J., 1979, Properties of Wilm's tumour line (Tu Wi) and pig kidney line (LLC–PK$_1$) typical of normal kidney tubular epithelium, *In Vitro* **15:**446–454.

Philo, R. D., and Eddy, A. A., 1978, Equilibrium and steady-state models of the coupling between the amino acid gradient and the sodium electrochemical gradient in mouse ascites-tumour cells, *Biochem. J.* **174:**811–817.

Rabito, C. A., 1981, Localisation of the Na$^+$-sugar cotransport system in a kidney epithelial cell line (LLC–PK$_1$), *Biochim. Biophys. Acta* **649:**286–296.

Rabito, C. A., and Ausiello, D. A., 1980, Na$^+$-dependent sugar transport in a cultured epithelial cell line from pig kidney, *J. Membr. Biol.* **54:**31–38.

Rabito, C. A., and Karish, M. V., 1982, Polarized amino acid transport by an epithelial cell line of renal origin (LLC–PK$_1$), *J. Biol. Chem.* **257:**6802–6808.

Rabito, C. A., Kreisberg, J. I., and Wight, D., 1984, Alkaline phosphatase and γ-Glutamyltranspeptidase as polarization markers during the organization of LLC–PK$_1$ cells into an epithelial membrane, *J. Biol. Chem.* **259:**574–582.

Reid, M., Gibb, L. E., and Eddy, A. A., 1974, Ionophore mediated coupling between ion fluxes and amino acid absorption in mouse ascites-tumour cells. Restoration of the physiological gradients by valinomycin in the absence of adenosine triphosphate, *Biochem. J.* **140:**383–393.

Rodriguez–Boulan, E., Paskiet, K. T., and Sabatini, D. D., 1983, Assembly of enveloped viruses in Madin–Darby canine kidney cells: polarized budding from single attached cells and from clusters of cells in suspension, *J. Cell Biol.* **96:**866–874.

Rosenberg, L. E., Blair, A., and Segal, S., 1961, The transport of amino acids by rat kidney cortex slices, *Biochim. Biophys. Acta* **54:**479–488.

Samaržija, I., and Frömter, E., 1982, Electrophysiological analysis of rat renal sugar and amino acid transport. III. Neutral amino acids. *Pflügers Arch.*, **393**:199–209.

Schafer, J. A., and Barfuss, D. W., 1980, Membrane mechanisms for transepithelial amino acid absorption and secretion, *Am. J. Physiol.* **238**:F335–F346.

Schultz, S. G., and Curran, P. F., 1970, Coupled transport of sodium and organic solutes, *Physiol. Rev.* **50**:637–718.

Sepúlveda, F. V., and Burton, K. A., 1982, γ-Glutamyl transferase activity in the pig proximal colon during early postnatal development, *FEBS Lett.* **139**:171–173.

Sepúlveda, F. V., and Pearson, J. D., 1982, Characterisation of neutral amino acid uptake by cultured epithelial cells from pig kidney, *J. Cell. Physiol.* **112**:182–188.

Sepúlveda, F. V., and Pearson, J. D., 1984a, Deficiency in intercellular communication in two established renal epithelial cell lines (LLC–PK$_1$ and MDCK). *J. Cell. Sci.* **66**:81–93.

Sepúlveda, F. V. and Pearson, J. D., 1984b, Localization of alanine uptake by cultured renal epithelial cells (LLC–PK$_1$) to the basolateral membrane. *J. Cell Physiol.* **118**:211–217.

Sepúlveda, F. V., and Smith, M. W., 1978, Discrimination between different entry mechanisms for neutral amino acids in rabbit ileal mucosa, *J. Physiol. (London)* **282**:73–90.

Sepúlveda, F. V., Burton, K. A., and Pearson, J. D., 1982, The development of γ-glutamyltransferase in a pig renal-epithelila-cell line *in vitro, Biochem. J.* **208**:509–512.

Sigrist-Nelson, K., Murer, H., and Hopfer, U., 1975, Active alanine transport in isolated brush border membranes, *J. Biol. Chem.* **250**:5674–5680.

Silbernagl, S., 1979, Renal transport of amino acids, *Klin. Wocheschr.* **57**: 1009–1019.

Silbernagl, S., 1982, Metabolism and absorption of glutathione (GSH) in the tubule lumen. An *in vivo* microperfusion study in rat kidney, *J. Physiol. (London)* **325**:59P.

Silbernagl, S., Foulkes, E. C., and Detjeen, P., 1975, Renal transport of amino acids, *Rev. Physiol. Biochem. Pharmacol.* **74**:105–167.

Slack, E. N., Liang, C.–C. T., and Sacktor, B., 1977, Transport of L-proline and D-glucose in luminal (brush border) and contraluminal (basal–lateral) membrane vesicles from the renal cortex, *Biochem. Biophys. Res. Comm.* **77**:891–897.

Smith, M. W., Sepúlveda, F. V., and Paterson, J. Y. F., 1983, Cellular aspects of amino acid transport, in: *Intestinal Transport: Fundamental and Comparative Aspects* (M. Gilles-Baillien and R. Gilles, eds.), Springer-Verlag, Berlin, pp. 46–63.

Stevens, B. R., Ross, H. J., and Wright, E. M., 1982, Multiple transport pathways for neutral amino acids in rabbit jejunal brush border vesicles, *J. Membr. Biol.* **66**:213–225.

Tate, S. S., and Meister, A., 1975, Identity of maleate-stimulated glutaminase with γ-glutamyl transpeptidase in rat kidney, *J. Biol. Chem.* **250**:4619–4627.

Thomas, E. L., and Christensen, H. N., 1971, Nature of the cosubstrate action of Na$^+$ in a transport system, *J. Biol. Chem.* **246**:1682–1688.

Tsao, B., and Curthoys, N. P., 1980, The absolute asymmetry of orientation of γ-glutamyl transpeptidase and aminopeptidase on the external surface of the rat renal brush border membrane, *J. Biol. Chem.* **255**:7708–7711.

Turner, R. J., and Silverman, M., 1977, Sugar uptake into brushborder vesicles from normal human kidney, *Proc. Natl. Acad. Sci. USA* **74**:2825–2829.

Ullrich, K. J., 1979, Sugar, amino acid and Na$^+$ cotransport in the proximal tubule, *Ann. Rev. Physiol.* **41**:181–195.

White, M. F., and Christensen, H. N., 1982, The two-way flux of cationic amino acids across the plasma membrane of mammalian cells is largely explained by a single transport system, *J. Biol. Chem.* **257**:10069–10080.

White, M. F., Gazzola, G. C., and Christensen, H. N., 1982, Cationic amino acid transport into cultured animal cells. I. Influx into cultured human fibroblasts, *J. Biol. Chem.* **257**:4443–4449.

7

Transepithelial Transport in Cell Culture
Mechanism and Bioenergetics of Na^+, D-Glucose Cotransport

DAYTON S. MISFELDT and MARTIN J. SANDERS

1. INTRODUCTION

1.1. Expression of Transporting Epithelial Phenotype in Cell Culture

This chapter is a general review of our laboratory's effort to utilize epithelial cell culture to explore questions of epithelial transport that are less amenable to other experimental approaches. It is remarkable that cultured transporting epithelial cells express their phenotype when dissociated from the mesenchymal scaffolding of the tissue of origin (Misfeldt *et al.*, 1976). Our experimental insight was to culture the epithelial cells on a permeable support, such as a membrane filter, that would allow the cell layer to be experimentally manipulated (Misfeldt *et al.*, 1975). The cultured cells could then be mounted in an Ussing chamber, which provided access to, and isolation of, the fluid bathing either side of the epithelial sheet. Thus, by this experimental technique there is formed in culture a functioning epithelial tissue from dissociated individual cells. Experimental questions could thus be directed at cellular and subcellular processes that require tissue level function for solution.

1.2. Advantages of Cell Culture for the Study of Transepithelial Transport

The inherent technical advantages of cultured cells for the study of epithelial transport include their ability to be controlled and their simplification as a model

DAYTON S. MISFELDT and MARTIN J. SANDERS • Palo Alto Veterans Medical Center, Palo Alto, California 94304; and Stanford University, Stanford, California 94305. *Present address for M.J.S.:* Center for Ulcer Research and Education, VA Wadsworth Medical Center, Los Angeles, California 90073.

system. For those cells that can be clonally cultured, the greatest possible cellular homogeneity can be achieved. In addition, the experimenter can control the external physical and chemical environment, e.g., the gaseous or organic metabolic substrates or the temperature, as well as the type and concentration of the transport substrates of interest. It is even possible to modify the lipid composition of the plasma membranes by manipulation of the culture media. Then the effect of such a manipulation on transport function can be analyzed (Ryan and Simoni, 1980). From the dissociation of epithelial tissues to individual cells, complex epithelial organ architecture can be reduced to a simple functioning monolayer sheet. This is readily appreciated in the study of transport function by pulmonary alveolar cells. Cell cultures were derived from dissociated pulmonary tissue, which was enriched for type II pneumocytes. The epithelial transport functions of such pneumocytes could then be identified. Their epithelial transport functions were thus first appreciated and studied *in vitro* (Mason *et al.*, 1982; Goodman *et al.*, 1983). The full power of genetic manipulation of epithelial cells as microorganisms for the analysis of epithelial transport has not been realized. The problems of genetic manipulation are related to the difficulty of selection and to subsequent amplification of "mutant" cell populations that have altered transport properties. Such alterations in transport properties may not influence cell proliferation and thus preclude the use of changes in proliferation for selection. The use of fluorescent-activated cell sorting can overcome this limitation; as appropriate antibodies are isolated or other fluorescent labels are developed, cells of interest may then be identified. The other equally difficult problem has been the resistance of mammalian epithelial cells to continuous proliferation and passage, let alone the requirement for clonal growth, a necessity for genetic analysis. This, too, may be a problem that is finally yielding (Ehmann and Misfeldt, 1984).

Our laboratory specifically proceeded to identify an organic solute (1) that was vectorially transported across the cell layer of an epithelial cell line without chemical modification, (2) that was ideally available with radioisotopic labels, and (3) whose transport is reflected as an electrical signal. Na^+-coupled D-glucose transport by the LLC–PK_1 pig renal cell line fulfilled all three of these criteria. This transepithelial transport system was first identified, and validated as authentic, by a comparison with the results of studies *in vivo* concerning this system. Subsequently, we sought, in a series of experiments, (1) to characterize the mechanism of Na^+-dependent uptake of D-glucose, (2) to test the Koefoed–Johnsen and Ussing hypothesis of transepithelial Na^+ transport, and (3) to characterize the cellular bioenergetic pattern associated with coupled transport by cultured cells.

Table I.   Transepithelial Electrical Properties of
LLC–PK$_1$ Monolayer

Potential difference (mV)	-2.7 ± 0.16 (32)[a]
Short-circuit current (μ A cm^{-2})	13.0 ± 0.7 (32)
Resistance (ohm·cm^2)	211.0 ± 13.0 (44)

[a] $\bar{\chi} \pm$ SEM(n) measured at 37°C in Hanks' salt solution (HSS) 5.5 mM glucose.

2. TRANSEPITHELIAL NA$^+$, d-GLUCOSE COTRANSPORT BY THE RENAL CELL LINE LLC–PK$_1$

The Na$^+$-dependent d-glucose uptake system was described for the LLC–PK$_1$ in monolayer culture (Mullin *et al.*, 1980; Rabito and Ausiello, 1980). The nonmetabolized d-glucose analog alpha-methyl-d-glucoside was demonstrated to be concentrated intracellularly by a Na$^+$-dependent, phlorizin-inhibitable, carrier-mediated transport process. Whereas d-glucose and specific analogs are taken up by all cells, the Na$^+$-dependent process, whereby there is cellular concentration and transepithelial transport, is limited to the proximal tubule of the nephron, the small intestine, and the choroid plexus. This Na$^+$-dependent transport process is specifically inhibited by the drug phlorizin. When the dissociated cells were plated onto collagen-coated membrane filters, the individual cells became uniformly polarized, joined one to another by juxtaapical tight junctions, and formed a monolayer epithelial tissue sheet (Misfeldt and Sanders, 1981). The membrane filter support allowed handling of the cell monolayer and placement in an Ussing chamber for the measurement of electrophysiological parameters and the fluxes of isotopically labeled molecules. This technique (Misfeldt *et al.*, 1976; Cereijido *et al.*, 1978) was applied to the analysis of Na$^+$, d-glucose cotransport by the renal cell line LLC–PK$_1$.

2.1. Analysis of Cotransport by I_{sc} (Short-Circuit Current)

When the LLC–PK$_1$ monolayer was placed in an Ussing chamber, in which each side of the monolayer was bathed by identical d-glucose-free solutions, there was a transepithelial resistance of over 200 ohms · cm^2. The addition of d-glucose resulted in a prompt increase in both the transepithelial electrical potential (PD) and the I_{sc} without a significant change in the resistance. The electrophysiological parameters are summarized in Table I. The addition of phlorizin caused a rapid and nearly complete inhibition of the PD and I_{sc} as illustrated in

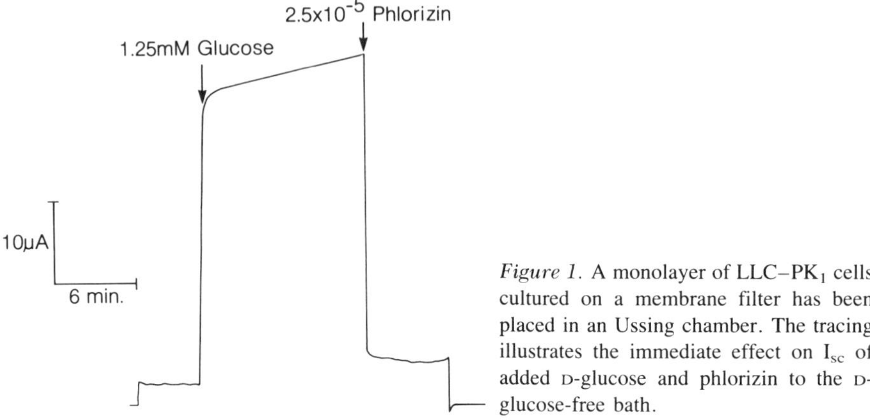

Figure 1. A monolayer of LLC–PK$_1$ cells cultured on a membrane filter has been placed in an Ussing chamber. The tracing illustrates the immediate effect on I_{sc} of added D-glucose and phlorizin to the D-glucose-free bath.

Fig. 1. The phlorizin effect occurred only when the drug was added to the apical bath, suggesting that the I_{sc} was a reflection of apical membrane interactions between the cotransporter, Na$^+$ and D-glucose. Studies to be described subsequently support this interpretation.

2.1.1. K_m and V_{max} of Cotransport

If the I_{sc} is a true reflection of the interaction between the apical cotransporter and glucose, then the Michaelis constants of this sugar transport system may be determined in a Ussing chamber. The I_{sc} was measured as the concentration of D-glucose in the bathing solutions was progressively increased. The resultant I_{sc}s were analyzed by an Eadie–Hofstee plot of the I_{sc} versus I_{sc}/(D-glucose). In this manner the V_{max}, the maximum transport rate, and the K_m, the concentration of D-glucose required for half-maximal I_{sc}, were determined. Similar studies were done with other sugar analogs, which then permitted the relative affinities of these compounds for the cotransport system to be compared (Fig. 2).

2.1.2. Specificity of Cotransport

From the relative affinity, K_m, the relation between sugar structure and cotransporter interaction was established for comparison with that observed in the small intestine and proximal tubule. In Table II D-glucose is compared with five analogs. The hierarchy of effectiveness of these agents in stimulating I_{sc} is similar to that observed in the small intestine (Barry *et al.*, 1965); the same pattern of structural effectiveness has been associated with apical membrane depolarization (Frömter, 1979) and steady-state concentration differences across

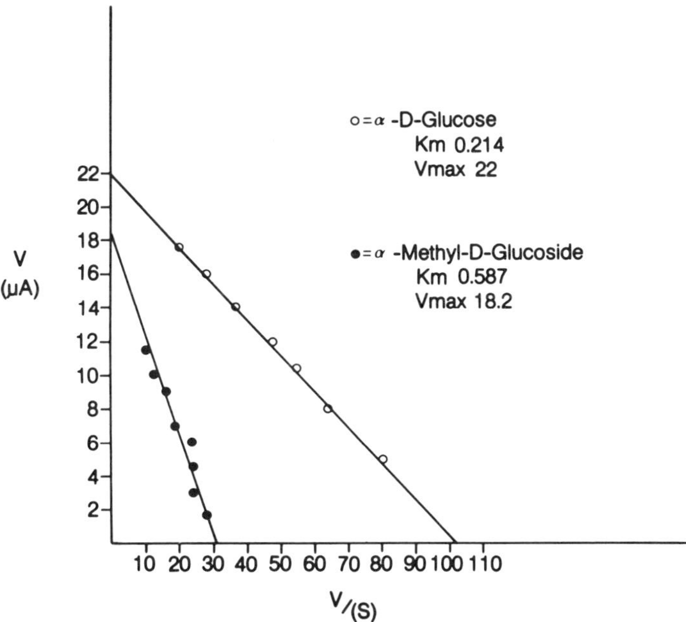

Figure 2. An Eade–Hofstee plot of an experiment illustrating the method for comparing D-glucose with other sugars in the cotransport mechanism. The effect of concentration on I_{sc} was used to estimate the relative affinity, K_m, between the sugar and cotransporter and apparent number of cotransport sites, V_{max}. The slopes and intercepts were calculated by least-squares linear regression.

Table II.   Concentration of Sugars to Stimulate I_{SC}[a]

	$K_{1/2}$(mM)	I_{SC} $V_{max}(\mu$ A cm^{-2})
β-Methyl-D-glucoside	0.131 ± 0.01 (7)[b]	4.9 ± 1.1 (7)[b]
D-Glucose	0.284 ± 0.03 (17)	11.7 ± 1.9 (17)
α-Methyl-D-glucoside	0.653 ± 0.04 (7)[b]	8.4 ± 2.1 (7)[c]
6-Deoxy-D-glucose	0.772 ± 0.08 (8)[b]	2.3 ± 0.14 (8)[b]
D-Galactose	4.76 ± 1.4 (4)[b]	15.1 ± 3.8 (4)[c]
3-O-Methyl-D-glucoside	29.5 ± 3.0 (4)[b]	17.1 ± 6.4 (4)[c]

[a]Paied t test to D-glucose. $\bar{\chi} \pm$ SEM (*n*).
[b]p < 0.05.
[c]p > 0.05.

Table III. ^{14}C + ^{3}H *Bidirectional Simultaneous*
D-*Glucose Flux* (μ *mol* cm^{-2} h^{-1})[a]

Direction	Control	Phlorizin
$J_{A \to Bl}$	0.259 ± 0.02 (10)	0.130 ± 0.02 (4)
$J_{Bl \to A}$	0.150 ± 0.03 (10)	0.133 ± 0.04 (4)

[a] $\bar{\chi}$ ± SEM(n) at 37°C in HSS (5.5 mM glucose).

proximal tubule (Ullrich *et al.*, 1974). The cotransporter strictly requires a D-glucopyranoside with an equatorial position of C2 hydroxyl and prefers a hydroxyl at C6 or methyl in the alpha or beta plane at C1. Only Li^+ at 3–5% the effectiveness in generating I_{sc} could replace Na^+. When the [Na^+] was varied there was a decrease in both the K_m and V_{max}, but more pronounced on the K_m.

2.2 Analysis of Cotransport by Isotope Fluxes

Validation of the assumption that I_{sc} is representative of Na^+, D-glucose cotransport was provided by direct correlation of I_{sc} with the fluxes of Na^+ and D-glucose by determination of the flux of $^{22}Na^+$ and ^{3}H and ^{14}C labeled D-glucose. Net D-glucose transport occurred from the apical-to-basolateral bath (Table III). Since D-glucose can also be metabolized, the transported radioactive material was analyzed by chromatography and compared to a D-glucose standard; the results were identical. Thus, the transported radioactivity was indeed labeled D-glucose (Misfeldt and Sanders, 1981). $^{22}Na^+$ fluxes were measured in both the apical-to-basolateral and basolateral-to-apical directions under short-circuited conditions and the net apical-to-basolateral flux compared to the I_{sc}; there was no significant difference (Misfeldt and Sanders, 1982). Concurrent determination of both $^{22}Na^+$ and ^{3}H-D-glucose unidirectional fluxes at concentrations of D-glucose of 5.5 and 0.3 mM exhibited a fixed ratio of net apical-to-basolateral transport, J^{Na^+}/J^{D-glc}, of about 2 (Table IV).

3. MECHANISM OF COUPLED NA$^+$, D-GLUCOSE APICAL MEMBRANE UPTAKE

Although the I_{sc} is a result of transepithelial coupled D-glucose and Na^+ transport, the I_{sc} may also be a sensitive measure of the interaction between the cotransporter (CTP) at the apical membrane, Na^+ and D-glucose. The I_{sc} may reflect such interactions if the rate-limiting step for I_{sc} is due to processes occurring at the apical membrane. This possibility is supported by the following results: The sugar analogs which stimulate I_{sc} are the same as those which (1) are

Table IV. Transepithelial Unidirectional Fluxes

D-glucose in bath (mM)	D-glucose (μ moles cm^{-2}hr^{-1})			Sodium (μ moles cm^{-2}hr^{-1})			$J_{net}^{Na}/J_{net}^{D\text{-}glc}$
	$J_{ap\to bl}$	$J_{bl\to ap}$	J_{net}	$J_{ap\to bl}$	$J_{bl\to ap}$	J_{net}	
5.5 ($n=6$)	0.364 ± 0.035^a	0.102 ± 0.013	0.262 ± 0.03	1.44 ± 0.16	0.85 ± 0.08	0.58 ± 0.08	2.2
0.3 ($n=3$)	0.189 ± 0.040	0.015 ± 0.004	0.174 ± 0.04	1.24 ± 0.18	0.81 ± 0.08	0.41 ± 0.13	2.3

[a] $\bar{\chi} \pm$ SD.

Table V. *Coupled Cellular Uptake of ^{22}Na and ^{3}H-D-Glucose across the Apical Membrane*

	A. Control[a]	B. Control with phlorizin (10^{-4} M) (nmoles min^{-1} mg protein^{-1})	(A$^{\Delta}$ − B)	Δ Na/ Δ D-glucose
^{3}H-D-Glucose	2.53 ± 0.24(8)[b]	1.90 ± 0.14(8)	0.63 ± 0.28	2.2 ± 0.57[c]
^{22}Na	4.86 ± 0.25(8)	3.44 ± 0.28(8)	1.42 ± 0.38	

[a]Medium composition: 30 mM Na, 110 mM Tris, 2.5 mM D-glucose.
[b]$\bar{\chi}$ ± SEM (n).
[c]$\bar{\chi}$ ± SD calculated by the "jacknife" method (Mosteller and Tukey, 1977).

actively transported (Ullrich *et al.*, 1974) and (2) are associated with apical membrane depolarization (Frömter and Luer, 1973). The analogs have an identical hierarchy of effectiveness. The K_m of our I_{sc} response for D-glucose at a [Na$^+$] of 140 mM is 0.27 mM (Misfeldt and Sanders, 1981). This value lies within the range of 0.2–1.0 mM. The K_m for D-glucose uptake by brush border membrane vesicles (Turner and Silverman, 1978; Fairclough *et al.*, 1979) and the K_m for proximal tubule depolarization (Frömter and Luer, 1973) also lie within this concentration range. An inhibition of I_{sc} occurs either when Na$^+$ is removed, or when phlorizin is added only to the solution bathing the apical surface.

3.1. Apical Uptake of Na$^+$:D-Glucose Is 2:1

As presented, the ratio of *net* transepithelial Na$^+$/D-glucose transport was 2:1 (Table IV). If indeed I_{sc} reflects the rate-limiting step of apical uptake, then the stoichiometry of uptake should have certain characteristics. It was determined, using ^{22}Na$^+$ and ^{3}H-D-glucose, that the initial rate of uptake across the apical membrane is linear over the first 5 min. In order to determine the cotransport uptake stoichiometry, the uptake of both Na$^+$ and D-glucose was determined simultaneously; uptake studies were done in both the presence and absence of phlorizin. The phlorizin-sensitive component of uptake of these two molecules, which was then calculated, was presumably uptake by the cotransport mechanism (Table V) (Misfeldt and Sanders, 1982). The ratio of Na$^+$:D-glucose apical uptake by the cotransport system was 2.2 by this method. Thus, when considering either the net transepithelial transport, or the apical uptake of Na$^+$ and D-glucose, a stoichiometry of 2:1 for Na$^+$ and D-glucose was obtained. These observations thus suggest that Na$^+$ and D-glucose maintain parallel pathways of uptake and egress which the cotransport process is functioning.

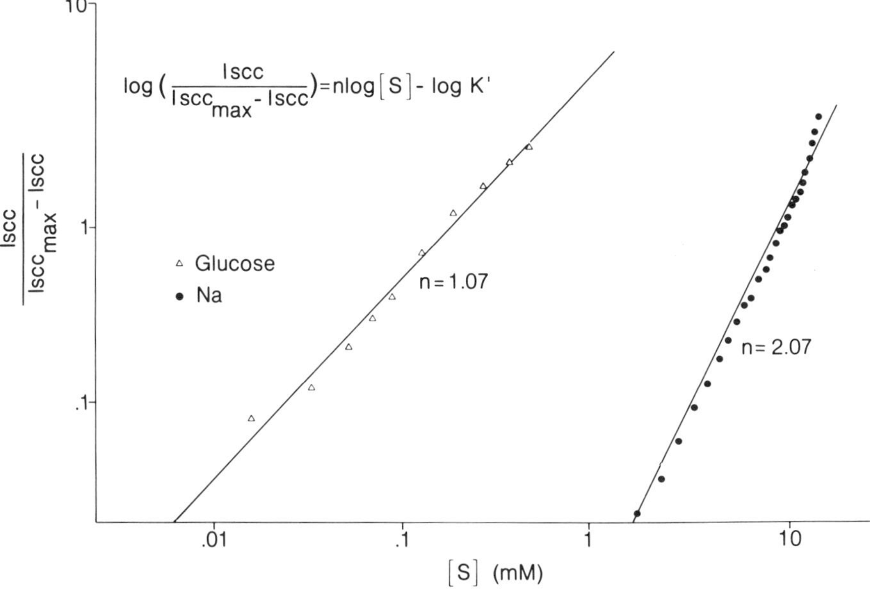

Figure 3. Hill plots of I_{sc} as the concentration of D-glucose or Na$^+$ was varied while the concentration of the other was held constant, [Na$^+$] at 140 mM and [D-glucose] at 20 mM. The slope was calculated by least-squares regression of a log–log plot. The slope indicates the least number of molecules translocated in the generation of I_{sc}.

3.2. I_{sc} as a Function of [Na$^+$] Is 2:1, as a Function of [D-Glucose] Is 1:1

The stoichiometry of ligand [Na$^+$ or D-glucose] interaction with the CTP was further analyzed. The I_{sc} was determined as a function of ligand concentration. In order to determine if more than one of the same species of the ligand interacts with the cotransporter for an I_{sc} response, the data were analyzed by means of a Hill plot (i.e., a plot of the I_{sc} versus Na$^+$ and of the I_{sc} versus D-glucose). The Hill plot allows an estimate of the minimum number of ligands interacting with the CTP receptor. Figure 3 illustrates the results of a typical experiment. In such an experiment, the D-glucose concentration was increased, while keeping the Na$^+$ concentration constant at 140 nM. When the I_{sc} was measured under these conditions, a slope of 1.02 was calculated by linear regression. However, when Na$^+$ varied and the concentration of D-glucose held constant, the slope from the Hill plot was 2.08. Thus, these studies confirmed the stoichiometry of apical membrane uptake (i.e., that the generation of I_{sc} is the result of two sodiums and a single D-glucose interacting with the CTP at the apical membrane). This stoichiometry for Na$^+$, D-glucose coupled uptake has been reported for other systems, which are present in chicken enterocytes (Kimmich and Randles, 1980), intestinal brush border vesicles (Kaunitz *et al.*, 1982), and the skin of the proglottids in two species of cestodes (Fisher and Read, 1971; Read *et al.*, 1974).

The electrochemical gradient that is generated across the apical membrane, due to the cotransport of Na^+ with glucose, is dramatically affected by the 2:1 uptake ratio between Na^+ and D-glucose. Indeed, when the apical uptake stoichiometry between Na^+ and D-glucose is 2:1, the electrochemical potential due to the Na^+ gradient will be twice as large, as in the case where the uptake stoichiometry between Na^+ and D-glucose is 1:1. (The force operating on the cotransport process is a function of the gradient on each Na^+ ion.) Doubling of the Na^+ electrochemical potential, which occurs as a consequence (at least theoretically), increases the maximum concentration gradient for D-glucose from outside to within the cell to the second exponential power.

3.3. Phlorizin Binding as a Function of $[Na^+]$ Is 1:1

The glycoside phlorizin is a specific and potent inhibitor of Na^+, D-glucose transport. The binding of phlorizin to the CTP is Na^+ dependent, and increase in the $[Na^+]$ is associated with a decrease in the apparent phlorizin binding constant (Frasch *et al.*, 1979; Glossman and Neville, 1972; Chesney *et al.*, 1974; Turner and Silverman, 1981). The Na^+-independent and weak inhibitory potency ($< 5\%$) of phloretin, the aglycone of phlorizin, suggests that the Na^+-dependent binding of phlorizin at the CTP is by the sugar moiety of phlorizin. The stoichiometry of Na^+:phlorizin binding was examined by a Hill plot analysis of ^3H-phlorizin binding as a function of $[Na^+]$ (Fig. 4). The minimum number of Na^+ ions bound per phlorizin molecule was estimated as 1:1, which was the slope of the Hill plot. Thus, whereas the uptake has a Na^+:D-glucose stoichiometry of 2:1, the Na^+:Na^+-dependent phlorizin-binding ratio is 1:1. These findings indicate that the Na^+, D-glucose CTP interaction that results in translocation is not the same as Na^+-dependent CTP binding of phlorizin, although both processes occur at the same membrane site.

3.4. Two-Step, Two-Sodium Model of Na^+, D-Glucose Cotransport

From our electrophysiological analysis of Na^+, D-glucose cotransport by the LLC–PK$_1$ cells, the apical uptake was rate limiting. Studies of the cotransport mechanism in brush border membrane vesicles provide additional insight. Sodium was required for phlorizin-inhibitable D-glucose uptake (Aronson and Sacktor, 1975; Kinne *et al.*, 1975; Toggenburger *et al.*, 1978; Turner and Silverman, 1978). D-glucose uptake and the number of phlorizin-binding sites were determined in the presence of sodium, but absence of a sodium chemical gradient across apical membrane vesicles. Under these conditions, increased intravesicular negativity was associated with enhanced D-glucose up-

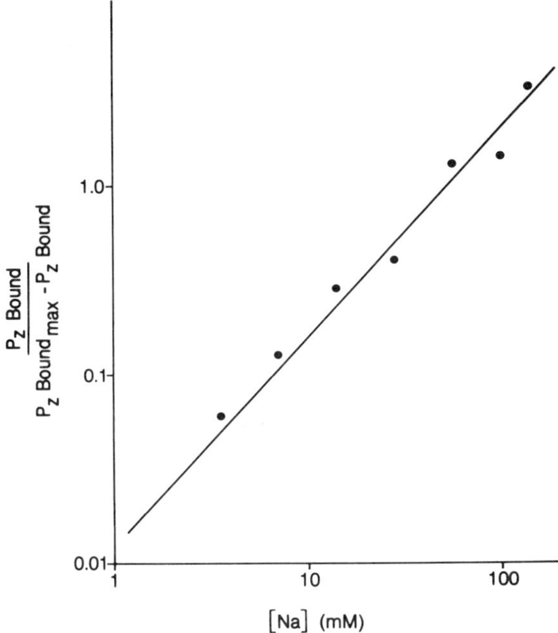

Figure 4. Hill plot analysis of Na$^+$-dependent ^{3}H-phlorizin binding as a function of [Na]. The slope was calculated by least-squares regression to be 1.1 indicating the minimum sodium binding for each phlorizin molecule.

take (Murer and Hopfer, 1974; Beck and Sacktor, 1978) as well as an increased *number* of exteriorly located phlorizin-binding sites. However, the dissociation of bound phlorizin was not observed (Aronson, 1978; Toggenburger *et al.,* 1978). These effects of intravesicular negativity (i.e., an increase in the number of exterior exposed CTPs and an enhancement of cotransport uptake) suggest that the empty, or unloaded, CTP carries a net negative charge and the loaded CTP a net positive charge. The reversal of net CTP charge from empty to loaded is possible with the 2:1, Na$^+$:D-glucose stoichiometry of our results. In addition, the 1:1 stoichiometry of Na$^+$:Na$^+$-dependent phlorizin binding is consistent with a stepwise process, whereby the two sodium ions interact with the CTP-D-glucose complex.

We propose a two-step, two-sodium model for the cotransport mechanism (Fig. 5). First, there is a random (Crane and Dorando, 1979; Turner and Silverman, 1981) binding of Na$^+$ and D-glucose to the negatively charged empty CTP, CTP$^-$, forming a ternary complex without a net charge. To this complex a second Na$^+$ binds in what necessarily must be an ordered process (Hopfer and Groseclose, 1980). The net charge is now positive. There is an apparent translocation of the complex, followed by dissociation, resulting in the movement of

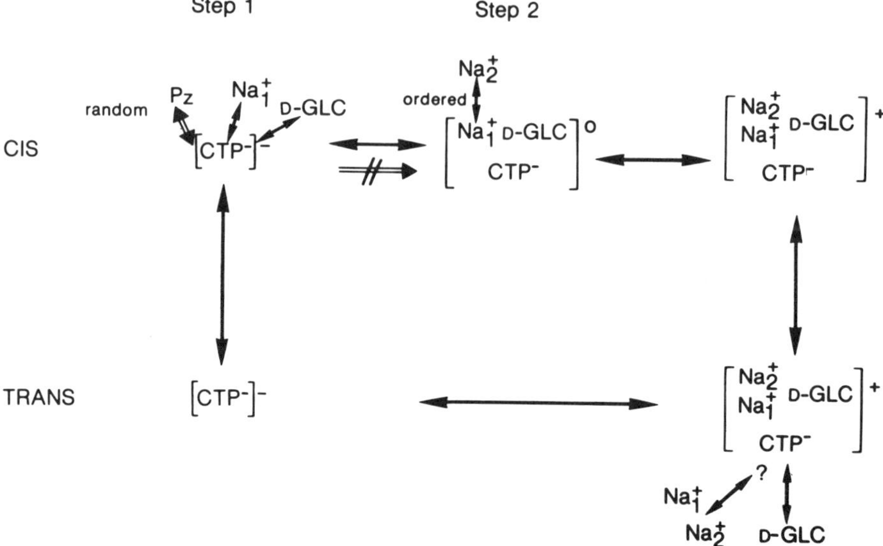

Figure 5. Diagram of a sequential, two-step, two-sodium mechanism of D-glucose binding and translocation. The net negatively charged empty carrier and net positively charged translocating complex are consistent with reported experiments of others. See text for details.

two Na$^+$ molecules and a single molecule of D-glucose into the cell. Remaining in the membrane is a negatively charged CTP. The negative cell interior may then enhance the translocation of the empty CTP back to an exterior exposure. In the presence of phlorizin there is binding to the CTP with a single Na$^+$ with an inability of further Na$^+$ binding or translocation. Since the affinity of phlorizin for one CTP is greater than for D-glucose, cotransport is competitively inhibited.

4. ROLE OF Na$^+$/K$^+$ ATPase IN NET TRANSEPITHELIAL Na$^+$ TRANSPORT

The Na, D-glucose cotransport by the pig kidney cell line LLC–PK$_1$ provided an experimental system to examine another question concerning epithelial transport, that is, to examine whether the Na$^+$/K$^+$ ATPase plays the role proposed by Koefoed–Johnsen and Ussing (1958) in facilitating transepithelial Na$^+$ transport. According to their model of transepithelial Na$^+$ transport, Na$^+$ enters the cell across the apical membrane and exits via the exchange of intracellular Na$^+$ for extracellular K$^+$ by means of the basolateral-membrane-associated ATPase. The coupling between transepithelial Na$^+$ transport and basolateral K$^+$ uptake has been difficult to confirm, despite continued interest (DeLong and Civan, 1978; Curran and Cereijido, 1965: Nellans and Schultz,

Table VI. Initial Rates of Influx of ^{3}H-D-Glucose and ^{86}Rb into Disrupted LLC–PK$_1$ Monolayers

| | Influx (nmole·min^{-1}·mg protein^{-1}) | | | |
Conditions	^{3}H-D-Glucose		^{86}Rb	
1. Control (HSS)[a]	4.07 ± 0.214(18)[b]		13.8 ± 0.52(18)	
+ Phlorizin (10^{-4} M)	2.56 ± 0.10(18)		11.7 ± 0.38(18)	
− Phlorizin	1.51 ± 0.26	$p < 0.05$[c]	2.18 ± 0.64	$p < 0.05$
2. D-Glucose-free HSS	—		7.69 ± 0.36(8)	
+ Phlorizin (10^{-4} M)	—		7.51 ± 0.39(8)	$p > 0.05$
3. HSS + ouabain (10^{-4} M)	2.76 ± 0.12(12)		2.29 ± 0.14(12)	
+ Phlorizin (10^{-4} M)	2.60 ± 0.17(12)	$p > 0.05$	3.34 ± 0.09(12)	$p > 0.05$

[a]HSS, Hanks' salt solution.
[b]x̄ ± S.E. (n).
[c]Student's t test.

1976); however, results supporting an alternative model have not been compelling. Among the experimental difficulties, a diverse and relatively large intracellular potassium pool may have prevented the detection of K^+ uptake specifically coupled to transepithelial Na^+ transport. The intracellular potassium pool may be reduced to the smallest possible size by using LLC–PK$_1$. Such monolayers are composed only of a uniform population of transporting epithelial cells.

Since the phlorizin-inhibitable apical Na^+ uptake is coupled to the uptake of D-glucose, the cellular entry of ^{3}H-D-glucose was used as a marker for the entry of Na^+ by means of a cotransport mechanism. ^{86}Rb was used to determine K^+ entry. The uptake experiments were performed on monolayers of LLC–PK$_1$ cells loosened from the culture dish to facilitate medium access to both sides. To determine if there was a linkage between transepithelial sodium transport and K^+ uptake, simultaneous initial rates of uptake of ^{3}H-D-glucose and ^{86}Rb were determined, in the presence and absence of phlorizin.

Table VI summarizes the results of these experiments. In the presence of glucose, phlorizin inhibited Rb^+ uptake. As a control, Rb^+ uptake was also examined in the absence of D-glucose. Under such conditions, phlorizin had no effect on Rb^+ uptake. The addition of ouabain reduced the uptake of Rb^+ and D-glucose to the phlorizin-inhibited level; however, ouabain did not further inhibit either Rb^+ or D-glucose uptake in the presence of phlorizin.

The stoichiometry of Na^+:Rb(K)$^+$ exchange was calculated with the knowledge of the 2:1 stoichiometry of Na^+:D-glucose uptake and of transepithelial transport. A ratio of 1.2 ± 0.32 (x̄ ± SD) further suggests that Na^+ for K^+ exchange occurs via the Na^+,K^+ ATPase with a coupling ratio of 3 Na^+ to 2 K^+. This coupling ratio has similarly been observed in other epithelial tissue (Zeuthen and Wright, 1978; Kirk *et al.*, 1980; Nielsen, 1979).

5. BIOENERGETICS OF Na^+, D-GLUCOSE COTRANSPORT BY CULTURED CELLS

The study of the cellular metabolic substrate utilization that occurs during the use of specific physiologic functions can be done particularly well in the experimentally controlled conditions of renal cell cultures (Sanders, Simon, and Misfeldt, 1983). In tissue cultures that have cellular homogeneity, each cell is functionally equivalent, which provides an accurate cellular and subcellular denominator. Given the control of the environment, which is possible under cell culture conditions, the metabolic organic substrates and oxygen can be precisely regulated.

As presented earlier in this chapter, the LLC–PK$_1$ cells form a simplified epithelial tissue with Na^+, D-glucose cotransport accounting for nearly all of the net Na^+ transport and the I_{sc} (Misfeldt and Sanders, 1981, 1982). The mechanism of coupled-Na^+ transepithelial transport is consistent with Na^+ extrusion by basolateral Na^+/K^+ ATPase (Sanders and Misfeldt, 1982). Thus, the consumption of ATP linked to transepithelial Na^+ transport can be completely and reversibly regulated by phlorizin without toxicity.

The cells were cultured on membrane filters. Transport function was monitored by I_{sc} when the filter was placed in an Ussing chamber; oxygen consumption was measured when the filter was incubated under identical conditions in a cuvette sealed airtight by placement of a Clarke-type oxygen electrode; lactate production was chemically measured in the incubation medium after the appropriate manipulations.

5.1. *The Relation between Glycolysis and I_{sc}*

The formation of lactate is a measure of glycolysis and was determined after stimulation of cotransport by alpha-methyl-D-glucoside (AMG), a metabolically inert Na^+ dependent substrate (Crane, 1960; Barry *et al.*, 1965; Misfeldt and Sanders, 1981). After preincubation for 48 hr in carbohydrate-free Hanks' salt solution supplemented by L-glutamine (2 mM), maximal cotransport was stimulated; lactate was measured before and after the addition of a noncotransported glycolytic substrate, D-mannose. Lactate production was undetectable until D-mannose was added; at the time it increased to 0.148 ± 0.054 μmol h^{-1} 10^6 cells^{-1}. In a similar study, which monitored the I_{sc}, a twofold increase in the I_{sc} was observed immediately after the addition of D-mannose and reached a maximum after 15 min. Thus, when metabolic substrate is a limiting factor, an increased I_{sc} response is still observed, although with a markedly slower time course. The I_{sc} nevertheless eventually reaches the same maximal level under such limiting conditions. These experiments also clearly show that 2mM L-glutamine does sustain sufficient ATP production for maximal Na^+ transport.

5.2. O_2 and I_{sc} Vary According to Glycolysis

Cells cultured on a membrane filter under standard conditions were placed in Hank's salt solution containing 5.5 mM D-glucose and 140 mM Na^+, concentrations that stimulate maximal cotransport (Misfeldt and Sanders, 1981). O_2 consumption, which was measured before and after cotransport, was inhibited by the addition of phlorizin (Table VII). Phlorizin was observed to reduce the O_2 consumption by half; this inhibitory effect of phlorizin was reversible. A comparable decrease in O_2 consumption was also observed after the removal of Na^+. Our studies that demonstrate the function of Na^+,K^+ ATPase in the transepithelial cotransport of Na^+ (presented earlier in this chapter and in Sanders and Misfeldt, 1982) would not only explain the observed effect of cotransport on O_2 consumption, but would predict a relationship of cotransport to ATP production.

In order to examine the dependence of cotransport and oxygen consumption on ATP production, $LLC-PK_1$ cells were preincubated in the absence of glycolytic substrates. O_2 consumption was then measured in response to cotransport stimulation by AMG. There was a prompt increase by 3.5 fold (Table VII), which remained steady. There was also a prompt increase in the I_{sc}. However, in contrast to the case with O_2 consumption, the prompt increase in I_{sc} was followed by either (1) a decrease to a lower steady state, which was still greater than prestimulation (Fig. 6), or (2) a slower rise to a new steady state. Despite this variability in the initial I_{sc} response, O_2 consumption reached a maximal rate during the first 5 min and continued constant at this maximal rate into the 5- to 10-min period. When a noncotransported glycolytic substrate was then added, e.g., D-mannose or D-fructose, the I_{sc} increased over a 15-min period to a new maximum, twofold over the I_{sc} supported by oxidative phosphorylation alone (Fig. 6). The twofold increase in I_{sc} caused by the addition of the glycolytic substrate was associated with a 30% decrease in O_2 consumption. The further addition of oxidative phosphorylation substrates such as citrate, diacetate, triacetate, or succinate, however, did not further increase the I_{sc} (Fig. 6).

The experiments provide information for a qualitative understanding of the cotransport bioenergetics of cultured cells. The stimulation of cotransport, I_{sc}, is associated with increased O_2 consumption and lactate production. The relation between I_{sc} and O_2 consumption is variable and depends on the rate of glycolysis. The observation that oxidative phosphorylation alone did not sustain maximal cotransport may be explained by two possible constraints—either the amount of ATP from the mitochondria is insufficient, or an adequate amount of ATP is produced but is unavailable to the Na^+,K^+ ATPase. In contrast, glycolysis can support maximum cotransport, even when oxidative phosphorylation and O_2 consumption are completely inhibited by cyanide (10^{-3} M).

A higher O_2 consumption was observed after I_{sc} stimulation in the absence

Table VII.   Oxygen Consumption: Effect of Sequential Additions after Preincubation in Absence of Glycolytic Substrates

Conditions		O_2 consumption (μmol h^{-1} cm^{-2})
Control (GF-HSS),[a] no cotransport		0.042 ± 0.006 (6)[b]
Alpha-methyl-D-glucoside (3.75 mM)	0–5 min	0.148 ± 0.028 (4)
	5–10 min	0.148 ± 0.028 (4)
D-Mannose (2.5 mM) added		0.105 ± 0.021 (4)
Phlorizin (0.025 mM)		0.053 ± 0.007 (4)

[a]GF-HSS, D-glucose-free Hanks' salt solution, 2 mM L-glutamine added.
[b]$\bar{x} \pm$ SD (n).

of glycolysis. Nevertheless, the maximum I_{sc} that was sustained by oxidative phosphorylation under these conditions was only half that obtained in the presence of glycolysis. The maximal I_{sc} response was not limited by the concentration of cotransport substrates, but rather the energy available for the Na$^+$,K$^+$ ATPase. Thus, we may tentatively conclude that only glycolysis is capable of providing adequate ATP to sustain maximal transport. Alternatively, only the

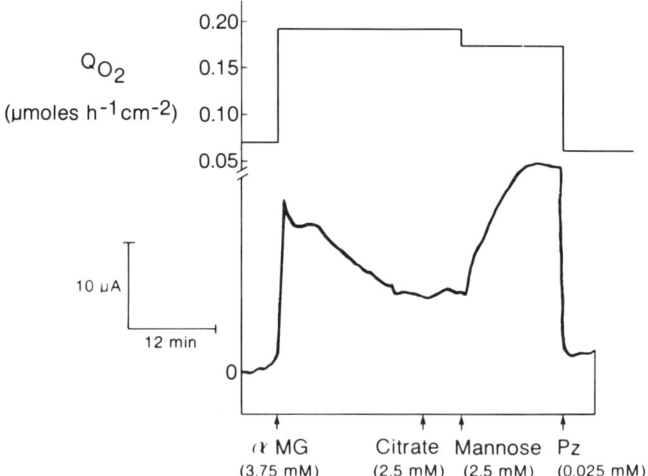

Figure 6. The upper tracing indicates oxygen consumption over the time course parallel with the I_{sc} response in the lower panel. The filter with cell layer was preincubated in glucose-free Hanks' salt solution (2 mM L-glutamine added) for 24 hr prior to the experiment. On addition of alpha-methyl-D-glycoside there was a prompt increase in both O_2 consumption and I_{sc}. The rate of oxygen consumption was constant, but I_{sc} gradually assumed a new steady state that was unresponsive on the further addition of mitochondrial oxidative substrates. In contrast, the addition of glycolytic substrates increased I_{sc} and caused a decrease in oxygen consumption. See text for discussion.

ATP from glycolysis is available for the Na^+,K^+ ATPase, whereas the ATP from oxidative phosphorylation is not.

Whereas the metabolic demands of epithelial transport by LLC–PK$_1$ cells are only adequately met through utilization of ATP formed through glycolysis, the bioenergetic pattern of cultured cells during nonspecific function or growth can be supported by other means. For example, if D-glucose is replaced by D-fructose, the metabolic pathways can support cultured cell growth (Reitzer *et al.*, 1979). The need for D-glucose by cultured cells appears to be anabolic; cell growth may readily occur in the absence of D-glucose if purine and pyrimidine nucleosides are supplied (Zielke *et al.*, 1976). In contrast, for epithelial transport, the contribution of glycolytic energy may be considerable; the glycolytic energy provision for Na^+ transport by the toad and turtle urinary bladder has been measured (Schwartz and Steinmetz, 1977; Steinmetz *et al.*, 1981). In the freshwater turtle adapted to anaerobiosis, near maximal urinary bladder Na^+ transport was sustained by the glycolytic energy provision alone (Klahr and Bricker, 1964).

The relationship between bioenergetics and transport may be formulated in terms of transport regulation: is the provision of energy the stimulus for epithelial transport, or is metabolism stimulated by transport and the consumption of ATP? Handler *et al.* (1969) noted that aldosterone-stimulated Na^+ transport by the toad urinary bladder was associated with increased glycolysis; however, when Na^+ transport was inhibited by ouabain, there was no increase in glycolysis. Our cell culture studies demonstrate that this relationship can be manipulated. Under conditions where the energy provision is the apparent limiting factor, and in particular when there is only oxidative phosphorylation, the addition of glycolytic metabolic substrates stimulates Na^+ transport. In the presence of glycolytic substrates, the apical membrane transport mechanism is the saturable limiting step in transepithelial cotransport. Thus, the regulation of I_{sc} by LLC–PK$_1$ cells may depend on glycolysis. The corollary is that there is a variable relationship between the energy contributed by oxidative phosphorylation and transport; the stoichiometry of bioenergetics must include the measurement of both glycolysis and O_2 consumption.

6. CONCLUSION

The validity of the cell culture approach to problems in epithelial transport is established only by the demonstration that the transport functions are identical to those observed in either *in vivo* or *in vitro* organ studies. Yet the importance of this technique is only supported to the extent that new problems are identified and old problems are solved. It is of little advantage to repeat in cell culture what has or could be done by other approaches. Our studies of Na^+ D-glucose cotransport in LLC–PK$_1$ cells illustrate the descriptive validating phase and the problem-solving application of the cell culture approach to questions concerning the cotransport uptake mechanism, transepithelial Na^+ transport via K^+ exchange, and the energy provision for cotransport. These are but examples of such applications; the extent to which others are demonstrated will determine the actual usefulness of cell culture for the study of epithelial transport.

REFERENCES

Aronson, P. S., 1978, Energy-dependence of phlorizin binding to isolated renal microvillus membranes. Evidence concerning the mechanism of coupling between the electrochemical Na^+ gradient and sugar transport, *J. Membrane Biol.* **42**:81–98.

Aronson, P. S., and Sacktor, B., 1975, The Na^+ gradient-dependent transport of D-glucose in renal brush border membranes, *J. Biol. Chem.* **250**:6032–6039.

Barry, R. J. C., Smyth, D. H., and Wright, E. M., 1965, Short-circuit and solute transfer by rat jejunum, *J. Physiol.* **181**:410–431.

Beck, J. C., and Sacktor, B., 1978, The sodium electrochemical potential mediated uphill transport of D-glucose in renal brush border membrane vesicles, *J. Biol. Chem.* **253**:5531–5535.

Cereijido, M., Robbins, E. S., Dolan, W. J., Rotunno, C. A., and Sabatini, D. D., 1978, Polarized monolayers formed by epithelial cells on a permeable and translucent support, *J. Cell Biol.* **77**:853–880.

Chesney, R., Sacktor, B., and Kleinzeller, A., 1974, The binding of phlorizin to the isolated luminal membranes of the renal proximal tubule, *Biochem. Biophys. Acta* **332**:263–277.

Crane, R. K., 1960, Intestinal absorption of sugars, *Physiol. Rev.* **40**:789–825.

Crane, R. K., and Dorando, F. C., 1979, On the mechanism of Na^+-dependent glucose transport, in: *Function and Molecular Aspects of Biomembrane Transport* (E. Quagaliariello, E., Palmieri, F., Papa, S., and Klingenberg, M., eds.), Elsevier North Holland Biomedical Press, New York, pp. 271–278.

Curran, P. F., and Cereijido, M., 1965, K fluxes in frog skin, *J. Gen. Physiol.* **48**:1101–1033.

DeLong, J., and Civan, M. M., 1978, Dissociation of cellular K^+ accumulation from net Na^+ transport by toad urinary bladder, *J. Membrane Biol.* **42**:19–43.

Ehmann, U. K., Peterson, W. D., Jr., Misfeldt, D. S., 1948, To grow mouse mammary cells in culture, *J. Cell Biol.* **98**:1026–1032.

Fairclough, P., Malathi, H., Preiser, H., and Crane, R. K., 1979, Reconstitution into liposomes of glucose active transport from the rabbit renal proximal tubule. Characteristics of the system, *Biochem. Biophys. Acta* **552**:295–306.

Fisher, F. M., Jr., and Read, C. P., 1971, Transport of sugar in the tape worm *Calliobothrium verticillatum. Biol. Bull. (Woods Hole, Massachusetts)* **140**:46–62.

Frasch, W., Frohnert, P. P., Bode, F., Baumann, K., and Kinne, R., 1970, Competitive inhibition of phlorizin binding by D-glucose and influence of sodium: A study on brush border membrane of rat kidney, *Pfluegers Arch.* **320**:265–284.

Frömter, E., 1979, Solute transport across epithelia: What can we learn from micropuncture studies on kidney tubules? *J. Physiol.* **288**:1–38.

Frömter, E., and Luer, K., 1973, Electrical studies on sugar transport kinetics of rat proximal tubule, *Pfluegers Arch.* **343**:R47.

Glossman, H., and Neville, D. M., Jr., 1972, Phlorizin receptors in isolated kidney brushborder membrane, *J. Biol. Chem.* **247**:7779–7789.

Goodman, B. E., Sleischer, R. E., and Crandall, E. D., 1983, Evidence for active sodium transport by monolayers of pulmonary alveolar epithelial cells, *Am. J. Physiol.* **245**:C78–C83.

Handler, J. S., Preston, A. G., and Orloff, J., 1969, The effect of aldosterone on glycolysis in the urinary bladder of the toad, *J. Biol. Chem.* **244**:3194–3199.

Hopfer, U., and Groseclose, R., 1980, The mechanism of Na^+-dependent D-glucose transport, *J. Biol. Chem.* **225**:4453–4462.

Kanuitz, J. D., Gunther, R., and Wright, E. M., 1982, Involvement of multiple Na^+ ions in intestinal D-glucose transport, *Proc. Natl. Acad. Sci. USA* **79**:2315–2318.

Kimmich, G. A., and Randles, J., 1980, Evidence for an intestinal Na^+:sugar transport coupling stoichiometry of 2.0, *Biochem. Biophys. Acta* **596**:439–444.

Kinne, R., Murer, H., Kinne–Saffran, E., Thees, M., and Sachs, G., 1975, Sugar transport by renal plasma membrane vesicles. Characterization of the systems in the brush border microvilli and basal-lateral plasma membranes, *J. Membrane Biol.* **21**:375–395.

Kirk, K., Halm, D., and Dawson, D. C., 1980, Active sodium transport by turtle colon via electrogenic Na–K exchange pump, *Nature* **287**:237–239.

Klahr, S., and Bricker, N. S., 1964, Na transport by isolated turtle bladder during anaerobiosis and exposure to KCN, *Am. J. Physiol.* **206**:1333–1339.

Koefoed–Johnsen, V., and Ussing, H. H., 1958, The nature of frog skin potential, *Acta Physiol. Scand.* **42**:298–308.

Mason, R. J., Williams, M. C., Widdicombe, J. H., Sanders, M. J., Misfeldt, D. S., and Berry, L., 1982, Transepithelial transport by pulmonary alveolar type II cells in primary cultures, *Proc. Natl. Acad. Sci. USA* **79**:6033–6037.

Misfeldt, D. S., and Sangers, M. J., 1981, Transepithelial transport in cell culture: D-glucose transport by a pig kidney cell line (LLC–PK₁), *J. Membrane Biol.* **59**:13–18.

Misfeldt, D. S., and Sanders, M. J., 1982, Transepithelial transport in cell culture: Stiochiometry of Na/phlorizin binding and Na/D-glucose cotransport. A two-step, two sodium model of binding and translocation, *J. Membrane Biol.* **70**:191–198.

Misfeldt, D. S., Hamamoto, S. T., and Pitelka, D. R., 1976, Transepithelial transport in cell culture, *Proc. Natl. Acad. Sci. USA* **73**:1212–1216.

Mosteller, F., and Tukey, J. W., 1977, *Data Analysis and Regression,* Addison Wesley, Menlo Park, California, pp. 133–164.

Mullin, J. M., Weibel, J., Diamond, L., and Kleinzellar, A., 1980, Sugar transport in the LLC–PK₁ renal epithelial cell line similarity to to mammalian kidney and the influence of cell density, *J. Cell Physiol.* **104**:375–379.

Murer, H., and Hopfer, U., 1974, Demonstration of electrogenic Na^+-dependent D-glucose transport in intestinal brush border membranes, *Proc. Natl. Acad. Sci. USA* **71**:484–488.

Nellans, H., and Schultz, S. G., 1976, Relations among transepithelial sodium transport, potassium exchange and cell volume in rabbit ileum, *J. Gen. Physiol.* **68**:441–463.

Neilsen, R., 1979, Coupled transepithelial sodium and potassium transport across isolated frog skin: Effects of ouabain, amiloride and the polyene antibiotic filipin, *J. Membrane Biol.* **51**:161–184.

Rabito, C. A., and Ausiello, D. A., 1980, Na^+-dependent sugar transport in a cultured epithelial cell line (LLC–PK₁), *J. Membrane Biol.* **54**:31–38.

Read, C. P., Stewart, G. L., and Pappas, P. W., 1974, Glucose and sodium fluxes across the brush border of *Hymenolepis dimunata* (cestoda), *Biol. Bull. (Woods Hole, Massachusetts)* **147**:146–162.

Reitzer, L. J., Wice, B. M., and Kennell, D., 1979, Evidence that glutamine, not sugar, is the major energy source for cultured HeLa cells, *J. Biol. Chem.* **254**:2669–2676.

Robinson, B. A., and MacKnight, A. D. C., 1976, Relationship between serosal medium potassium concentration and sodium transport in toad urinary bladder: III Exchangeability of epithelial cellular potassium, *J. Membrane Biol.* **26:**269–286.

Ryan, J., and Simoni, R. D., 1980, Alanine transport by Chinese hamster ovary cells with altered phospholipid acyl chain composition, *Biochim. Biophys. Acta* **598:**606–615.

Sanders, M. J., and Misfeldt, D. S., 1982, Ouabain-sensitive ^{86}Rb(K) influx is linked to trans-epithelial Na transport in pig kidney cell line, *Biochim. Biophys. Acta* **685:**383–385.

Sanders, M. J., Simon, L. M., and Misfeldt, D. S., 1983, Transepithelial transport in cell culture: Bioenergetics of Na-, D-glucose-coupled transport, *J. Cell. Physiol.* **114:**263–266.

Schwartz, J. H., and Steinmetz, P. R., 1977, Metabolic energy and pCO_2 as determinants of H$^+$ secretion by turtle urinary bladder, *Am. J. Physiol.* **233:**F145–F149.

Steinmetz, P. R., Husted, R. F., Mueller, A., and Beauwens, R., 1981, Coupling between H$^+$ transport and anaerobic glycolysis in turtle urinary bladder. Effect of inhibitors of H$^+$ ATPase, *J. Membr. Biol.* **59:**27–34.

Toggenburger, G., Kessler, M., Rothstein, A., and Semenza, G., 1978, Similarity in effects of Na$^+$ gradients and membranes potentials on D-glucose transport by, and phlorizin binding to, vesicles derived from brush borders of rabbit intestinal mucosal cells, *J. Membrane Biol.* **40:**269–290.

Turner, R. J., and Silverman, M., 1978, Sugar uptake into brush border vesicles from dog kidney. I. Specificity, *Biochim. Biophys. Acta* **507:**305–321.

Turner, R. J., and Silverman, M., 1981, Interaction of phlorizin and sodium with the renal brush-border membrane D-glucose transporter: Stoichiometry and order of binding, *J. Membrane Biol.* **58:**43–55.

Ullrich, K. H., Rumrich, G., and Kloss, S., 1974, Specificity and sodium dependence of the active sugar transport in the proximal convolution of the rat kidney, *Pfluegers Arch.* **351:**35–48.

Zuethen, T., and Wright, E. M., 1978, An electrogenic Na$^+$/K$^+$ pump in the choroid plexus, *Biochim. Biophys. Acta* **511:**517–522.

Zielke, H. R., Ozand, P. T., Tildon, J. T., Sevdalian, D. A., and Cornblath, M., 1976, Growth of human diploid fibroblasts in the absence of glucose utilization, *Proc. Natl. Acad. Sci. USA* **73:**4110–4114.

IV

MEMBRANE PROTEINS AND HORMONE RECEPTORS

Transporting epithelial cells exhibit a structural and functional polarity *in vivo*. The cells transport solutes from one tissue compartment to another. In order for such transepithelial transport to occur, the cells must possess two different plasma membrane domains, which are enriched with respect to particular plasma membrane proteins. For example, in proximal tubule cells the sodium-dependent sugar transport system is localized in the apical membrane which faces the lumen of the nephron. The Na^+/K^+ ATPase is localized to the basolateral membrane, which faces the blood side of the tubule. Hormone receptors coupled to adenylate cyclase are also localized at the basolateral membrane, which is affected by the hormones in the blood. The polarity of epithelial cells is maintained in the culture situation. The apical surface faces the culture medium. The basolateral surface faces the dish.

The polarity of MDCK plasma membranes has been examined extensively using enveloped viruses as a probe. Some viruses (VSV and Sendai) bud exclusively from the basolateral surface. The hormone receptors and responsiveness of such polarized cells have been examined in detail. MDCK cells respond to numerous hormones that act via cyclic AMP, including arginine vasopressin, prostaglandin E_1, glucagon, and norepinephrine. Although LLC–PK_1 cells have several proximal tubule transport systems, these cells possess a high density of receptors for arginine vasopressin, a distal tubule marker. Thus LLC–PK_1 has been examined extensively with regard to its vasopressin receptors. Monoclonal antibodies should prove to be an important means to characterize the integral membrane transport and receptor proteins of cultured epithelial cells. Not only are monoclonal antibodies available for beta-adrenergic receptors, but monoclonals are also available for the Na^+-dependent glucose transport system of transporting epithelial cells. Such monoclonal antibodies will be of use in the study of MDCK cells, LLC–PK_1 cells, and appropriate primary kidney cell cultures.

8

Viruses in the Study of the Polarity of Epithelial Membranes

MARY TAUB

1. INTRODUCTION

Transporting epithelial cells are present in many organs *in vivo*. These epithelial cells play an important role in regulating the passage of solutes from one tissue compartment to another. In order to perform such functions, transporting epithelial cells must possess a remarkable structural and functional polarity. Their plasma membrane is characterized by two distinct domains, which are separated by tight junctions: an apical (or mucosal) domain with microvilli, which often faces a lumenal compartment, and a basolateral (or serosal) domain, which is associated with another tissue compartment (Fig. 1). These two plasma membrane domains possess different transport systems, membrane receptors, and enzymes. For example, distal tubule cells of the kidney readsorb salt and water from the lumen of the tubule back into the blood. This vectorial transport is facilitated by the polarized structure of the distal tubule cells (Fig. 2). The apical surface (which faces the lumen of the tubule) possesses an amiloride-sensitive Na^+ channel, as well as proteolytic enzymes (leucine aminopeptidase and alkaline phosphatase). In contrast, the basolateral surface possesses high Na^+/K^+ ATPase activity. Hormone-sensitive transepithelial solute transport is facilitated by the presence in the basolateral surface of arginine vasopressin receptors cou-

MARY TAUB • Department of Biochemistry, State University of New York at Buffalo, School of Medicine, Buffalo, New York 14214.   Work funded by National Institutes of Health Grant 1 RO1 CA28111-04 and National Institutes of Health Research Career Development Award 1 KO4 CA/Am 00888-01.

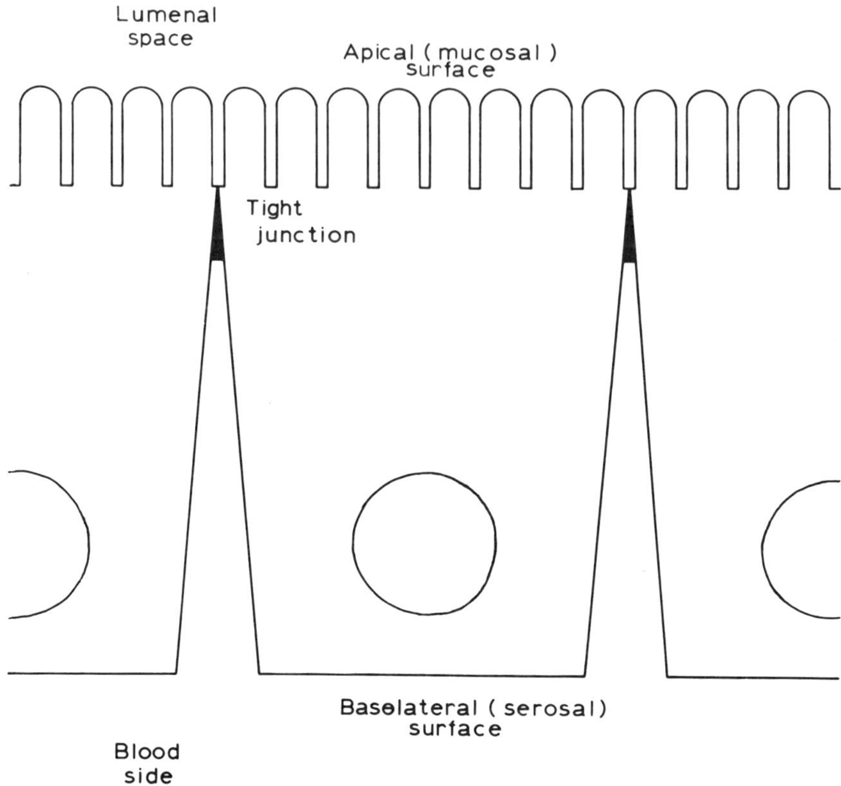

Figure 1. Polarized structure of transporting epithelial cell.

pled to adenylate cyclase. So, the polarized structure of the epithelial cell is essential to the appropriate expression of its differentiated transport functions.

Thus, an important question for our understanding of epithelial transport is ''What are the mechanisms responsible for the establishment and maintenance of epithelial membrane polarity?'' The organelles involved in the biosynthesis of the membrane components are arranged in a highly ordered manner inside epithelial cells. Membrane glycoproteins are generally assembled and glycosylated during their biosynthesis in the rough endoplasmic reticulum (RER). These proteins are subsequently transported to the Golgi complex, where they are further modified. The glycoproteins are finally transported, perhaps by endocytotic vesicles, to the appropriate plasma membrane domain (Fig. 3).

The molecular mechanisms involved in the sorting out of membrane proteins to the two domains are *not* well defined. A number of determinants may be critical in the insertion of proteins into one of the two plasma membrane do-

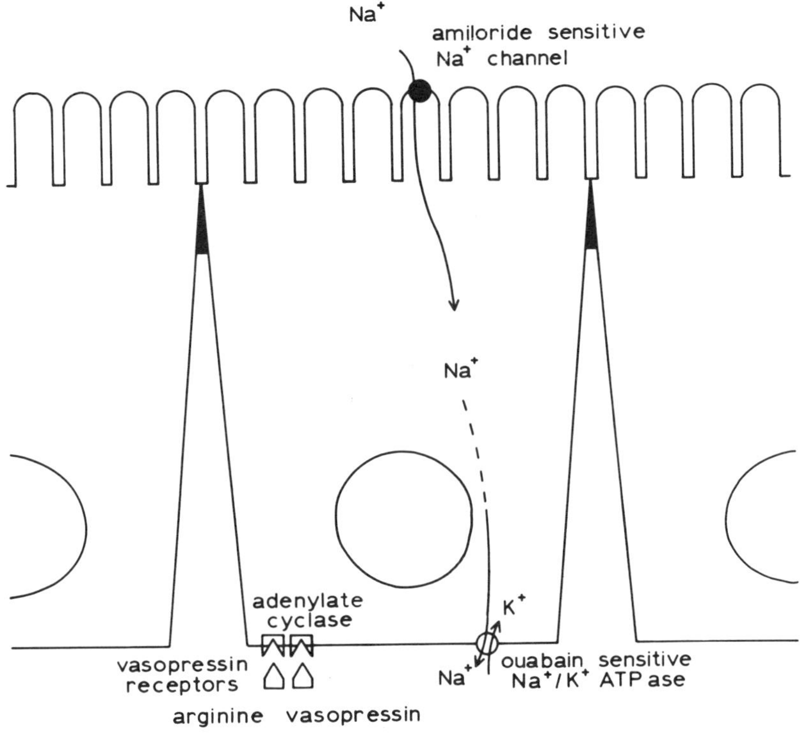

Figure 2. Distal tubule cells.

mains. These determinants could include the locations of the biosynthetic organelle for the membrane protein, the processing pathway for the protein, the structural characteristics of the membrane protein (e.g., its glycosylation, its signal sequence), as well as the state of the plasma membrane itself. The preexistence of two plasma membrane domains as well as tight junctions may be critical for the polarized insertion of membrane proteins to occur. These parameters may also be important in the maintenance of epithelial polarity.

This paper describes the use of the Madin–Darby canine kidney (MDCK) cell line and enveloped RNA viruses in the study of the polarity problem. The MDCK cell line is a well-characterized kidney epithelial cell line (McRoberts *et al.*, 1981; Taub, 1984) that has been shown to retain *in vitro* both the structural and functional polarity of kidney tubule cells (Cereijido *et al.*, 1978; Leighton *et al.*, 1969; Misfeldt *et al.*, 1976). Enveloped RNA viruses mature by budding from the host cell plasma membrane. In MDCK cells these viruses have been shown to bud asymmetrically, from either the apical or the basolateral surface, depending on the class of RNA virus. The enveloped RNA viruses are an excel-

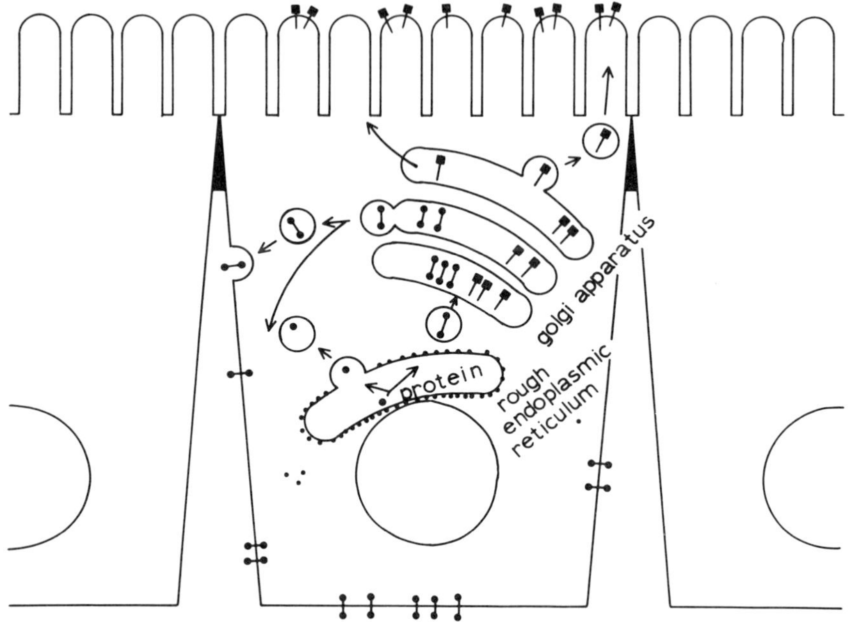

Figure 3. Transport pathways for membrane proteins.

lent system for the study of epithelial polarity, as they code for only a small number of proteins, and these proteins are made in great abundance. Furthermore, unlike membrane proteins encoded by the host genome, the synthesis of viral proteins may be initiated at a particular time, and thus the sequence of events that occurs during viral protein synthesis may be easily studied.

2. INITIAL STUDIES CONCERNING VIRUSES AND MDCK CELLS

The MDCK cell line was established from the kidney of a normal male cocker spaniel in 1958 (Gausch *et al.,* 1966). This kidney epithelial cell line, like most others, was originally used by virologists for viral production. Indeed, a number of viruses were reported to grow in MDCK cells including influenza (PR8), vesicular stomatitus virus (VSV), and vaccinia (Gausch and Smith, 1968; Green, 1962). The abortive infection of MDCK cells by herpes simplex virus was also examined (Roizman and Aurelian, 1965). However, detailed studies concerning viruses and MDCK cells were only done after the MDCK cell line was extensively characterized. The structure initially thought to be viral foci by the virologist Gausch (Leighton, 1981) were later identified by Leighton as being multicellular domes or hemicysts, structures indicative of vectorial fluid transport across MDCK monolayers (Leighton *et al.*, 1970). In order for this vectorial transport to occur, the cells had to retain the structural and functional polarity of transporting epithelial cells *in vivo*. The desire to understand the biogenesis of such polarity ultimately led to initiation of a number of studies concerning viruses and MDCK cells.

The possibility that enveloped viruses could be used in the study of the membrane polarity of MDCK cells was first reported by Rodriguez–Boulan and Sabatini (1978). These investigators observed asymmetrical budding of enveloped viruses from MDCK cells by electron microscopy. Influenza virus (WSN strain), simian virus, and Sendai virus were observed to bud exclusively from the basolateral surface (Fig. 4). Similar observations were obtained with the MDBK epithelial cell line, whereas polarized budding was not observed from primary chick embryo fibroblasts, even from the surface attached to the plastic dish.

The observation of asymmetrical viral budding from MDCK cells is consistent with previous reports of polarized viral budding in natural epithelial monolayers. For example, the budding of Newcastle disease virus from the apical surface of the chorioallantoic membrane of chick embryos was reported as early as 1952 (Murphy and Bang, 1952). Rodriguez-Boulan and Sabatini (1978) extended these observations to a well-characterized, readily manipulated kidney cell line. These investigators also demonstrated that the direction of budding is a specific property of each particular enveloped virus.

3. POSSIBLE ROLE OF VIRAL ENVELOPE PROTEINS IN DETERMINING THE MEMBRANE DOMAIN FOR VIRAL BUDDING

The enveloped RNA viruses that bud from MDCK cells have a complex structure (Fig. 5). The RNA (which is covered with nucleoprotein) is encased

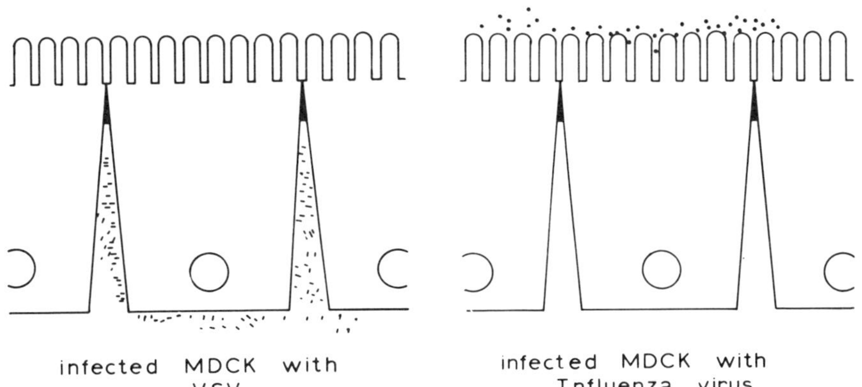

Figure 4. Polarized budding of viruses from MDCK cells.

with a lipoprotein envelope. The inside of the lipoprotein envelope consists of a layer of membrane protein (M protein), whereas the outside of the envelope is composed of lipids. Glycoproteins often called peplomers project out from the lipid envelope. The peplomers of influenza include both hemagglutinin and neuraminidase, whereas the peplomers of VSV are polymers of glycoprotein G (Fenner *et al.*, 1974). The overall structures of the virions used to infect MDCK cells are compared in Fig. 6.

The viral glycoproteins (or peplomers) have been proposed to initiate viral budding, as they are the first components of RNA viruses to enter the plasma membrane. Subsequently, as summarized in Table I, the other components of the virus also become associated with the plasma membrane, and budding occurs. Thus the initial interaction of the viral glycoproteins with the plasma membrane may be a critical factor in determining the plasma membrane domain for the budding of viruses from epithelial cells.

The possible role of the viral envelope proteins (glycoproteins) in determining the plasma membrane domain for budding has been studied. Rodriguez-Boulan and Pendergast (1980) have observed the surface distribution of the envelope glycoproteins of influenza, Sendai, and VSV viruses by means of both immunofluorescence and immunoelectron microscopy. The glycoproteins of in-

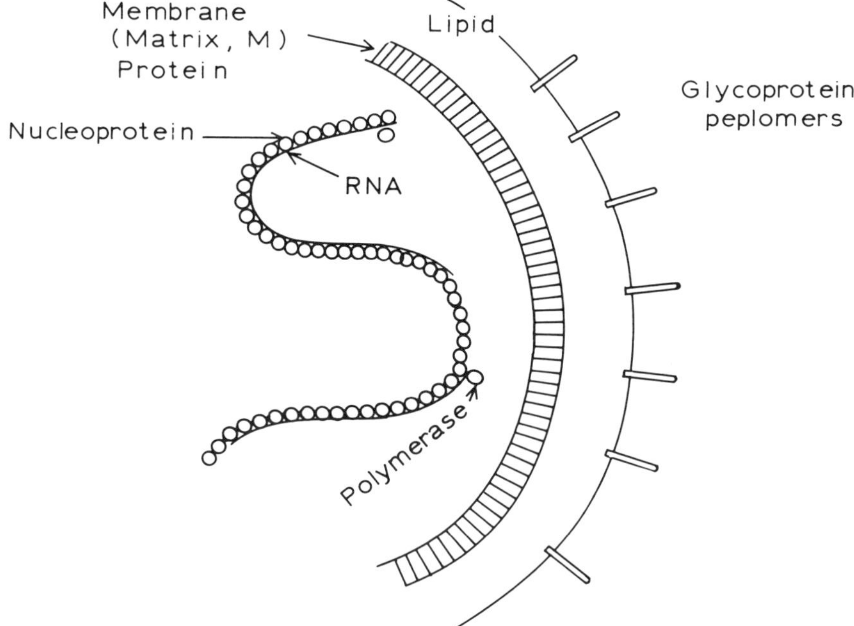

Figure 5. Comparison of structures of different enveloped RNA viruses.

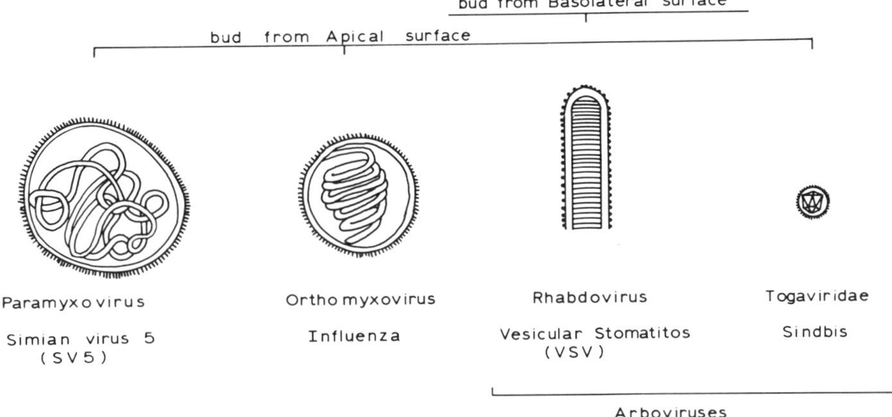

Figure 6. Detailed structure of enveloped RNA virus.

fluenza and Sendai were observed only at the apical surface and the G protein of VSV predominantly at the basolateral surface, prior to the onset of viral budding. (Note that Rindler *et al.,* 1982, have recently observed the presence of some G protein in the apical surface.) On the other hand, the Sendai virus nucleocapsides were observed throughout the cell prior to budding. Thus, a model was proposed in which the asymmetrical distribution of viral envelope glycoproteins, rather than the nucleocapsides, directs the polarity of viral budding.

The viral glycoproteins may be selectively integrated into particular plasma membrane domains as a result of their unique structural properties. Indeed, the investigations of Dragsten *et al.* (1981) indicate that the structural properties of the membrane components of MDCK cells can determine their polarized distribution in the plasma membrane. Fluorescently conjugated membrane probes were used in these investigations. Alternatively, the transport pathway of the glycoproteins through the Golgi, following their biosynthesis, may determine their integration into a particular plasma membrane domain. In either case, the oligosaccharide residues of the viral envelope proteins may ultimately be critical

Table I.   Sequence of Events That Occur during the Process of Viral Budding

1. The viral glycoproteins enter the plasma membrane; other cellular proteins are displaced
2. The viral membrane protein (M, or matrix protein) forms a layer on the cytoplasmic side of the modified plasma membrane
3. The viral nucleocapside then becomes associated with the membrane protein; viral peplomers appear on the outside of the plasma membrane
4. Budding of a mature viricn then occurs

in determining whether they are distributed into a particular plasma membrane domain.

The possible role of the carbohydrate residues of the viral glycoproteins in determining the asymmetrical distribution of membrane proteins was studied in two ways. First, enveloped viruses were grown in MDCK cells in the presence of a glycosylation inhibitor, tunicamycin (Roth *et al.*, 1979). Tunicamycin had no effect on the polarity of VSV and influenza budding, even at tunicamycin concentrations that completely inhibited the glycosylation of viral proteins. Second, viral budding from lectin-resistant variants of MDCK cells was studied. These variants exhibit altered patterns of glycosylation of their glycoproteins. However, the enveloped viruses still bud from the usual surface in the lectin-resistant variants (Meiss *et al.*, 1979). These observations indicate that the domains for budding of VSV and influenza do not depend on the glycosylation of the viral envelope proteins.

These studies still do *not* rule out the possibility that the insertion of the viral envelope proteins into the plasma membrane (either the apical or basolateral domain) is *the* determining factor in polarized budding. The initial association of the influenza envelope protein with the apical surface would presumably then be followed by the assembly of the remainder of the viral components within the viral envelope (as described in Table I).

This possibility could be examined by viral mixing studies. MDCK cells were coinfected with VSV and influenza virus. If the location of the viral envelope in a particular membrane alone dictates the membrane domain of budding, then mixed virions composed of the influenza envelope and VSV RNA would bud from the apical surface; virions composed of the VSV envelope and influenza RNA would bud from the basolateral surface.

The results of the mixing studies were not consistent with these expectations, however. Rindler *et al.* (1982) observed that when MDCK cells were coinfected with VSV and influenza, mature VSV continued to bud from the basolateral surface and often contained some of the HA glycoproteins of influenza. Furthermore, influenza continued to bud from the apical surface. The influenza budding from the apical surface was *not* associated either with VSV RNA or with VSV G protein (significant amounts of G protein were observed in the apical membrane, however). Thus viral components in addition to the envelope glycoproteins probably also play a role in the final assembly and budding of mature virions.

4. THE TRANSPORT PATHWAYS FOR VIRAL GLYCOPROTEINS FROM THE RER TO THE PLASMA MEMBRANE: THEIR ROLE IN THE BIOGENESIS OF POLARITY

In order to study membrane polarity further using MDCK cells, a simpler phenomenon than viral budding was needed. The domain of viral budding may be determined by many other viral components than just the glycoproteins. The polarized distribution of viral glycoproteins in the MDCK plasma membrane is a simpler phenomenon to study. Thus, the mechanism underlying the polarized distribution of the hemagglutinin of influenza virus (which becomes integrated into the apical surface), and the G protein of VSV (which becomes integrated predominantly in the basolateral surface) has been studied.

The possibility has been examined that the polarized distribution of these viral glycoproteins in the plasma membrane is determined by their transport pathways from the RER, through the Golgi, to the plasma membrane. Particular glycoproteins may be transported by different routes from the RER to the plasma membrane (Fig. 3). These glycoproteins may undergo different types of structural modifications in the various cellular compartments that they enter, on their route to the plasma membrane, as summarized in Table II. The glycoproteins may be directly integrated into the plasma membrane from such a compartment or may be routed out to the appropriate plasma membrane domain by a particular class of cytoplasmic vacuoles. Any site in this pathway may prove to be important in determining the membrane domain into which the glycoprotein is ultimately integrated. Thus an examination of the transport pathways for viral glycoproteins may lead to an identification of the cellular-sorting-out processes for these glycoproteins.

Matlin and Simons (1983) have thus examined the kinetics of transport of influenza virus hemagglutinin (HA) to the apical membrane of MDCK cells. The presence of hemagglutinin on the basolateral surface was determined by a sensitive tryptic assay. When MDCK monolayers contained HA on their apical surface, trypsin treatment of the monolayers caused the HA to be converted into two smaller-molecular-weight polypeptides. Their presence could be determined by SDS gel electrophoresis. The investigations using this assay indicated that HA first appeared on the apical surface 15 min after viral infection. Glycosylation of HA had already occurred by this time interval. Reduction of the temperature to

Table II. Structural Modifications That Occur to Glycoproteins en Route to the Plasma Membrane

---

A. RER
    1. Assembly of the glycoprotein
    2. Addition of core oligosaccharide residues (Tabas *et al.*, 1978)
B. Transport to the Golgi, perhaps by clathrin-coated vesicles (Jamieson and Palade, 1971)
C. Golgi
    3. Trimming of core oligosaccharide residues (Leblond and Bennett, 1977)
    4. Terminal glycosylation (i.e., addition of sugar residues to oligosaccharide core; Lebond and Bennett, 1977)
D. Golgi or later steps
    5. Proteins may also be altered by proteolysis, sulfation, or fatty acid acylation (Lazarowitz and Choppin, 1975; Nakamura and Compans, 1977; Schmidt and Schlesinger, 1980)

20°C prevented the transfer of the influenza virus HA to the apical membrane. However, terminal glycosylation of the HA was not prevented by the temperature drop. The block in the transfer of HA to the apical membrane at 20°C was thus attributed to the failure of endosomes at this temperature (Marsh *et al.*, 1983).

Alonzo and Compans (1981, 1983) were able to prevent the integration of viral glycoproteins into the plasma membrane by another means, by the use of ionophores. Monensin (10^{-6} M), a Na$^+$ ionophore, was observed to completely inhibit the budding of VSV, as well as the appearance of VSV G protein on the basolateral membrane. In contrast, 10^{-6} M monensin had no effect on the budding of influenza from the apical membrane. Monensin altered the glycosylation of both the VSV G protein and the influenza HA in these studies.

The mechanism by which monensin specifically blocks the insertion of the VSV G protein into the basolateral membrane of MDCK is of interest. Monensin has been observed by other investigators (Griffiths *et al.*, 1983) to inhibit the transport of the glycoproteins of Semliki Forest virus from the medial to the *trans*-Golgi cisternae. Thus, this transport step through the Golgi may be an essential step in the pathway resulting in the integration of VSV G protein into the basolateral surface of MDCK cells. In contrast, the HA glycoprotein apparently does not require this transport step for its eventual integration into the apical membrane of MDCK cells. So the transport pathway of G protein to the MDCK cell surface may indeed be different from the pathway used by the influenza virus HA glycoprotein. Furthermore, this transport pathway is apparently an important determinant for the insertion of the VSV G protein into the basolateral membrane of MDCK cells.

5. ENDOCYTOSIS OF VIRAL GLYCOPROTEINS: ITS ROLE IN THE MAINTENANCE OF MEMBRANE POLARITY

The transport pathways involved in the recycling of plasma membrane components are also important to examine in the study of membrane polarity. Recycling of components of the plasma membrane is constantly occurring in animal cells. The plasma membrane proteins are continuously removed from the membrane by endocytosis and are subsequently replaced by membrane components in the cytosol. The recycling of membrane components that have been removed by endocytosis is an important means for the cells to maintain the proper balance of proteins in the plasma membrane. The recycling of plasma membrane components in epithelial cells must also be involved in maintaining the asymmetrical distribution of proteins in the two plasma membrane domains.

The membrane polarity may be maintained by several possible types of recycling (Fig. 7). After endocytosis apical membrane proteins may be recycled by a pathway distinct from the pathway used for recycling of basolateral membrane proteins. As a consequence membrane polarity is preserved. Alternatively, after endocytosis apical membrane proteins may be recycled by a pathway that converges with the pathway for recycling of basolateral membrane proteins. In this latter case, sorting-out mechanisms must exist, so that appropriate membrane proteins are subsequently routed out to the appropriate membrane (either apical or basolateral). The sorting-out mechanisms could be either the same or distinct from those involved in the initial insertion of a newly synthesized protein into a particular plasma membrane domain.

Matlin *et al.* (1983) have used viral glycoproteins and MDCK cells as a means to study this problem. VSV G protein was implanted into the apical membrane of MDCK cells by low-pH-dependent fusion. (Note that VSV G protein is normally inserted in the basolateral domain.) The location of the G protein on the MDCK plasma membrane was then followed by means of immunofluorescence. Pesonen and Simons (1983) were also able to quantitate the fate of implanted G protein by means of a protein-A-binding assay. After 60 min the amount of G protein on the apical membrane was dramatically reduced. Only 25–35% of the implanted G protein remained on the apical membrane. A significant fraction of the G protein (25–48%) had appeared on the basolateral membrane during this time period. Some G protein was also observed in the intracellular vacuoles.

These studies suggest, then, that following the endocytosis of virals, glycoproteins from either the apical or basolateral domains, the proteins are routed to a common site (e.g., an endosome), where sorting-out processes occur. So investigations concerning the recycling of the viral glycoproteins do indeed provide important clues about how normal plasma membrane proteins in epithelial cells are maintained in their appropriate membrane domain.

An example of such a normal plasma membrane protein is leucine aminopeptidase, a proteolytic enzyme of kidney epithelial cells. Louvard (1980) has observed that leucine aminopeptidase is localized to the apical membrane of MDCK cells, by means of immunofluorescence. Treatment of MDCK monolayers with antiaminopeptidase antibody stimulated internalization of the aminopeptidase conjugated to the antibody (the fluorescence gradually disappeared from the cell surface with time). Subsequently, antibody, presumably still bound to aminopeptidase, was shown to reappear on the apical membrane of MDCK cells. The specific recycling of antibody conjugated to aminopeptidase may occur by mechanisms similar to those which have been elucidated by the use of enveloped viruses.

6. THE ROLE OF TIGHT JUNCTIONS

The tight junctions of transporting epithelial cells have long been recognized as playing an important role in regulating the passage of diffusible sub-

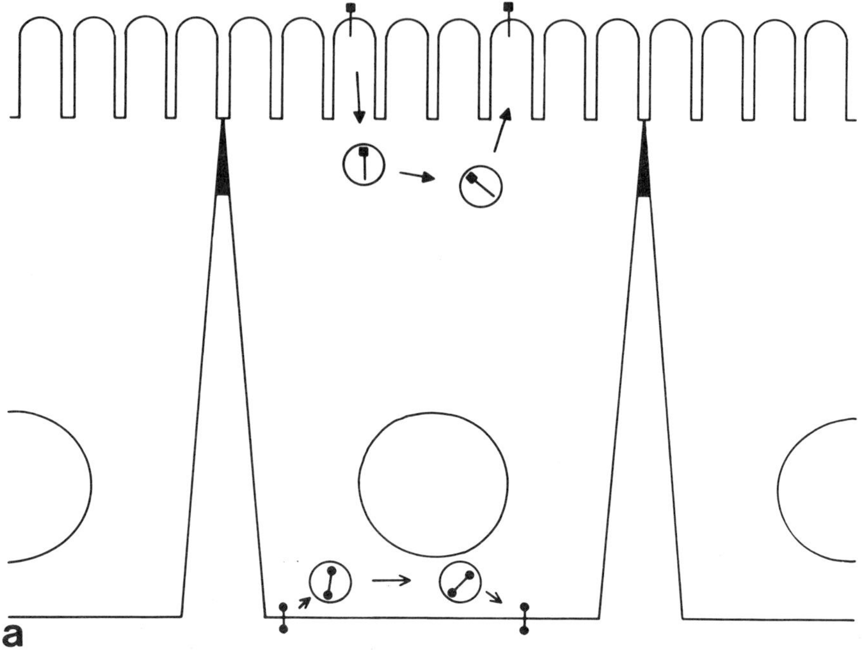

Figure 7. Two models for recycling of plasma membrane proteins in polarized cells: (a) recycled independently, and (b) recycled through pool.

stances across epithelial monolayers (Frömter and Diamond, 1972). Junctions have also been proposed to play an important role in maintaining the two plasma membrane domains of epithelial cells and have even been suggested as being essential in maintaining polarity. The tight junction presumably maintains such membrane asymmetry by acting as a barrier to the lateral diffusion of integral membrane constituents between the apical and basolateral domains. Indeed, the diffusion of integral membrane proteins in the plane of the plasma membrane has been demonstrated in many cells, although the diffusion constants are often smaller than predicted for free diffusion.

The role of junctions in maintaining the membrane asymmetry of MDCK cells has been examined by the use of viruses. Viral budding has been examined in MDCK cells that have been maintained in suspension culture for 12–20 hr (Rodriguez–Boulan *et al.*, 1983). The cell suspension contained not only many

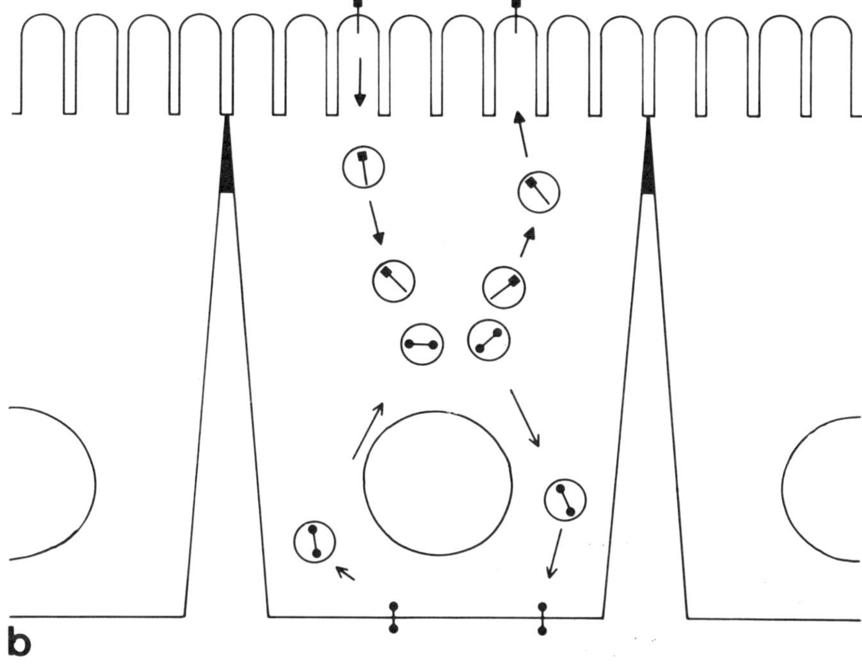

Figure 7. (Continued)

single, rounded cells, but also many cell aggregates. The cells in these aggregates displayed structural polarity as well as tight junctions. After infection with influenza and VSV viruses, these viruses emerged from the single cells in a nonpolarized manner (that is from the entire surface of the single cells). However, in the cell aggregates polarity of viral budding was observed. Influenza and VSV preferentially emerged from either the cells' apical or basolateral surfaces, respectively.

The dependence of polarity on junctions was also examined by studying sparsely plated MDCK cells. The MDCK cells in suspension were plated at a low density into culture dishes, so that cell–cell contact generally did *not* occur. Not only did attached single cells possess apical microvilli, but polarity of viral budding was observed. Influenza and VSV virions predominantly (although not stringently) emerged from the "apical" and "basolateral" surfaces, respectively, of single MDCK cells.

These studies indicate that either tight junction formation alone or attachment to a substratum can cause viruses to bud in a polarized manner from MDCK cells. However, a stringent polarity of viral budding is only expressed in MDCK cells that both possess tight junctions and are attached to a substratum. A stringent sorting out of plasma membrane proteins to the apical and basolateral domains may furthermore be critical in maintaining the functional polarization of MDCK cells.

The role of junctions in maintaining the membrane polarity of MDCK cells has similarly been examined by Roth and Compans (1981), also using enveloped viruses. MDCK monolayers were subjected to a mixed infection with VSV (which buds from the basolateral surface) and influenza (which buds from the apical surface). Initially, normal VSV and influenza virions budded from MDCK cells in a polarized manner. Later on in infection phenotypically mixed virions (with glycoproteins from both viruses) and VSV (WSN) pseudotypes were formed. VSV/WSN pseudotypes containing the VSV genome enveloped the WSN influenza coat.

The formation of mixed virions and pseudotypes was closely associated by Roth and Compans with the opening of the tight junctions in the MDCK monolayer. Presumably, diffusion of membrane glycoproteins to different domains could then occur. When the junctions were completely disrupted either very late on in infection, or by EGTA treatment, directional budding of the virions continued, although strict polarity of budding was lost. These observations concerning the role of junctions in polarized budding are consistent with those of Rodriguez–Boulan *et al.* (1983) with sparsely plated MDCK cells.

However, Roth and Compans (1981) have also obtained results unlike those of Rodriguez–Boulan *et al.* (1983). Confluent MDCK cells were infected with virus, trypsinized, and then replated at a very low density. Under these conditions, Roth and Compans observed a complete loss of polarity of viral budding. Thus, attachment to a substratum alone was *not* sufficient for the polarized budding of viruses from MDCK cells to occur. Rodriguez–Boulan *et al.* (1983) claim that Roth and Compans do not observe polarized budding only because they use freshly trypsinized MDCK cells. The development of polarity in freshly trypsinized MDCK cells can only occur after a protein synthetic phage (Griepp *et al.*, 1979), which may not have occurred in the Roth and Compans experiments.

The role of junctions in maintaining epithelial membrane polarity has also been examined without the use of viruses. Cellular plasma membrane markers have been studied both in natural monolayers as well as in cultured MDCK cells. For example, two apical membrane proteins, leucine aminopeptidase and alkaline phosphatase, have been studied in mouse intestinal epithelial cells. After the cells are dissociated into a single cell suspension, leucine aminopeptidase and alkaline phosphatase (normally on the apical surface) redistribute. Finally, both enzymes become uniformly distributed over the entire cell surface (Ziomek *et al.*, 1980). Similar observations have been made with the cell surface markers of MDCK cells. U *et al.* (1979, 1980) have observed the polarized distribution of intramembranous particles in the junctional complex by means of freeze–fracture electron microscopy. The asymmetrical distribution of these particles is lost simultaneously with the disruption of the junctions by EGTA and is regained when the junctions reform.

Dragsten *et al.* (1981) have made similar observations concerning the polarized distribution of fluorescently conjugated lectin and lipid probes. When either the apical or basolateral surface of MDCK cells is labeled with fluorescently conjugated wheat germ agglutinin or peanut agglutinin, these probes do *not* migrate to the other membrane domain. However, if monolayers are first labeled with lectin in one domain and are then dispersed into single-cell suspensions, then label will spread over the entire MDCK cell surface. Consequently, the junctional complex may indeed be a barrier to the diffusion of membrane components throughout the plasma membrane.

Rodriguez–Boulan *et al.* (1983) have criticized the interpretations of these experiments, which involve either cell dissociation or EGTA treatment. Any loss of polarity observed in such studies may be attributed to the loss of normal cell–cell and cell–substrate interactions, as well as to the loss of junctions.

7. SUMMARY

Transporting epithelial cells possess a remarkable structural and functional polarity. In order for this polarity to exist, membrane proteins must be selectively integrated into either the apical or basolateral domain of the plasma membrane. Not only must the polarized distribution of plasma membrane proteins be established after their biosynthesis, but this polarized distribution must be maintained.

The MDCK cell line and enveloped RNA viruses have been used to study the polarity problem. Some viruses (influenza, Sendai, SV5) bud from the apical surface of MDCK cells, whereas other viruses (VSV) bud from the basolateral surface. The viral glycoproteins may play a role in determining the direction of viral budding from MDCK cells. The HA of influenza becomes integrated into the apical surface prior to budding, whereas the G protein of VSV becomes integrated predominantly into the basolateral surface. The role of the sugar moieties on the glycoproteins in determining polarity has been considered. Polarized viral budding continued to occur in MDCK cells treated with tunicamycin and in lectin-resistant MDCK cells.

The viral glycoproteins have been examined with regards to the biogenesis of their asymmetrical distribution in the plasma membrane of MDCK. A sensitive tryptic assay was developed to determine the kinetics of appearance of the influenza HA in the apical membrane after its biosynthesis. The glycosylation of HA preceded its integration into the apical membrane. A temperature reduction to 20°C prevented the integration of the HA into the plasma membrane, but not glycosylation. The lack of integration of HA could be attributed to a failure of endosomes. The ionophore monensin was observed to specifically inhibit the appearance of newly synthesized VSV G protein into the basolateral membrane but did not affect the appearance of the influenza HA to the apical membrane domain. Monensin may block the transport of G protein from the *cis-* to the *trans*-cisternae.

Viruses have also been used to study the mechanisms responsible for the maintenance of membrane polarity. Epithelial membrane polarity is maintained despite the constant endocytosis and recycling of plasma membrane proteins. In order to study these phenomena and polarity, the VSV G protein was inserted into the apical surface of MDCK cells by low pH fusion (normally G protein is integrated into the basolateral surface). The implanted G protein was observed to disappear from the apical membrane within 60 min. A significant fraction subsequently reappeared on the basolateral membrane.

The role of tight junctions in maintaining membrane polarity has been studied. When MDCK cells are in suspension, viruses bud uniformly from all surfaces of MDCK. When MDCK cells attach to the dish, polarized budding is again observed, even without tight junction formation. However, in the absence of junctions, the strict polarity of the budding is lost. This observation was confirmed in viral mixing studies.

To summarize, MDCK cells and viruses have proved to be important in the study of membrane polarity of epithelial cells. Several important conclusions have been made. First, the glycosylation of membrane proteins is not a determinant of their ultimate plasma membrane domain. Second, after glycoprotein biosynthesis, the transport pathway of the glycoprotein to the plasma membrane does play a role in determining its asymmetrical distribution. Third, when the apical and basolateral glycoproteins are recycled, they enter a common pool. Sorting-out processes then occur. Finally, the viral studies indicate that tight junctions are not necessary for the maintenance of polarity, although junctions are important for a strict sorting out of ''apical'' and ''basolateral''plasma membrane proteins to their respective domains.

ACKNOWLEDGMENTS.   The author acknowledges Esther Olczak for preparation of the manuscript and Terrance O'Keefe for illustrations.

REFERENCES

Alonzo, F. V., and Compans, R. W., 1981, Differential effect of monensin on enveloped viruses that form at distinct plasma membrane domains, *J. Cell Biol.* **89:**700–705.

Alonzo–Caplen, F. V., and Compans, R. W., 1983, Modulation of glycosylation and transport of viral membrane glycoproteins by a sodium ionophore, *J. Cell Biol.* **97:**659–668.

Cereijido, M., Robbins, E. S., Dolan, J. W., Rotunno, C. A., and Sabatini, D. D., 1978, Polarized monolayers formed by epithelial cells on a permeable and translucent support, *J. Cell Biol.* **77:**853–880.

Dragsten, P. M., Blumenthal, R., and Handler, J. S., 1981, Membrane asymmetry in epithelia: Is the tight junction a barrier to diffusion in the plasma membrane? *Nature* **294:**718–722.

Fenner, F., McAuslan, B. R., Mims, C. A., Sambrook, J., and White, D. O., 1974, *The Biology of Animal Viruses,* 2nd ed., Academic Press, New York.

Frömter, E., and Diamond, J., 1972, Route of passive ion permeation in epithelia, *Nature* **35:**9–13.

Gausch, C. R., and Smith, T. F., 1968, Replication and plaque assay of influenza virus in an established line of canine kidney cells, *Applied Micro.* **16:**588–594.

Gausch, C. R., Hard, W. L., and Smith, T. F., 1966, Characterization of an established line of canine kidney cells (MDCK), *Proc. Soc. Exp. Biol. Med.* **122:**931–935.

Green, I. J., 1962, Serial propagation of influence B (Lee) virus in a transmissible line of canine kidney cells, *Science* **138:**42–43.

Griepp, E. B., Robbins, E., Malamet, S., Dolan, W., and Sabatini, D. D., 1979, Studies on the role of glycoproteins in tight junction formatijon, *J. Cell Biol.* **83:**88a.

Griffiths, G., Quinn, P., and Warren, G., 1983, Dissection of the Golgi complex. I. Monensin inhibits the transport of viral membrane proteins from medial to trans Golgi cisternae in baby hamster kidney cells infected with Semliki Forest virus, *J. Cell Biol.* **96:**835–850.

Jamieson, J. D., and Palade, G. E., 1971, Synthesis, intracellular transport and discharge of secretory proteins in stimulated pancreatic exocrine cells, *J. Cell Biol.* **50:**135–158.

Lazarowitz, S. G., and Choppin, P. W., 1975, Enhancement of the infectivity of influenza A and B viruses by proteolytic cleavage of the hemagglutinin polypeptide, *Virology* **68:**440–454.

Leblond, C. P., and Bennett, G., 1977, Role of the Golgi apparatus in terminal glycosylation, in: *Internatitonal Cell Biology, 1976–1977* (B. R. Brinkley and K. R. Porter, eds.), Rockefeller University Press, New York, pp. 326–336.

Leighton, J., 1981, Brief history of active transport in culture, *Ann. N.Y. Acad. Sci.* **373:**351–353.

Leighton, J., Brada, Z., Estes, L. W., and Justh, G., 1969, Secretory activity and oncogenicity of a cell line (MDCK) derived from canine kidney, *Science* **158:**472–473.

Leighton, J., Estes, L. W., Mansukhani, S., and Brada, Z., 1970, A cell line derived from normal dog kidney (MDCK) exhibiting qualities of papillary adenocarcinoma and of renal tubular epithelium, *Cancer* **26:**1022–1028.

Louvard, D., 1980, Apical membrane aminopeptidase appears at the site of cell–cell contact in cultured kidney epithelial cells, *Proc. Natl. Acad. Sci. USA* **77:**4132–4136.

Marsh, M., Bolzau, E., and Helenius, A., 1983, Penetration of Semliki Forest virus from acidic prelysosomal vacuoles, *Cell* **32:**931–940.

Matlin, K. S., and Simons, K., 1983, Reduced temperature prevents transfer of a membrane glycoprotein to the cell surface but does not prevent terminal glycosylation, *Cell* **34:**233–243.

Matlin, K., Bainton, D. F., Pesonen, M., Louvard, D., Genty, N., and Simons, K., 1983, Transepithelial transport of a viral membrane glycoprotein implanted into the apical plasma membrane of Madin–Darby canine kidney cells. I. Morphological evidence, *J. Cell Biol.* **97:**627–637.

McRoberts, J. A., Taub, M., and Saier, M. H., Jr., 1981, The Madin–Darby canine kidney (MDCK) cell line, in: *Functionally Differentiated Cell Lines* (G. Sato, ed.), Liss, New York.

Meiss, H., Green, R., and Rodriguez-Boulan, E., 1979, Lectin resistant mutants of polarized epithelial cells, *J. Cell Biol.* **83:**61a.

Misfeldt, D. S., Hamamoto, J. J., and Pitelka, D. R., 1976, Transepithelial transport in culture, *Proc. Natl. Acad. Sci. USA* **73:**1212–1216.

Murphy, J. S., and Bang, F. B., 1952, Observations with the electron microscope on cells of the chick chorio-allantoic membrane infected with influenza virus, *J. Exp. Med.* **95:**259–268.

Nakamura, K., and Compans, R. W., 1977, The cellular site of sulfation of influenza viral glycoproteins, *Virology* **79:**381–392.

Pesonen, M., and Simons, K., 1983, Transepithelial transport of a viral membrane glycoprotein implanted into the apical membrane of MDCK cells. II. Immunological quantitation, *J. Cell Biol.* **97:**638–643.

Rindler, M. J., Ivanov, I. E., and Sabatini, D. D., 1982, Membrane localization of viral glycoproteins in polarized epithelial cells, *J. Cell Biol.* **95:**262a.

Rodriguez–Boulan, E., and Pendergast, M., 1980, Polarized distribution of viral envelope proteins in the plasma membrane of infected epithelial cells, *Cell* **20:**45–54.

Rodriguez–Boulan, E., and Sabatini, D. D., 1978, Asymmetric budding of viruses in epithelial monolayers: A model system for study of epithelial polarity, *Proc. Natl. Acad. Sci. USA* **75:**5071–5075.

Rodriguez–Boulan, E., Paskier, K. T., and Sabatini, D. D., 1983, Assembly of enveloped viruses in Madin–Darby canine kidney cells: Polarized budding from single attached cells and from clusters of cells in suspension, *J. Cell Biol.* **96:**866–874.

Roizman, B., and Aurelian, L., 1965, Abortive infection of canine cells by herpes simplex virus, *J. Mol. Biol.* **11:**528–538.

Roth, M. G., and Compans, R. W., 1981, Delayed appearance of pseudotypes between vesicular stomatitus virus and influenza virus doing mixed infection of MDCK cells, *J. Virol.* **40:**848–860.

Roth, M. G., Fitzpatrick, J. P., and Compans, R. W., 1979, Polarity of influenza and vesicular stomatitis virus maturation in MDCK cells: Lack of a requirement for glycosylation of viral glycoproteins, *Proc. Natl. Acad. Sci. USA* **76:**6430–6434.

Schmidt, M. F. G., and Schlesinger, M. J., 1980, Relation of fatty acid attachment to the translation and maturation of vesicular stomatitis virus and sindbis virus membrane glycoproteins, *J. Biol. Chem.* **255:**3334–3339.

Tabes, I., Schlesinger, S., and Kornfeld, S., 1978, Processing of high mannose oligo saccharides on the newly-synthesized polypeptides of the vesicular stomatitis virus 6 protein and the IgG heavy chain, *J. Biol. Chem.* **253:**716–722.

Taub, M. L., 1984, Kidney cell cultures in hormonally defined serum-free medium, in: *Mammalian Cell Culture: The Use of Serum-Free Hormone-Supplemented Media* (J. Mather, ed.), Plenum Press, New York, pp. 129–150.

U, H. S., Saier, M. H., Jr., and Ellisman, M. H., 1979, Tight junction formation is closely linked to the polar redistribution of intramembranous particles in aggregating MDCK epithelia, *Exp. Cell Res.* **122:**384–385.

U, H. S., Saier, M. H., Jr., and Ellisman, M. H., 1980, Tight junction formation in the establishment of intramembraneous particle polarity in aggregating MDCK cells. Effect of drug treatment, *Exp. Cell Res.* **128:**223–235.

Ziomek, C. A., Schulman, S., and Edidin, M., 1980, Redistribution of membrane proteins in isolated mouse intestinal epithelial cells, *J. Cell Biol.* **86:**849–857.

9

Hormone Receptors and Response in Cultured Renal Epithelial Cell Lines

KATHRYN E. MEIER and PAUL A. INSEL

> [Philosophy, science, and engineering] are three schools of magic. . . .
> Record many facts. Try to see a pattern. Then make a wrong guess at
> the next fact; that's science.
> Heinlein, 1980

1. INTRODUCTION

Although hormones, neurotransmitters, and growth factors can activate renal transport and metabolic functions, a detailed understanding of the role of these agents in the regulation of renal homeostasis has been difficult to ascertain. In large part, this is due to the complexity of the kidney. With its large variety of different cell types, the kidney presents a veritable myriad of potential target cells. One solution to this problem has been to attempt to isolate and study glomeruli, tubular segments, vascular tissue, or other anatomically distinct portions of the kidney. However, studies using these structures have largely ignored the heterogeneity of cell types within anatomically distinct regions. Thus, major questions remain unanswered regarding the sites and mechanisms of hormone responses within the kidney. These questions are readily amenable to examina-

KATHRYN E. MEIER and PAUL A. INSEL • Department of Medicine, Division of Pharmacology, M-013 H, University of California, San Diego, La Jolla, California 92093. Work in the authors' laboratory has been supported by grants from the American Cancer Society (BC 405), the National Institutes of Health (HL 25457 and GM 31987), and a postdoctoral fellowship (GM 08183-01) to K. E. M. from the National Institutes of Health.

tion by using cultured renal epithelial cells in which one can study hormone action in homogeneous cell populations (Handler *et al.*, 1980). Although such cells lack the anatomical relationships and cell–cell interplay that exist *in vivo,* the advantages of being able to isolate variables, to control the cells' ionic and hormonal milieu, and to draw "clean" conclusions regarding properties of a particular cell type have been powerful "drawing cards" for investigators interested in how hormones and neurotransmitters regulate the kidney.

The goals of studies of hormone action in kidney epithelia have shifted in recent years; hence, new experimental approaches have emerged. Initially, hormonal responsiveness in renal cell lines was used primarily as a marker of differentiated function. By defining the pattern of response to different hormones and relating this to information available on responses occurring *in vivo,* one could pinpoint the region (and, in some cases, the cell type) of origin of a particular cell line. More recently, the hormone responses themselves have become the targets of study, and cell culture systems have led to new insights into the mechanism of hormone action. Of particular interest to us has been the utility of cultured renal epithelial cells as model systems to study receptors, the initial recognition sites for hormones and neurotransmitters. These studies have helped provide further information on the regulation of response and the events modulating signal transduction in hormonally responsive renal cells. Thus, in terms of Heinlein's definition of science, early work in this field has resulted in the recording of many facts and the emergence of patterns of hormone responsiveness in cultured renal cells. To test the "guess" that these hormones bind to specific receptors and in turn to connect receptor occupation to alterations in cellular response, one must study both receptors and response in these model systems.

In this article, we review the literature concerning hormone response and hormone receptors in several established renal epithelial cell lines (Table I). These cell lines are "continuous" in that they have been maintained in culture for many years. They are readily available and their cellular properties are well

Table I.   Origins and Properties of Established Renal Epithelial Cell Lines

Cell line	Year of origin	Species of origin	Segment-specific properties[a]
A6	1965	South African clawed toad	ND
LLC–PK$_1$	1958	Hampshire pig	Proximal tubule
MDCK	1958	Cocker spaniel dog	Distal tubule/collecting duct
MDBK	1957	Cow	ND
JTC-12.P3	1961	Cynomolgus monkey	ND

[a]ND, not determined.

documented. It is important to remember that these models are seldom as simple as one might like to believe; for example, cell lines that were not clonally derived (i.e., from a single cell) are unlikely to be homegeneous, nor are cultured cells identical in all respects to their *in vivo* counterparts. Nonetheless, these cell lines are extremely useful, in part because once hormone responses are defined they tend to remain relatively stable with time in culture.

One factor that has an impact on this consistency of hormone response of epithelial cells is the culture conditions in which the cells are grown. In particular, because serum present in growth medium contains hormones, studies of hormone receptors and response are complicated by the possible presence of the hormone of interest in the growth medium. Serum-derived hormones can modulate the expression of both hormone receptors and response through a variety of mechanisms. This complication can be avoided, or at least minimized, by growing cells in serum-free media that contain a defined mixture of growth-promoting factors and in which levels of the hormone of interest are known. Serum-free media have been developed for most of the cell lines discussed in this review. In addition, the culture substrate may be important in regulating epithelial cell hormone receptors and response. Epithelial cells in culture are polarized such that their apical, or luminal, surface faces the media and their basolateral, or antiluminal, surface faces the culture substrate. Tight junctions form boundaries between these two types of membrane. *In vivo,* the basolateral surface of tubular epithelial cells is the membrane at which nutrients and hormonal signals are received from the plasma and at which the cells interact with basal lamina components. Since in cultured cells the basolateral surface is attached to the plastic substrate, the access of media to this surface may be limited in confluent monolayers. New cell culture techniques, which eliminate some of these problems, will be discussed in this review.

Our discussion is organized by cell line and then further subdivided into sections for each class of hormone. Background material is provided for each cell line regarding known cellular functions that may be modulated by hormones. The literature search was completed in February 1983. Our goal in this review is to draw attention to developments in the field and to suggest further uses for these (and other) epithelial cell lines.

2. A6 CELLS

2.1. Background

The A6 cell line was derived by Rafferty from a mince of whole kidney from a normal male South African clawed toad, *Xenopus laevis* (Rafferty, 1969). These cells are available from the American Type Culture Collection (ATCC

CCL 102). The growth and morphology of A6 cells have been described by
Handler and co-workers (Perkins and Handler, 1981). A6 cells are epithelioid in
appearance and can be grown as monolayers on either a plastic or collagen
substrate. At confluence, A6 cells grown on plastic have tight junctions and
apical microvilli and form domes (indicative of apical-to-basolateral transport;
see Fig. 1). Motile cilia have recently been observed on the apical surface of A6
cells, but only when the cells are grown on porous collagen-coated filters such
that media is freely accessible to the basolateral cell surface (Handler *et al.*,
1984). Although they are not clonally derived, A6 cells are morphologically
homogeneous (Perkins and Handler, 1981).

The transport properties of A6 cells have been studied in detail (Perkins and
Handler, 1981). Confluent monolayers of A6 cells exhibit very high electrical
resistance ($\sim$5000 Ω-cm^2). When the cells are seeded at greater than confluent
density and the resistance is monitored over 1 wk, an increasing resistance is
measured during the first 6 days, after which time values for R, I_{sc}, and PD*
become stable. I_{sc}, which is equivalent to the net Na$^+$ flux, is decreased by
apical application of amiloride, an inhibitor of passive Na$^+$ transport.

2.2. Aldosterone

Ion transport in the distal tubule of the kidney and in amphibian urinary
bladder is regulated by the adrenal mineralocorticoid aldosterone. Because A6
cells are responsive to aldosterone, these cells provide a useful model system in
which to study aldosterone action. Cell lines derived from toad urinary bladder
also show aldosterone-mediated regulation of ion transport (Johnson *et al.*,
1981). However, these latter cell lines, which form multilayers consisting of
cells having more than one morphology, are less ideal models than the more
homogeneous A6 cell line.

The initial observations of Perkins and Handler (1981) described a fourfold
increase in an amiloride-inhibitable I_{sc} in cells that had been incubated overnight
with 100 nM aldosterone. Further studies by the same investigators (Handler *et
al.*, 1981) showed that treatment of A6 cells with aldosterone increased the
incorporation of amino acids into proteins of several molecular weights. The
function of these proteins has not been determined. Aldosterone treatment of A6
cells also resulted in a 40% increase in the number of [^{3}H]ouabain binding sites.
This increase in ouabain binding sites was apparently secondary to the stimulation of sodium transport since the effect was eliminated by simultaneous administration of amiloride. Although induction of citrate synthetase activity by aldosterone has been observed with epithelial cells scraped from toad bladder, A6
cells failed to respond similarly.

*R, transepithelial resistance; I_{sc}, short-circuit current, PD, potential difference.

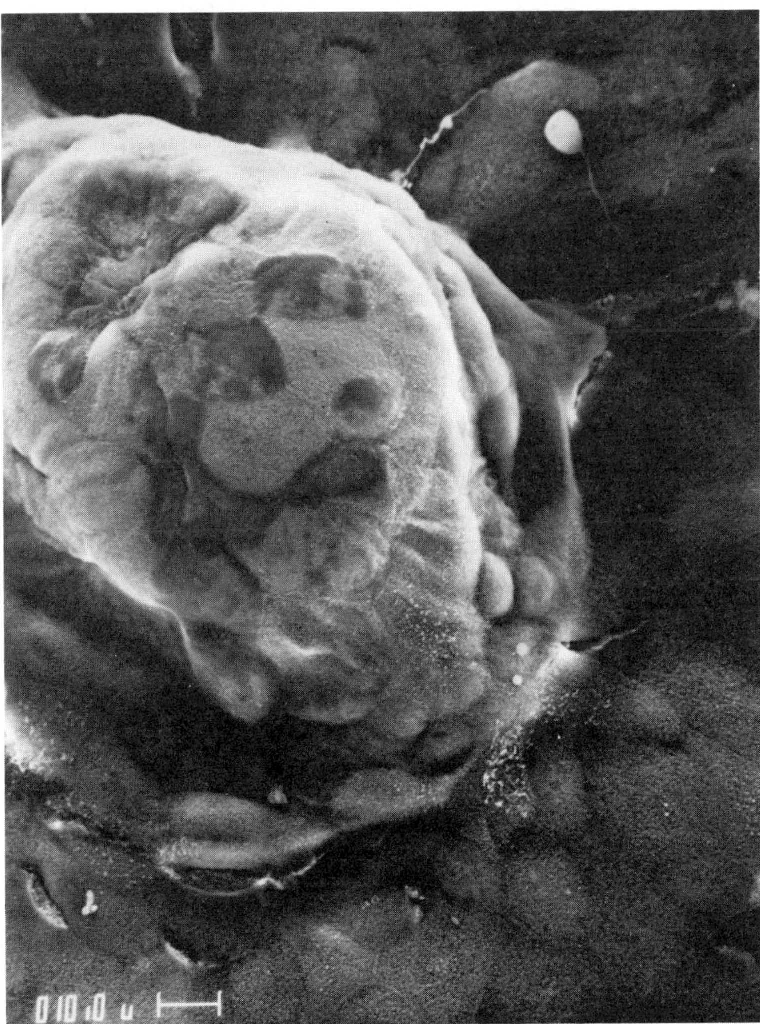

Figure 1. Dome formation by epithelial cells in culture. MDCK cells (clone C) were grown to confluence in serum-containing media on a Costar plastic cover slip. The cells were fixed with glutaraldehyde followed by osmium tetroxide, dried by the critical-point method, sputter-coated with gold/palladium, and photographed using a scanning electron microscope. On the left side of this field is a large, multicellular dome (also called a ''hemicyst'' or ''blister'') which is elevated above the substrate. Dome formation occurs when fluid, transported vectorially from the apical to basolateral cell surface, accumulates beneath the cell monolayer.

Aldosterone binding sites have been identified in A6 cells with radioligand binding techniques (Handler *et al.*, 1981). Nuclear binding of [³H]aldosterone was assessed by incubation of cells with radioligand, followed by hypotonic lysis and filtration of the cell lysate to collect the bound radioactivity. [³H]Aldosterone bound to two sites with different affinities (site 1, K_D = 86 pM and B_{max} = 18 fmol/mg cell protein; site 2, K_D = 16 nM, B_{max} = 280 fmol/mg cell protein*). Binding of various steroids to the site with a high K_D (as measured through competitive binding assays) correlated well with their potencies in stimulating sodium transport. Interestingly, corticosterone and other steroids classified as "glucocorticoids" in mammals were more potent than aldosterone in stimulating sodium transport. The distinction between "glucocorticoids" and "mineralocorticoids" in amphibians has not been clearly established.

A detailed report of [³H]aldosterone binding has recently appeared (Watlington *et al.*, 1982). This study is particularly elegant in that "LIGAND," a weighted, nonlinear, least-squares, curve-fitting computer program (Munson and Rodbard, 1980), was used to model the binding data, allowing for delineation of the two types of steroid binding sites. The binding of [³H]aldosterone was compared with that of [³H]corticosterone. Each steroid bound to two sites with differing affinities. Competition for the radioligand binding sites by a series of unlabeled steroids indicated that the sites with high K_Ds for [³H]aldosterone and [³H]corticosterone were identical, and that sites with lower K_Ds for [³H]aldosterone and [³H]corticosterone represented a second binding site. Receptor occupancy by aldosterone and corticosterone at the lower affinity steroid-binding site correlated well with activation of sodium transport (Fig. 2). The functional role of the high-affinity site is unknown.

2.3. Insulin

Insulin alters renal transport, but until recently effects of insulin had not been pursued in epithelial cell cultures. It may be difficult to assess effects of insulin on cultured cells, because insulin is present in serum and is a required component of serum-free media used for cell culture. Nevertheless, Fidelman and co-workers (1982), using A6 cells grown on a porous collagen-coated filter substrate, found that basolateral application of insulin to cells (previously incubated for 3 hr in the absence of insulin) increased I_{sc} sixfold and PD twofold. These effects of insulin, which were maximal at 3000 μU/ml and peaked within 1 hr, are consistent with the sodium retention produced by insulin *in vivo*. In the same study, these investigators examined the metabolic effects of insulin. Insulin (10,000 μU/ml) increased the incorporation of glucose into glycogen by 56–70%. Thus, A6 cells may provide a useful model for studies of insulin effects on

---

*K_D, equilibrium dissociation constant; B_{max}, maximum number of binding sites.

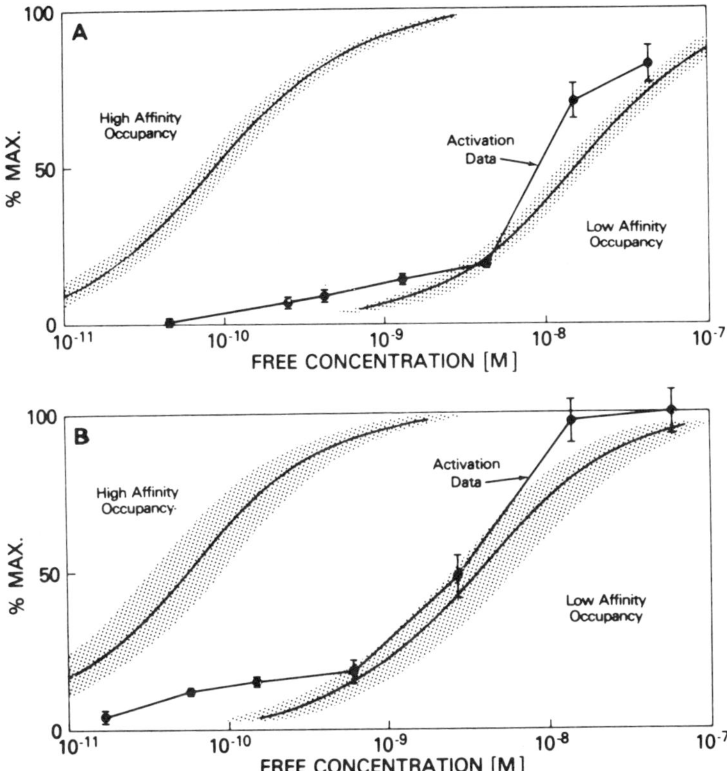

Figure 2. Aldosterone and corticosterone receptor occupancy and response in A6 cells. Occupancy of high- and low-affinity receptor sites in A6 cells by aldosterone (A) and corticosterone (B) is shown as smooth curves, with the shaded area representing the standard error. These curves were constructed (using the LIGAND computer program) from affinity constant values determined by two-site analysis of equilibrium binding data. Also shown are the activation curves for the increase in I_{sc} in response to the steroids, where the error bars represent the standard error. (From Watlington *et al.*, 1982; reproduced with permission.)

renal function. However, the concentrations of insulin required to produce effects on these cells are 10- to 100-fold higher than circulating levels generally achieved *in vivo*. This may reflect changes in the cells' insulin receptors or insulin response pathway produced by culture *in vitro*.

2.4. Vasopressin

Initial experiments by Perkins and Handler (1981) showed no effect of arginine vasopressin (AVP) on PD or I_{sc} in A6 monolayers grown on plastic. AVP also did not activate adenylate cyclase activity in a preparation of broken

A6 cells. By contrast, when these investigators tested membranes prepared from A6 cells that had been grown on porous collagen-coated filters, AVP stimulated adenylate cyclase activity (Handler *et al.*, 1984). Thus it appears that the substrate or culture condition used for growing A6 cells can determine responsiveness to vasopressin. This system may prove useful for studying how renal cells develop responsiveness to vasopressin.

2.5. Other Hormones

Lang and co-workers (1982) reported that adenosine stimulates cAMP* accumulation in A6 cells grown on a porous substrate. The results of tests with adenosine analogs indicated that the effect was mediated via ribose-specific stimulatory receptors. Stimulation of cAMP accumulation by the β-adrenergic agonist isoproterenol has also been noted (Handler *et al.*, 1983). Until the effects of other hormones on A6 cells are evaluated, it is not possible to define the overall receptor "fingerprint" of these cells. Such studies should be forthcoming now that culture methods have been developed that seem to facilitate more complete expression of the differentiated functions of A6 cells. It will be interesting to learn which nutrients or hormones must have access to the basolateral surface of A6 cells in order to induce differentiation.

3. LLC–PK$_1$ CELLS

3.1. Background

The LLC–PK$_1$ cell line was developed in 1958 by Hull and co-workers (1976) from a mince of whole kidney from a normal male Hampshire pig. LLC–PK$_1$ cells are available from the American Type Culture Collection (ATTC CRL 1392). LLC–PK$_1$ cells are epithelioid and nontumorigenic, form domes, and produce plasminogen activator but not renin. LLC–PK$_1$ cells were not clonally derived and may be heterogeneous since a subline of LLC–PK$_1$ produces higher levels of plasminogen activator than does the parent strain (Hull *et al.*, 1976).

A more detailed description of the differentiated properties of LLC–PK$_1$ was published by Perantoni and Berman (1979). These investigators used histochemical techniques to identify cellular enzyme activities. They found that the cells had γ-glutamyl-transpeptidase activity, but that the level of activity was quite variable. Most of the cells in the population had no leucine aminopeptidase activity, but some cells had high activity. These results provide further evidence for the heterogeneity of the population. LLC–PK$_1$ cells also had acid phos-

*cAMP, adenosine-3′,5′-cyclic monophosphate.

phatase activity but not glucose-6-phosphatase, 5'-nucleotidase, or adenosine triphosphatase activities. The latter result is somewhat difficult to resolve with the findings of Mills *et al.* (1979), who identified ouabain-binding sites on the surface of $LLC-PK_1$ facing the culture dish (basolateral surface) and presented evidence that these sites were associated with the Na^+/K^+ ATPase. This basolateral localization may have prevented access of substrate in the histochemical studies and is a limitation to the use of such techniques to define cellular constituents. Thus, some data on enzyme activities of $LLC-PK_1$ correspond to results found in proximal tubular cells, suggesting that $LLC-PK_1$ is derived from proximal renal tubular epithelium. In contrast, Goldring *et al.* (1978) studied the hormone responsiveness of $LLC-PK_1$ and concluded that the cells might be derived from the thick ascending limb of the loop of Henle. Those workers also noted the possibility of heterogeneity, based on their observation that long-term passage of the cells resulted in altered hormone response. Since the cells were not cloned, it is possible that different subpopulations of cells are preferentially expressed after different periods in culture.

A serum-free defined medium has been developed that supports continuous growth of $LLC-PK_1$ cells (Chuman *et al.*, 1982). This medium contains transferrin, insulin, selenium, hydrocortisone, triiodothyronine, vasopressin, and cholesterol. Presumably, the cells possess receptors to each of the hormones in this medium, although, as indicated in the next sections, detailed studies have only been reported for certain of these hormones. Hull *et al.* (1976) developed an unusual subline, $LLC-PK_{1L}$, which could be grown in medium 199 without supplementation with serum or growth factors. The use of serum-free, hormonally defined media may allow for more precise studies of hormone action in this cell line, as has been possible for certain other cells grown in defined media (Dibner and Insel, 1981; Darfler *et al.*, 1981).

Considerable work, which will not be reviewed here, has been devoted to the description of the properties and development of sodium-dependent hexose transport in LLC/PK_1 (Amsler and Cook, 1982; Mullin *et al.*, 1980; Moran *et al.*, 1982; Misfeldt and Sanders, 1981). The characteristics of sugar transport in confluent $LLC-PK_1$ cells closely resemble those in proximal tubule. Rabito and Karish (1982, 1983) have also described amino acid transport systems in $LLC-PK_1$.

$LLC-PK_1$ cells do not synthesize prostaglandins (Hassid, 1981; Goldring *et al.*, 1978; Lifschitz, 1982). Lifschitz (1982) showed that this is due to a defect in the cyclooxygenase step of the synthetic pathway. This feature may prove useful in hormonal studies, since secondary cellular effects due to endogenous prostaglandin release are nonexistent. $LLC-PK_1$ cells also fail to synthesize glucose from lactate (Mullin *et al.*, 1982) despite the evidence that gluconeogenesis is localized to the proximal tubule in some species (Maleque *et al.*, 1980; Schmidt and Guder, 1976).

3.2. Vitamin D_3

Metabolism of 25-hydroxyvitamin D_3 by a 24-hydroxylase to 24,25-$(OH)_2D_3$ has been demonstrated in LLC–PK_1 cells (Colston and Feldman, 1982). The 24,25-$(OH)_2$ D_3 metabolite is less active than the major active metabolite, 1,25-$(OH)_2D_3$ (which is also made by the kidney), but appears to be part of a negative-feedback loop of D_3 metabolism. This regulatory pathway was shown by Colston and Feldman (1982) in LLC–PK_1 cells, in which 1,25-$(OH)_2D_3$ induced 24-hydroxylase activity (peak at ~9 hr). In addition, Colston and Feldman identified specific cytosolic binding sites (K_D = 0.12 nM; B_{max} = 26 fmol/mg protein) for 1,25-$(OH)_2D_3$ and found that the EC_{50} for enzyme induction (0.13 nM) was virtually identical to the K_D of 1,25-$(OH)_2D_3$. Induction of the 24-hydroxylase activity was inhibited by actinomycin D and thus was considered a transcriptional event.

3.3. Vasopressin

Ausiello and co-workers have extensively characterized the vasopressin receptors of LLC–PK_1 cells and receptor interaction with adenylate cyclase. Goldring *et al.* (1978) initially showed that vasopressin increased cAMP levels in LLC–PK_1 cells. Further studies (Ausiello *et al.*, 1980) demonstrated that vasopressin increased cAMP-dependent protein kinase activity. These investigators found that rat and guinea pig medulla contained predominantly the soluble, type II form of protein kinase, rather than the type I form. The kinase activated by vasopressin in LLC–PK_1 cells was also a type II kinase.

In subsequent studies Roy and Ausiello (1981) described the specific binding of [^{3}H](8-lysine) vasopressin to intact LLC–PK_1 cells. Specific binding of the radioligand was saturable, attained equilibrium rapidly, and was partly reversible, and the binding sites were stereoselective (shown using structural analogs of vasopressin). Scatchard plots* of the saturation binding isotherms, using cells either grown in monolayers or maintained in suspension for 24 hr, were curvilinear, indicating a heterogeneity of binding sites. In contrast, when binding assays were performed using cells freshly suspended with EDTA, the Scatchard plots were linear, indicating a single class of noncooperative binding sites (B_{max} = 20-25 fmol/10^6 cells). Mg^{2+} ion increased the affinity of vasopressin for its receptor. Using computer-aided, curve-fitting methods, the authors showed that a ''receptor transition'' model would account for the observed heterogeneous nature of the binding sites in the cells grown in monolayer or maintained in

---

*Scatchard plots of equilibrium binding data are obtained by plotting the amount of bound radioligand (abscissa) versus the amount of bound radioligand divided by the free radioligand (ordinate). The negative slope of the line equals the K_D for the radioligand and the intercept on the abscissa equals the B_{max} (Scatchard, 1949).

suspension (Vassent *et al.*, 1981; Fig. 3). Other models (single site; two independent sites) did not describe the binding data as well as did the transition model. The transition model describes the conversion of the receptor to a lower affinity form after hormone binding. The authors postulate that the receptor is a dimer with two high-affinity binding sites. When one site binds hormone, the other site converts to a low-affinity form. In EDTA-suspended cells, the transition to low affinity is postulated to occur slowly such that a homogeneous population is observed. Thus, metal ion, perhaps Mg^{2+}, is required for the "receptor transition."

Roy *et al.* (1981) also showed that occupancy of the vasopressin receptors correlated well with increased cAMP production. However, rapid desensitization of the response occurred in the presence of high concentrations of vasopressin. There was no loss of vasopressin binding sites accompanying desensitization. The desensitization did not occur with EDTA-treated cells and could be blocked by addition of hypertonic sodium chloride. On the basis of these and related data,

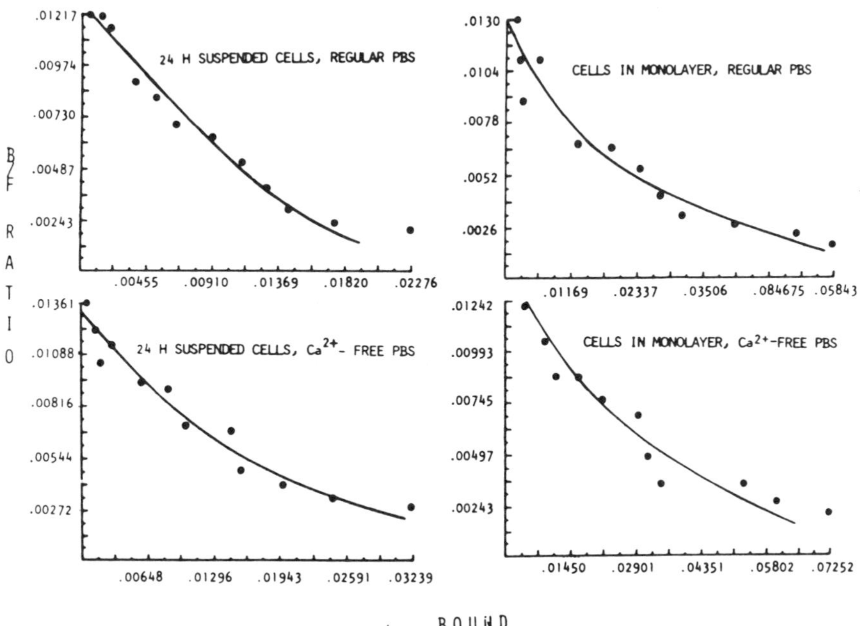

Figure 3. Vasopressin binding to receptors on LLC–PK$_1$ cells. Equilibrium binding data for arginine vasopressin binding to intact LLC–PK$_1$ cells are plotted according to the method of Scatchard. A receptor transition model, as described in the text, was used to fit the data by a nonlinear curve-fitting program. The lines represent the best fit obtained for each set of experimental data. (From Vassent *et al.*, 1981; reproduced with permission.)

the authors concluded that receptor transition is a required but not a rate-limiting step in desensitization, and that hypertonic sodium chloride blocks the transition. · Desensitization was correlated with occupancy of the low-affinity site. An intriguing concept derived from these studies is that, in the physiological concentration range of vasopressin, occupancy is directly coupled to response, whereas desensitization occurs preferentially with very high hormone concentrations. These studies suggest opportunities for further studies of the vasopressin receptor and its interaction with adenylate cyclase. For example, since these experiments were performed using intact cells, it should be possible to determine whether desensitization involves internalization and/or degradation of some components of the receptor–adenylate cyclase comples.

Ausiello and Hall (1981), using LLC–PK$_1$ membranes, found that calmodulin enhanced the vasopressin activation of adenylate cyclase up to 40% in the presence of calcium. This effect reflected an increase in V_{max}, rather than a change in K_m, for vasopressin activation.* Calmodulin activation was inhibited by trifluoperazine and was dependent on Mg^{2+} concentration. A careful delineation of this effect would be quite informative, as this is an excellent cultured cell system in which to explore effects of calmodulin on hormone-sensitive adenylate cyclase.

A novel study of the cellular regulation of vasopressin receptor expression in LLC–PK$_1$ was published by Roy et $al.$ (1980). In order to test the effects of growth factors, these investigators used the LLC–PK$_{1L}$ cell line developed by Hull et $al.$ (1976) for growth in serum-free media. Unlike the parent cells, LLC–PK$_{1L}$ cells have no calcitonin-stimulated adenylate cyclase activity, but do show isoproterenol-stimulated adenylate cyclase activity and have a 95% decrease in their number of vasopressin receptors. The binding properties of LLC–PK$_1$ and LLC–PK$_{1L}$ vasopressin receptors are identical. When LLC–PK$_{1L}$ cells were grown for 1 wk in medium containing 10% serum or 10 μg/ml insulin, the number of vasopressin binding sites was increased by approximately 80% and receptor affinity was unchanged. The effects of insulin and serum together were partly additive (130%). The increase in vasopressin receptors correlated closely with an increase in maximal stimulation of adenylate cyclase in response to vasopressin. Neither serum nor insulin had any marked effect on cell growth. Agents that had no effect on vasopressin receptors or response included hydrocortisone, dexamethasone, thyroid hormone, sodium butyrate, and dibutyryl cAMP. The LLC–PK$_{1L}$ cell line should therefore be a useful system in which to study the mechanisms whereby serum and insulin regulate the increased expression of vasopressin receptors: e.g., increased receptor synthesis, decreased receptor degradation, and altered receptor recycling.

As mentioned previously, vasopressin is a component of the LLC–PK$_1$ serum-free defined media developed by Chuman et $al.$ (1982). The requirement

---

*V_{max}, maximal velocity; K_m, Michaelis–Menten constant.

for vasopressin could be partly replaced by dibutyryl cAMP or isobutylmethyl xanthine, a phosphodiesterase inhibitor. Thus increases in cellular levels of cAMP appear to enhance LLC–PK$_1$ cell proliferation.

3.4. Calcitonin

Goldring *et al.* (1978) showed that salmon calcitonin increased the cAMP content of LLC–PK$_1$ cells approximately 35-fold with an apparent K_m of 2.7 nM. Ausiello *et al.* (1980) showed that calcitonin, like vasopressin, activated LLC–PK$_1$ protein kinase activity. However, calcitonin did not substitute for vasopressin in LLC–PK$_1$ hormone-supplemented medium (Chuman *et al.*, 1982). A possible explanation for this observation is that some sublines of LLC–PK$_1$ may not respond to calcitonin. Further studies of calcitonin receptors or response in LLC–PK$_1$ have not yet appeared, although this cell line would seem to be a very appropriate model system for characterization of calcitonin receptors.

3.5. Catecholamines

Goldring and co-workers (1978) presented data showing a 1.7-fold increase in LLC–PK$_1$ cAMP levels in the presence of epinephrine; however, the authors stated that this did not constitute an epinephrine response. The low response (compared to that elicited by calcitonin and vasopressin) may reflect a paucity of β-adrenergic receptors on these cells, or, alternatively, the cells might possess receptors but fail to show large hormonal responses for one or more reasons: (1) heterogeneity of the cell population, such that only a portion of the cells respond to catecholamines; (2) variations in the degree of coupling of β-adrenergic receptors to adenylate cyclase, or (3) coupling of both stimulatory (β-adrenergic) and inhibitory (α-adrenergic) receptors to adenylate cyclase. As mentioned previously, Roy *et al.* (1980) found that the LLC–PK$_{1L}$ subline (but not the parent line) responded to isoproterenol with a 50–80% increase in cAMP over basal levels when grown in serum-free media, but no response could be detected when these cells were grown in serum. These results suggest that both cellular heterogeneity and growth conditions may alter catecholamine response in LLC–PK$_1$. In preliminary radioligand binding studies, we have found that LLC–PK$_1$ cells obtained from ATCC and grown in 10% fetal bovine serum do not express α_1, α_2, or β-adrenergic receptors (Meier and Insel, unpublished data). Further studies of LLC–PK$_1$ cells may reveal whether catecholamine receptors can be expressed under other growth conditions, e.g., in the absence of the catecholamines present in fetal bovine serum (Dibner and Insel, 1981).

3.6. Other Hormones

As previously discussed, insulin increases the number of vasopressin receptors on LLC–PK$_{1L}$ (Roy *et al.*, 1980). Both insulin and triiodothyronine are

growth stimulatory for the parent cell line (Chuman *et al.*, 1982). Prostaglandin
E_2 and parathyroid hormone do not appear to increase cAMP levels significantly
in LLC–PK$_1$ (Goldring *et al.*, 1978). Since LLC–PK$_1$ cells do not synthesize
prostaglandins (Lifschitz, 1982), these cells will provide a very "clean" system
for the study of any prostaglandin action which can be demonstrated.

4. MDCK CELLS

4.1. Background

Madin–Darby canine kidney (MDCK) cells (ATCC CCL 34) were derived
in 1958 by Madin and Darby from a mince of kidney cortex obtained from a
normal female cocker spaniel and were first characterized by Gausch *et al.*
(1966). This popular cell line has been the subject of an extensive literature
which has been reviewed recently (McRoberts *et al.*, 1981; Saier, 1981; Taub
and Saier, 1979; Simmons, 1982). Features relevant to the current discussion are
as follows: Confluent MDCK cells form a polarized monolayer with domes (Fig.
1) (Leighton *et al.*, 1969, 1970). Dome formation is promoted by elevation of
intracellular cAMP (Valentich *et al.*, 1979), by inducers of differentiation
(Thomas *et al.*, 1982; Lever, 1982) and by reduction in cell–substrate adhesion
(Rabito *et al.*, 1980). Cloned MDCK cell lines exhibit varying tendencies toward
dome formation (Lever, 1981; Meier *et al.*, 1983a). The transport properties of
MDCK epithelia appear to vary between different sublines (Richardson *et al.*,
1981). Transepithelial resistance values range from approximately 100 Ω cm^{-2}
(Misfeldt *et al.*, 1976; Cereijido *et al.*, 1978) to 4000 Ω cm^{-2} (Simmons, 1981).
Ouabain-sensitive sodium pump sites are located predominantly on the basolateral cell surface (Cereijido *et al.*, 1980). Sodium transport pathways in MDCK,
including amiloride-sensitive, ATP-independent Na$^+$-H$^+$ antiport, have been
studied in detail and have been reviewed by McRoberts *et al.* (1981).
The morphological features of MDCK cells are dependent on the substrate
on which the cells are grown. When grown on a collagen substrate, MDCK cells
form two types of morphologically distinct, clonally stable colonies, as dis-
tinguished in scanning electron micrographs (Valentich, 1981). When grown on
collagen, one of these cell types forms basal lamina and is ciliated. However,
MDCK cells do not have cilia when grown on plastic. MDCK cells grown on
collagen and then overlaid with a collagen gel form tubulelike structures (Hall *et
al.*, 1982). When grown on carbon-coated Nucleopore filters, MDCK cells assemble basal lamina only on those portions of the cells which span holes in the
filter (Valentich, 1982). MDCK cells secrete glycosaminoglycans, which are
basal lamina components, when grown on either plastic or collagen (Chin *et al.*,
1982). Taken together, these results indicate that MDCK cells are capable of

synthesizing basal lamina components but assemble the components only under specific growth conditions.

MDCK cells are probably derived from distal tubule and collecting duct, as evidenced by the following data: (1) Rindler and co-workers (1979) showed that MDCK cells have high activities of ecto p-nitrophenylphosphate phosphatase, leucine aminopeptidase, and ouabain-sensitive Na^+/K^+ ATPase. MDCK cells lack several proximal tubule markers: maltase, trehalase, Na^+-dependent glucose uptake, and p-amino hippurate uptake. (2) Hormonal responsiveness of MDCK cells is similar to that of the distal tubule, as will be discussed later. (3) The ciliated and nonciliated cells described by Valentich (1981) are morphologically similar to the principal and intercalated cells of the mammalian collecting tubule. (4) Herzlinger et al. (1982) prepared monoclonal antibodies against MDCK cells and then used immunofluorescence staining techniques to localize antigenic sites in dog kidney sections that cross-reacted with MDCK cells. This elegant methodology revealed that one monoclonal antibody directed against MDCK recognized antigenic sites on the basolateral surface of the thick ascending limb of the loop of Henle and on distal convoluted tubule. All these studies point to a distal tubule/collecting duct origin for MDCK cells.

MDCK cells are nontumorigenic (McRoberts et al., 1981), despite earlier reports to the contrary (Leighton et al., 1969, 1970). MDCK cells can be transformed by Moloney sarcoma virus (MSV) (Taub et al., 1981), Harvey murine sarcoma virus (Lin et al., 1982), or chemical mutagenesis (Boerner and Saier, 1982); these transformed cells are tumorigenic.

As mentioned previously, data from several studies indicate that the MDCK cell line is not homogeneous, a fact that should be kept in mind when considering hormonal response of the cells. Evidence for this inhomogeneity includes the morphological variations between clonal MDCK cells (Valentich, 1981), the variation in electrical resistance between different MDCK sublines (Richardson et al., 1981), and differences in dome formation in various clonal MDCK cell lines (Lever, 1981). In addition, different MDCK cell lines synthesize different prostaglandins (Lewis and Spector, 1981). Recently we have confirmed that MDCK cells cloned from early-passage cultures obtained from ATCC have varying morphologies (Fig. 4) (Meier et al., 1983a). The observation that these clones are morphologically stable for more than 50 passages in culture favors the interpretation that the parental MDCK cell line consists of various cell types, rather than different cells having arisen by dedifferentiation or mutation. Several different clones from this series (shown in Fig. 4) all bind anti-MDCK monoclonal antibodies, as detected using immunofluorescence (G. Ojakian, unpublished observations). Since MDCK cells are easily cloned, such homogeneous sublines should prove extremely useful for investigations of hormonal response, transport, and cellular metabolism.

A serum-free, defined medium has been developed that supports long-term

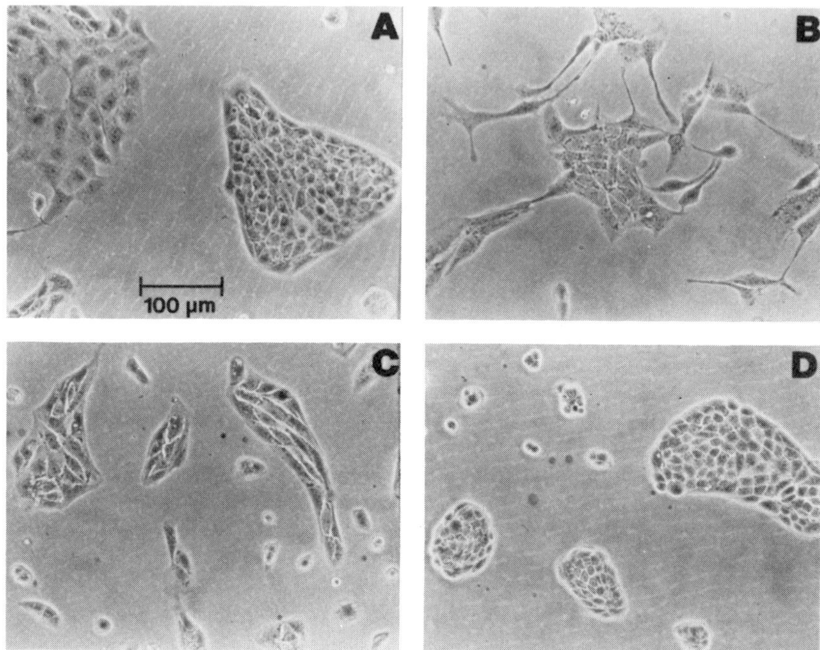

Figure 4. Morphologic variations in parental and clonal MDCK cell lines. Subconfluent cultures of MDCK cells, grown in serum-containing media on a plastic substrate, were photographed using phase-contrast optics. (A) shows several types of cell morphologies present in early-passage cultures of the parental line (obtained from ATCC). (B), (C), and (D) show clonal lines D, N, and L, respectively, which were all derived from early-passage parental MDCK cells (ATCC passage 50-60).

growth and dome formation in MDCK cells (Taub *et al.*, 1979). The growth factors present in this medium are insulin, transferrin, prostaglandin E_1, hydrocortisone, and triiodothyronine. Amino acid transport by MDCK cells grown in this medium has been described (Boerner and Saier, 1982).

MDCK cells synthesize prostaglandins (Levine and Hassid, 1977; Lin *et al.*, 1982; Lifschitz, 1982; Hassid, 1981; Lewis *et al.*, 1981; Levine and Moskowitz, 1979; Beaudry *et al.*, 1982; Lewis and Spector, 1981; Ohuchi and Levine, 1978). The predominant product is either PGE_2 or $PGF_{2\alpha}$, depending on the MDCK subline used (Lewis and Spector, 1981). The regulation of prostaglandin synthesis in MDCK cells is discussed further below. One virally transformed MDCK subline does not produce prostaglandins (Lin *et al.*, 1982). MDCK cells also synthesize sulfolipids, which are components of the mammalian kidney (Ishizuka *et al.*, 1978).

4.2. Aldosterone and Deoxycorticosterone

Studies designed to search for effects of mineralocorticoids on MDCK cells have yielded results that are both encouraging and disappointing. Ludens and co-workers (1978) were able to demonstrate specific binding of [^{3}H]deoxycortico-sterone ([^{3}H]DOC) to MDCK cytosol. The K_D of [^{3}H]DOC was 100 nM, and competitive binding assays indicated that the binding specificity was appropriate for a mineralocorticoid receptor. Aldosterone, which is secreted in dogs in much lower quantities than is DOC, had lower affinity for the receptor than did DOC. [^{3}H]DOC was concentrated in the nuclei of intact MDCK cells incubated with the radioligand. However, no cellular response to DOC was identified. Baker and Fanestil (1977) showed that [^{3}H]DOC binding to the receptor was inhibited by a broad class of protease inhibitors and substrates, indicating that the receptor binding site might have structural features similar to those of a protease.

Simmons (1978), in a preliminary report, described a reversal in sodium transport (from basolateral-to-apical to apical-to-basolateral) in MDCK cells treated for 4 hr with 200 nM aldosterone. Pretreatment with 10 μM DOC for 60 hr had a similar effect. Recently, Cobb *et al.* (1982) found that addition of 80 μM aldosterone 1 hr prior to addition of radiolabeled amino acids increased the incorporation of amino acids over the following 3–6 hr into at least two MDCK membrane proteins (molecular weights, 14,000 and 35,000 daltons). This oc-curred with MDCK cells grown in either serum-containing or serum-free medi-um, and similar results were obtained using DOC. These investigations are encouraging in that they document effects of mineralocorticoids on MDCK cells, yet they are incomplete since receptor occupancy has not been correlated with response and because the identity and function of the two aldosterone-induced membrane proteins are as yet unknown.

4.3. Bradykinin

Bradykinin stimulates prostaglandin synthesis in MDCK cells (Hassid, 1981; Lewis and Spector, 1981; Lewis *et al.*, 1981). In an MDCK subline that produced predominantly PGE$_2$, bradykinin increased the biosynthesis of both PGE$_2$ and 6-oxo-PGF$_{1\alpha}$ (as measured by radioimmunoassay) (Hassid, 1981); 6-oxo-PGF$_{1\alpha}$ is the principal degradation product of PGI$_2$ (prostacyclin). The EC$_{50}$* for bradykinin was approximately 1 ng/ml, and the ratio of PGE$_2$ to 6-oxo-PGF$_{1\alpha}$ was the same as that measured in unstimulated control cells. In the same study, when cells were prelabeled with [^{3}H]arachidonic acid, bradykinin increased release of arachidonic acid and also stimulated the synthesis of radi-olabeled PGE$_2$ and 6-oxo-PGF$_{1\alpha}$. Such results suggest that bradykinin acts to

---

*EC$_{50}$, concentration of hormone producing a response that is 50% of the maximum response.

stimulate prostaglandin formation by enhancing phospholipase activity and release of arachidonic acid from membrane phospholipids. Lewis and Spector (1981) found that bradykinin increased synthesis of both PGE_2 and $PGF_{2\alpha}$ in an MDCK subline making predominantly PGE_2, as well as in a subline making predominantly $PGF_{2\alpha}$. Maximal prostaglandin release was achieved after 5 min exposure to bradykinin, and the ratios of prostaglandin products were the same in control and bradykinin-treated cells. In the MDCK subline making predominantly PGE_2, supplementation of cells with linoleic acid (but not with oleic or arachidonic acid) enhanced the bradykinin-induced release of PGE_2 approximately fourfold (Lewis et al., 1981). Since the overall distribution of prostaglandin products was not altered by linoleic acid supplementation, the effect of linoleic acid appeared to be largely due to accumulation of arachidonic acid in cellular phospholipids. We are unaware of further characterization of bradykinin receptors on MDCK cells or of studies of lipoxygenase pathway products derived from arachidonic acid in these cells.

4.4. Catecholamines

The following discussion will be facilitated by a brief review of catecholamine receptor specificities. Receptors for catecholamines have been divided into several classes based on the differential sensitivity of responsive tissues to various adrenergic agonists and antagonists. Alpha-adrenergic responses are characterized by the order of agonist potency epinephrine > norepinephrine $\gg$ isoproterenol. Responses mediated by the α_1 receptor subtype are blocked by certain antagonists, such as prazosin, whereas the α_2 receptor subtype is blocked by other antagonists, such as yohimbine. The order of agonist potency for β-adrenergic receptors is isoproterenol > epinephrine $\geq$ norepinephrine. For β_1 receptors, epinephrine and norepinephrine are roughly equipotent, whereas for β_2 receptors, epinephrine is much more potent than norepinephrine. Dopamine receptors are more responsive to dopamine than to norepinephrine or epinephrine. Responses to α_1-adrenergic stimulation usually involve Ca^{2+} influx and alterations in phosphatidylinositol metabolism, whereas β-adrenergic and certain dopaminergic responses are mediated by increased intracellular cAMP, and α_2-adrenergic (and certain dopaminergic) stimulation results in decreased intracellular cAMP. In the kidney, there are two sources of adrenergic input: plasma catecholamines and renal sympathetic nerves (Moss, 1982).

Numerous reports have established that epinephrine or isoproterenol increases cAMP levels in MDCK cells (Ishizuka et al., 1978; Rindler et al., 1979; Lin et al., 1982; Rugg and Simmons, 1982; Meier et al., 1983a). In general, the response to isoproterenol is small in comparison to that elicited by other hormones (especially PGE_1 and vasopressin). In most studies the stimulation produced by isoproterenol represents only about a twofold increase over basal cAMP levels even when a phosphodiesterase inhibitor is added to inhibit cAMP degradation. The reason for the low level of response is not clear but may depend on the MDCK subline used or on the assay conditions, since Lin and co-workers (1982) reported a 15-fold elevation of cAMP levels by isoproterenol. We have found that levels of isoproterenol responsiveness are similar in several different MDCK clonal sublines (Meier *et al.*, 1983a). The following observations indicate that the increase in cAMP is mediated by β-adrenergic receptors: (1) the rank order of agonist potency is isoproterenol > epinephrine > norepinephrine (Rugg and Simmons, 1982), and (2) the response to epinephrine is totally blocked by concomitant addition of propranolol (Meier *et al.*, 1983a).

Effects of catecholamines on transport properties of MDCK cells have been documented by Simmons and co-workers. They showed that the I_{sc} in a high-resistance MDCK subline increased when 1 μM epinephrine was applied to the basolateral cell surface (Richardson *et al.*, 1981). This effect was not observed in a subline with low electrical resistance. Epinephrine (EC_{50} = 3 nM) enhanced net Cl^- secretion from the basolateral to apical surface (Brown and Simmons, 1981). The order of potency of adrenergic agonists in increasing I_{sc} was isoproterenol > epinephrine > norepinephrine, indicating mediation by β-adrenergic receptors. The Cl^- secretion induced by epinephrine was inhibited by furosemide and phloretin, but not by amiloride or 4-acetamido-4'-isothiocyano-2,2'disulphonic acid (SITS), and was dependent on the presence of sodium in the media on the basolateral surface.

These investigators also described an effect of epinephrine on K^+ transport in MDCK cells (Brown and Simmons, 1982). Epinephrine rapidly (within 3 min) and transiently increased K^+ efflux across both the basolateral and apical cell surfaces, accompanied by a net loss of cellular K^+. A similar effect was elicited by the calcium ionophore A23187, and no effect of either ionophore or epinephrine could be demonstrated in calcium-free medium. Epinephrine (EC_{50} = 0.91 μM) was effective only when present on the basolateral cell surface. Addition of epinephrine to cells exposed 12 min previously to the drug elicited no response, indicating that desensitization occurred. The epinephrine response could be shown only in an MDCK subline with high electrical resistance and not in a low-resistance line, although the latter cells did show K^+ efflux in response to ionophore. Phentolamine, an α-adrenergic antagonist, totally blocked the response to epinephrine. By contrast, K^+ efflux was not elicited by either isoproterenol or oxymetazoline, which are β-adrenergic and α_2- adrenergic agonists, respectively. Phenylephrine, an α_1 agonist, was not as efficacious as epinephrine, and addition of the the β blocker propranolol concomitantly with epinephrine elicited a greater response than that elicited by epinephrine alone. These results are most consistent with an epinephrine effect mediated by α_1-adrenergic receptors but also suggest some β-adrenergic modulation of the response. However, since the response was not altered by addition of a phosphodiesterase inhibitor (which would be expected to increase β-adrenergic-stimulated cAMP generation) concomitantly with epinephrine, the nature of any β-adrenergic effect is not clear.

Levine and Moskowitz (1979) showed that 6 μM norepinephrine produced a threefold increase in the synthesis of arachidonic acid metabolites in MDCK cells. These metabolites included $PGF_{2\alpha}$, 6-oxo-$PGF_{1\alpha}$, PGE_2, PGD_2, and thromboxane B_2, indicating that stimulation occurred at either the cyclooxygenase or phospholipase step. The response to norepinephrine required the continuous presence of agonist, continued for at least 24 hr, and was not dependent on protein synthesis (since it was not blocked by cycloheximide). The cytoskeletal disrupting drugs colchicine and cytochalasin B also did not interfere with catecholamine action. The EC_{50} for norepinephrine was in the micromolar range. Epinephrine was more potent than norepinephrine in eliciting the response, and isoproterenol had no effect at concentrations up to 20 μM. Dopamine also stimulated prostaglandin production. A series of α-adrenergic and dopaminergic antagonists were used to block the norepinephrine response, and their IC_{50}s* were determined. Since the response was neither blocked by propranolol nor elicited by isoproterenol, the effect appeared to be mediated through α-adrenergic receptors. However, the α-adrenergic subtype was not determined since the α antagonists tested did not clearly discriminate between α_1- and α_2-adrenergic receptors. The effect of dopamine, unlike that of norepinephrine, could not be blocked by the irreversible α antagonist phenoxybenzamine. Thus MDCK cells may have both dopaminergic and α-adrenergic receptors that stimulate prostaglandin production.

Taub *et al.* (1979) found that 10 μM norepinephrine was growth stimulatory for MDCK cells when added to a serum-free, defined medium lacking PGE_1, hydrocortisone, and triiodothyronine. Since prostaglandins of the E series were shown (in the same study) to stimulate growth of MDCK cells via elevation of cAMP levels, norepinephrine-induced prostaglandin release (Levine and Moskowitz, 1979) might account for the growth-stimulatory effect of norepinephrine. Additionally, the catecholamine might stimulate growth by increasing cAMP production via occupation of β-adrenergic receptors.

We have conducted a detailed characterization of the adrenergic receptors in MDCK cells using radioligand binding techniques (Meier *et al.*, 1982, 1983a). We prepared a crude membrane fraction from MDCK cells and assessed binding of several radioligands to this preparation. Specific binding of the selective adrenergic radioligands [^{125}I]ICYP† (β), [^{3}H]prazosin (α_1), and [^{125}I]IHEAT* (α_1) was demonstrated, but no specific binding of [^{3}H]yohimbine (α_2) could be

---

*IC_{50}, concentration of inhibitor blocking response by 50%.
†[^{125}I]ICYP, [^{125}I]iodocyanopindolol; [^{125}I]IHEAT, [^{125}I]iodo-2[β-(4-hydroxylphenyl)-ethylaminomethyl]tetralone.

detected. Further studies using competitive binding assays with a wide variety of subtype-selective adrenergic ligands showed that the β receptors were exclusively of the β_2 subtype (e.g., Fig. 6). Additional studies indicated that binding sites recognized by [^{3}H]prazosin and [^{125}I]IHEAT were identical and had the binding specificities typical of the α_1 subtype, as described in other tissues such as the rat renal cortex (Schmitz *et al.*, 1981; Snavely and Insel, 1982). These receptors were expressed by cells grown in either serum-containing or serum-free media. Epinephrine and isoproterenol increased cAMP levels in MDCK through β-adrenergic receptors (isoproterenol $EC_{50} = 0.1$ μM), and epinephrine increased phosphatidylinositol incorporation through α_1 receptors ($EC_{50} = 1.8$ μM). These responses are consistent with α_1- and β_2-adrenergic responses documented in other tissues. To date, we have been unable to demonstrate gluconeogenesis from lactate in MDCK clones bearing α_1 receptors, although α_1 receptors mediate renal gluconeogenesis *in vivo* (McPherson and Summers, 1982). Since the α_1 receptors mediating gluconeogenesis *in vivo* are thought to be located on the proximal tubule, the failure of MDCK cells (which are derived from distal tubule/collecting duct) to show α_1-stimulated gluconeogenesis is not unexpected (McPherson and Summers, 1982). Based on data discussed previously, it seems reasonable to suppose that α_1 and β_2 receptors on MDCK cells are preferentially located on the basolateral surface. It is interesting to note that the two adrenergic subtypes expressed on MDCK cells represent the less abundant of the two α- and β-adrenergic receptor subtypes observed in membranes prepared from kidney cortex, in which $\alpha_2 > \alpha_1$ and $\beta_1 > \beta_2$ (Snavely and Insel, 1982; Snavely *et al.*, 1982; Schmitz *et al.*, 1981).

In order to determine whether these two adrenergic receptor types were localized to the same cell type, we developed a number of clonal cell lines from single MDCK cells (Meier and Insel, 1982; Meier *et al.*, 1983a; Fig. 4). These clones varied in morphology, with at least two basic cell types present, a flattened, motile type and a more cuboidal, nonmotile type. All cell types possessed β_2-adrenergic receptors (1000–5000 receptors/cell), but only the motile clones had α_1-adrenergic receptors (7000–20,000 receptors/cell). These results show that a single renal tubule cell can express both α_1- and β_2-adrenergic receptors.

Further studies of α_1- and β_2-adrenergic receptors were carried out using a clone (MDCK-D) that possessed both receptor types (Meier *et al.*, 1983a). In the absence of added guanine nucleotide, agonist binding to the β-adrenergic receptor could be resolved (using the LIGAND program) into two sites with different affinities. Agonist affinity for the β receptor was reduced in the presence of guanine nucleotide such that only the low-affinity site was detected (Fig. 5). These results are consistent with results obtained in other β-adrenergic receptor systems in which the effect of guanine nucleotide has been linked to interactions between the receptor and a guanine nucleotide binding regulatory component, N, which mediates the coupling between receptor and adenylate cyclase (Stadel *et*

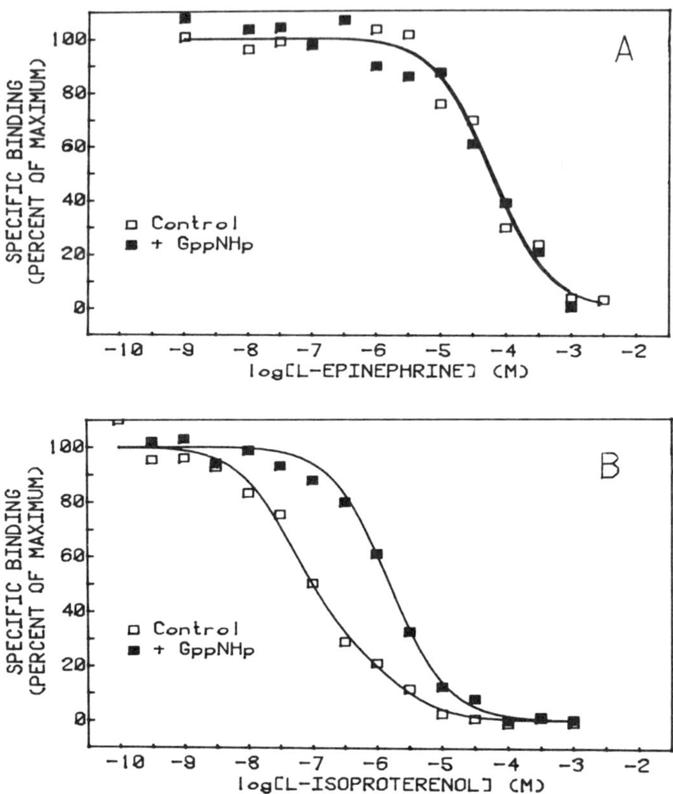

Figure 5. Effects of guanine nucleotide on agonist binding to MDCK alpha$_1$- and beta$_2$-adrenergic receptors. MDCK (clone D) membranes were incubated with [^{125}I]IHEAT and varying concentrations of (-)-epinephrine (A) or with [^{125}I]ICYP and varying concentrations of (-)-isoproterenol (B), in the absence (□) or presence (■) of 100 μM 5′-guanyl-imidodiphosphate [Gpp(NH)p]. The incubation buffer consisted of 2mM MgCl$_2$, 145 mM NaCl, and 20 mM tris–HCl, pH 7.5. Curve fitting for the competition binding data was performed using the LIGAND program. Data are shown as one-site fits except for one case [B, in the absence of Gpp(NH)$_p$] in which the two-site fit was statistically superior (p < 0.01). The K_D for epinephrine for the α_1-adrenergic receptor (A) was 15 μM in the absence of Gpp(NH)p and 16 μM in the presence of Gpp(NH)p. In B, the K_D of isoproterenol in the absence of Gpp(NH)p for the β-adrenergic receptor was 15 nM at the high-affinity site (76% of total sites) and 480 nM at the low-affinity site (24% of total sites). In the presence of Gpp(NH)p, the K_D of isoproterenol was 460 nM.

al., 1982). Agonist binding to the α_1-adrenergic receptor was not affected by guanine nucleotide (Fig. 5), which is consistent with the lack of interaction between this receptor and adenylate cyclase on MDCK cells; this result contrasts with recent data obtained in rat kidney cortical membranes in which guanine nucleotides alter agonist binding to α_1 receptors (Snavely and Insel, 1982).

Agonist affinities for both α_1- and β_2-adrenergic receptors in MDCK-D were higher in the absence of sodium than in the presence of isotonic sodium chloride (Meier *et al.*, 1983b). This suggests that catecholamine-mediated alterations in ion flux might lead to alterations in receptor binding of agonists.

In additional studies, we have taken advantage of the coexpression of α_1- and β_2-adrenergic receptors on MDCK-D cells to examine whether these two receptors are independent molecular entities (Meier *et al.*, 1984). Both α_1- and β_2-adrenergic receptors, prelabeled with radioligands, were solubilized from MDCK-D membranes with digitonin, a nonionic detergent. The solubilized receptors were incubated with a series of lectins immobilized on agarose and then were eluted with the sugars specific for those lectins. Both receptor types were retained by wheat germ agglutinin (WGA) from which they were eluted by *N*-acetyl-D-glucosamine or *N*-acetylneuraminic acid (sialic acid). Further experiments showed that the receptors were also retained by *Limax flavus* lectin (LFA), a sialic-acid-specific lectin, but not by lectins specific for terminal or internal *N*-acetylglucosamine residues. The receptors were eluted from LFA by sialic acid. These data indicated that both receptors were glycoproteins possessing terminal sialic acid residues. The molecular sizes of the digitonin-solubilized α_1- and β_2-adrenergic receptors were determined by gel exclusion chromatography to be 250,000 daltons (6.2 nm) for the α_1-adrenergic receptor and 150,000 daltons (5.1 nm) for the β_2-adrenergic receptors, showing that the two receptors are discrete membrane proteins. Lin *et al.* (1982) found that treatment of transformed MDCK cells with tunicamycin (an inhibitor of glycoprotein processing) for 24 hr decreased the isoproterenol-mediated stimulation of cAMP accumulation, consistent with our observation that the β-adrenergic receptor in MDCK cells is a glycoprotein. Interestingly, Phillips *et al.* (1982) have shown that MDCK cells will adhere to WGA-coated Petri dishes and can be maintained indefinitely on such surfaces, suggesting that sialic acid residues may be common to a number of basolateral membrane components on MDCK cells as well as to both types of adrenergic receptors.

4.5. Glucagon

Responsiveness to glucagon in MDCK cells was first documented by Rindler and co-workers (1979), who found a threefold increase in cAMP production in cells exposed to glucagon for 2–4 hr. Glucagon is growth stimulatory for MDCK cells grown in defined media lacking PGE_1, presumably because of its activation of adenylate cyclase (Taub *et al.*, 1979).

Lin and co-workers (1982) have examined glucagon receptors and response in MDCK cells in some detail. They measured a sixfold increase in intracellular cAMP in parental MDCK cells incubated with 2 μM glucagon. However, no glucagon response was detected in a cloned MDCK cell line transformed by

Harvey murine sarcoma virus. When the transformed cells were exposed to 4 mM butyrate (for 3 days) or to PGE_1 in serum-free medium (up to 7 days), glucagon responsiveness (as measured in intact cells or membranes) was restored. Butyrate did not affect glucagon response in the parental cell line. The EC_{50} for glucagon stimulation of cAMP formation was increased to 200 nM in butyrate-treated transformed cells, as compared to 20 nM for nontransformed cells. The reappearance of responsiveness correlated with a reappearance of specific binding sites for $[^{125}I]$glucagon, although these sites could not be fully characterized owing to a high level of nonspecific binding. The affinity of the ligand was slightly decreased in the butyrate-treated transformed cells as compared to parental cells. Simultaneous addition of butyrate and tunicamycin completely blocked the reappearance of the glucagon response, suggesting that the glucagon receptor is a glycoprotein whose *de novo* synthesis is induced by butyrate. One question raised by this intriguing study is whether the alterations in glucagon receptors are solely or directly due to viral transformation. Recently F. Darfler and M. C. Lin have found (personal communication), and we have confirmed (K. E. Meier and P. A. Insel, unpublished data), that the MDCK-D clonal subline, described previously (Meier *et al.*, 1983a), has no glucagon responsiveness (i.e., cAMP generation), whereas several other clonal sublines do respond to glucagon. Responsiveness to isoproterenol, vasopressin, and PGE_1 is retained by all these clones. Thus some of the cell types present in the parental MDCK population may lack glucagon receptors; further studies are required to establish this point.

4.6. Prostaglandins

Many of the effects of prostaglandins on MDCK cells have been mentioned in earlier sections of this discussion. These effects are particularly complicated since this cell line both produces and responds to prostaglandins. Levels of 0.1 μM PGE_2 have been measured in MDCK culture medium after only 3 days of culture and may contribute to the high basal cAMP values that have often been observed in these cells (Lin *et al.*, 1982).

PGE_1 stimulation of cAMP accumulation in MDCK cells was initially noted by Ishizuka *et al.* (1978), who recorded a 22-fold stimulation over basal levels. Rindler *et al.* (1979) found a similar stimulation and obtained an EC_{50} for PGE_1 of about 0.2 μM. These workers also showed varying efficacies of activation of cellular cAMP accumulation by different types of prostaglandins: $PGE_1 \simeq PGE_2 > PGA_1 > PGB, PGF_{2\alpha}$. Activation of adenylate cyclase by PGE_1 was also directly demonstrated in a membrane preparation. Direct evidence for the presence of prostaglandin receptors in MDCK cells has not yet been shown in radioligand binding studies.

Of further interest in this regard, data documenting PGE_1 response beyond the cAMP level has been provided by Richardson *et al.* (1981). They found that high-resistance MDCK cells, but not low-resistance cells, increased their I_{sc} when 1 μM PGE_1 was applied to the basolateral cell surface. We speculate that prostaglandin receptors, like those for catecholamines, are likely to be located on the basolateral surface of MDCK cells.

Taub and co-workers (1979) found that PGE_1 was a critical factor for stimulation of cell growth in the serum-free, defined media developed for MDCK cells. PGE_2 was as effective as PGE_1, but PGA_1 and $PGF_{2\alpha}$ did not stimulate cell proliferation. Dibutyryl cAMP could be substituted for PGE_1, indicating that the growth stimulation resulted from the increased cAMP levels. Dome formation by MDCK cells in defined medium is increased with increasing PGE_1 concentration (Taub, 1982). In subsequent studies, MDCK cells were chemically mutagenized and then selected for growth in defined medium lacking PGE_1 (Taub *et al.*, 1981). The growth of "PGE_1-independent" clones could be inhibited by exogenous PGE_1, although these cells responded to PGE_1 with even greater increases in cellular cAMP accumulation than did the parental line. It therefore appears that these cells may have abnormally low activities of cyclic nucleotide (or cAMP) phosphodiesterase. The variant clones were also tumorigenic. Whether the chemical transformation actually induced PGE_1 independence and whether this property is directly related to tumorigenicity have not yet been established.

4.7. *Vasopressin*

Arginine vasopressin induces a substantial increase in cAMP levels in MDCK cells (Ishizuka *et al.*, 1978; Rindler *et al.*, 1979; Lin *et al.*, 1982). In contrast to the large body of work devoted to study of the vasopressin response in the $LLC–PK_1$ cell line, this subject has not received much attention in the MDCK cell line. Richardson *et al.* (1981) showed that high-resistance MDCK cells responded to vasopressin with an increase in I_{sc}. Surprisingly, arginine vasopressin is not growth stimulatory for MDCK cells (Taub *et al.*, 1979); it also increases PGE_2 synthesis to only a very minor extent, if at all (Hassid, 1981).

4.8. *Other Hormones*

Simmons (1978) reported that angiotensin stimulated sodium transport in MDCK cells when present in low concentrations (10 pM) but not as high concentrations (1 μM). Rindler *et al.* (1979) found no effect of 10 μM angiotensin II on cAMP levels in MDCK. We have been unable to demonstrate any specific binding of [^{125}I]angiotensin II to membranes prepared from several different MDCK clonal lines (F. M. Chen, K. E. Meier, P. A. Insel, and M. Printz, unpublished observations).

Calcitonin does not increase cAMP in MDCK, whereas parathyroid hormone may increase cAMP levels slightly (Ishizuka *et al.*, 1978; Rindler *et al.*, 1979). Prolactin, ACTH, and adenosine also do not increase cAMP levels (Rindler *et al.*, 1979). Epidermal growth factor (EGF) appeared to increase cAMP levels very slightly after a 2–4 hr incubation with EGF (Rindler *et al.*, 1979); this increase in cAMP may be secondary to prostaglandin formation which is stimulated by EGF (Levine and Hassid, 1977).

The presence of insulin receptors on MDCK cells can be inferred from the growth-stimulatory effect of insulin (Taub *et al.*, 1979), although these could either be receptors for insulin or for insulinlike growth factors. An MDCK subline transformed by Moloney sarcoma virus requires a lower concentration of insulin for optimal growth than does the parental line (Taub *et al.*, 1981).

5. ADDITIONAL CELL LINES

Madin–Darby bovine kidney cells (Madin and Darby, 1958) (ATCC CCL 22) have not received the attention afforded to their "sibling" MDCK cells; this may be because the cell line was infected with bovine diarrhea virus. A virus-free strain is now available from the ATCC. The hormone responsiveness of MDBK cells has been described by Ishizuka *et al.* (1978) and by Rindler *et al.* (1979). The former investigators found that the basal cAMP level in MDBK was very high and no elevation could be detected in response to PTH, whereas calcitonin, vasopressin, isoproterenol, and PGE_1 produced small (but significant) increases in cAMP levels in the presence of isobutylmethylxanthine. Rindler *et al.* (1979) found that PGE_1 and isoproterenol elicited large (15- and 20-fold, respectively) increases in cAMP levels whereas vasopressin, PTH, and calcitonin produced only slight increases. We have found that MDBK cells express β_2-adrenergic receptors (Fig. 6) but not α_1-adrenergic receptors (Meier and Insel, unpublished data).

Ishizuka and co-workers (1978) also studied hormonal responsiveness in JTC-12.P3, a cell line derived from cynomolgus monkey kidney in 1961 (Katsuka and Takaoka, 1973). These cells had low basal cAMP levels and increased cAMP in response to parathyroid hormone (PTH), isoproterenol, and PGE_1, but not to calcitonin or vasopressin. The EC_{50} for PTH 1-34 was approximately 0.5 U/ml. The cAMP generation in response to PTH was greater in confluent than in nonconfluent JTC-12.P3 cells.

6. CONCLUSIONS AND PERSPECTIVE

At this time, this field resembles a partly finished patchwork quilt: the general pattern is evident but only a few of its sections have actually been "tied

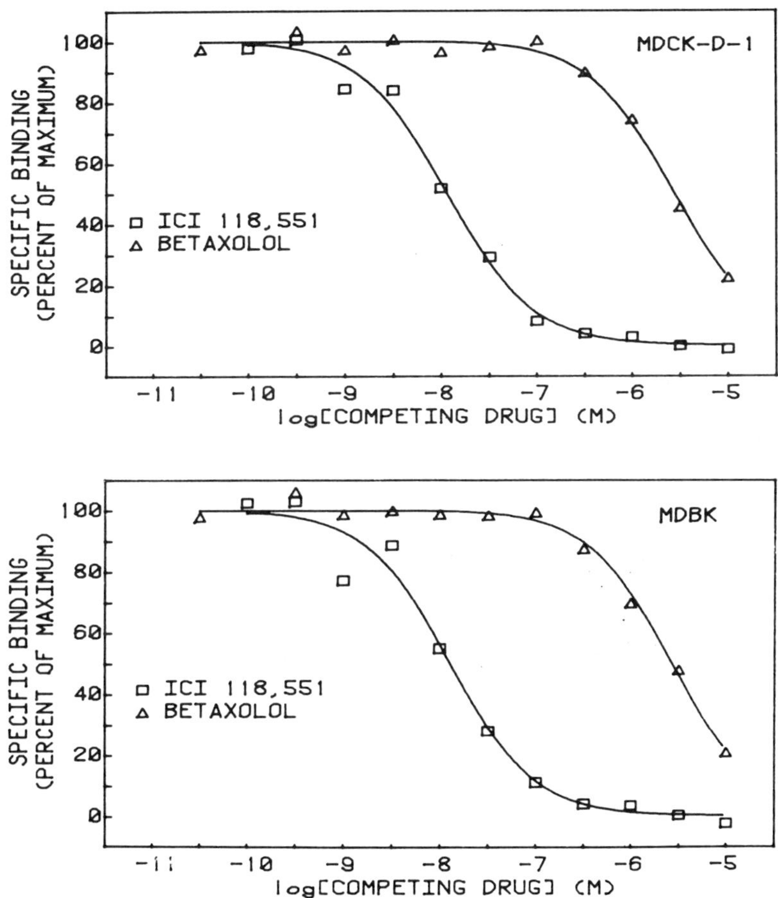

Figure 6. Competition for beta-adrenergic binding sites on membranes prepared from MDCK-D-1 and MDBK cells. MDCK (clone D-1) and MDBK membranes were prepared by hypotonic lysis from confluent cells grown in 7.5% horse serum and 2.5% fetal calf serum. In the same experiment, both membrane preparations (MDCK-D-1, upper panel; MDBK, lower panel) were incubated for 60 min at 37°C with 80 pM [^{125}I]iodocyanopindolol in the presence of varying concentrations of ICI 118,551 (a β_2-selective antagonist) or betaxalol (a β_1-selective antagonist). The competition binding data were analyzed using the LIGAND program to give the one-site fits shown. In both cell types, each ligand bound to a single class of sites; the affinities of the receptors for these ligands (ICI 118,551, K_D = 4–5 nM; betaxolol K_D = 1000 nM) were typical for the β_2-adrenergic-receptor subtype.

down.'' To this extent, the work is still at "stage 2" of Heinlein's definition of science, which we quoted at the outset. Responses to many hormones (Table II) have been identified in renal cell lines, and many others await description. These should provide model systems for the study of hormone action for those interested in renal physiology and pharmacology. In addition, these cells should be of interest to investigators wishing to study receptors for hormones such as catecholamines, insulin, and EGF, which are active on many other nonrenal tissues. In general, renal epithelial cells lines grow quite rapidly in relatively simple and inexpensive media, providing large amounts of material for biochemical studies. Since the renal cell lines described in this review are continuous (but not transformed), they may be useful for correlating *in vitro* binding to receptors and *in vivo* functional responses.

The molecular properties of receptors in epithelial cells have not been examined in detail, although these cells have a unique and fascinating cytoarchitecture which probably affects receptor function. Tight junctions between renal epithelial cells form a boundary between the apical (facing the tubular lumen) and the basolateral surfaces (facing the blood). This serves not only to limit the penetration of water and solutes, but also segregates the plasma membrane components such that different proteins and lipids are present in the apical and basolateral membranes (Dragsten *et al.*, 1981). Most renal hormone receptors appear to be localized on the basolateral surface where they are exposed to plasma-borne hormones. However, since small molecules (such as catecholamines) are present in the glomerular filtrate and since renal cells secrete certain hormones (such as prostaglandins) into the urine, the possibility that

Table II.   Hormone Responsiveness of Established Renal Epithelial Cell Lines

Hormone	A6	LLC–PK$_1$	MDCK	MDBK	JTC-12.P3
Adenosine	+	ND	−	ND	ND
Aldosterone	+	ND	+	ND	ND
Bradykinin	ND	ND	+	ND	ND
Calcitonin	ND	+	−	±	−
Catecholamines	+	±	+	+	+
Glucagon	ND	ND	+	ND	ND
Insulin	+	+	+	ND	ND
Parathyroid hormone	ND	+	±	−	+
Prostaglandins	ND	−	+	+	+
Vasopressin	+	+	+	±	−
Vitamin D	ND	+	ND	ND	ND

Cell line[a] spans the five cell-line columns.

[a] +, Response present; −, no response present; ±, conflicting data; ND, not determined.

some hormone receptors are located on the apical surface should not be over-looked. An interesting aspect of renal cell polarity is the interaction of the basolateral surface with the extracellular matrix; this interaction may modulate receptor response in numerous ways.

Another unresolved question is: what are the functional consequences of having different types of renal cells within a renal segment? In the distal nephron, for example, (1) Do the principal and intercalated cells receive different hormonal signals? (2) Do they respond differently to the same signal? (3) Do they modulate each other's functions by releasing locally active agents such as prostaglandins? Microtechniques have revealed much about the hormone responsiveness of various renal segments (Morel, 1981), but culture of individual cell types will permit further studies to be conducted on a larger scale and with greater biochemical sophistication.

Finally, there is every reason to develop new renal cell lines. Despite their numerous advantages, the MDCK and LLC–PK$_1$ cell lines are heterogeneous mixtures of cells whose site of origin can only be determined with much effort. Ideally, cell lines should be derived from dissected portions of the tubule or glomerulus where the cells of origin can be known and so that hormone response can be correlated directly to that of the parent tissue. The development of serum-free defined media (which inhibits fibroblast growth) has made it easier to select for epithelial cell lines without fibroblastic "contamination." Other advances in tissue culture methodology such as new types of substrates for growing epithelial cells should also help improve one's ability to correlate cell function *in vitro* and *in vivo*. Thus, investigators interested in hormonal response in the kidney should find cultured renal epithelial cells fruitful for further exploration and guesswork (Heinlein, 1980) for many years to come.

ACKNOWLEDGMENTS.   The authors wish to extend their thanks to Dr. Joseph Handler for providing preprints of data in press, to Jim Moberly for helpful editorial comments, and to Sandy Dutky for expert typing of this manuscript.

REFERENCES

Amsler, K., and Cook, J. S., 1982, Development of Na$^+$-dependent hexose transport in a cultured line of porcine kidney cells, *Am. J. Physiol.* **242**:C94–C101.

Ausiello, D. A., and Hall, D., 1981, Regulation of vasopressin-sensitive adenylate cyclase by calmodulin, *J. Biol. Chem.* **256**:9796–9798.

Ausiello, D. A., Hall, D. H., and Dayer, J.–M., 1980, Modulation of cAMP-dependent protein kinase by vasopressin and calcitonin in cultured porcine renal LLC–PK$_1$ cells, *Biochem. J.* **186**:773–780.

Baker, M. E., and Fanestil, D. D. 1977, Effect of protease inhibitors and substrates on deoxycorticosterone binding to its receptor in dog MDCK kidney cells, *Nature* **269**:810–812.

Beaudry, G. A., King, L., Daniel, L. W., and Waite, M., 1982, Stimulation of deacylation in Madin–Darby canine kidney cells: Specificity of deacylation and prostaglandin production in 12-0-tetradecanoyl-phorbol-13-acetate-treated cells, *J. Biol. Chem.* **257**:10973–10977.

Boerner, P., and Saier, M. H., 1982, Nutrient transport and growth regulation in kidney epithelial cells (MDCK) cultured in a defined medium, in: *Growth of Cells in Hormonally Defined Media,* Book A, Cold Spring Harbor Conferences on Cell Proliferation, Volume 9 (G. H. Sato, A. B. Pardee, and D. A. Sirbasku, eds.), Cold Spring Harbor Laboratory, pp. 555–565.

Brown, C. D. A., and Simmons, N. L., 1981, Catecholamine-stimulation of Cl^- secretion in MDCK cell epithelium, *Biochim. Biophys. Acta* **649**:427–435.

Brown, C. D. A., and Simmons, N. L., 1982, K^+ Transport in "tight" epithelial monolayers of MDCK cells: Evidence for a calcium-activated K^+ channel, *Biochim. Biophys. Acta* **690**:95–105.

Cereijido, M., Robbins, E. S., Dolan, W. J., Rotunno, C. A., and Sabatini, D. D., 1978, Polarized monolayers formed by epithelial cells on a permeable and translucent support, *J. Cell Biol.* **77**:853–880.

Cereijido, M., Ehrenfeld, J., Meza, I., and Martinez-Palomo, A., 1980, Structural and functional membrane polarity in cultured monolayers of MDCK cells, *J. Membr. Biol.* **52**:147–159.

Chin, S., Hall, H. G., and Bissell, M. J., 1982, Substratum affects production and distribution of glycosaminoglycans from epithelial cell lines, *J. Cell Biol.* **95**:124a.

Chuman, L., Fine, L. G., Cohen, A. H., and Saier, M. H., 1982, Continuous growth of proximal tubular kidney epithelial cells in hormone-supplemented serum-free medium, *J. Cell Biol.* **94**:506–510.

Cobb, M. H., Yang, C.–P. H., Jefferson, D. M., Pasnikowski, E., and Scott, W. N., 1982, Mineralocorticoid-induced membrane proteins in MDCK cells, *Mol. Cell Endocrinol.* **27**:129–137.

Colston, K., and Feldman, D., 1982, 1,25-Dihydroxyvitamin D_3 receptors and functions in cultured pig kidney cells ($LLC-PK_1$): Regulation of 24,25-dihydroxyvitamin D_3 production, *J. Biol. Chem.* **257**:2504–2508.

Darfler, F. D., Hughes, R. J., and Insel, P. A., 1981, Characterization of serum-induced alterations in the cyclic AMP pathway in S49 lymphoma cells, *J. Biol. Chem.* **256**:8422–8428.

Dibner, M. D., and Insel, P. A., 1981, Serum catecholamines desensitive β-adrenergic receptors of cultured C6 glioma cells, *J. Biol. Chem.* **256**:7343–7346.

Dragsten, P. R., Blumenthal, R., and Handler, J. S., 1981, Membrane asymmetry in epithelia: Is the tight junction a barrier to diffusion in the plasma membrane? *Nature* **294**:717–722.

Fidelman, M. L., May, J. M., Biber, T. U. L., and Watlington, C. O., 1982, Insulin stimulation of Na^+ transport and glucose metabolism in cultured kidney cells, *Am. J. Physiol.* **242**:C121–C123.

Gausch, C. R., Hard, W. L., and Smith, T. F., 1966, Characterization of an established line of canine kidney cells (MDCK), *Proc. Soc. Exp. Biol. Med.* **122**:931–935.

Goldring, S. R., Dayer, J.-M., Ausiello, D. A., and Krane, S. M., 1978, A cell strain cultured from porcine kidney increases cyclic AMP content upon exposure to calcitonin or vasopressin, *Biochem. Biophys. Res. Commun.* **83**:434–440.

Hall, H. G., Larson, D., and Bissell, M. J., 1982, Lumen formation by epithelial cell lines in response to collagen overlay: A morphogenetic model in culture, *Proc. Natl. Acad. Sci. USA* **79**:4672–4676.

Handler, J. S., Perkins, F. M., and Johnson, J. P., 1980, Studies of renal cell function using cell culture techniques, *Am. J. Physiol.* **238**:F1–F9.

Handler, J. S., Preston, A. S., Perkins, F. M., and Matsumura, M., 1981, The effect of adrenal steroid hormones on epithelia formed in culture by A6 cells, *Ann. N.Y. Acad. Sci.* **372**:442–454.

Handler, J. S., Preston, A. S., and Steele, R. E., 1984, Factors affecting the differentiation of epithelial transport and responsiveness to hormones, *Fed. Proc.* **43**:2221–2224.

Hassid, A., 1981, Transport-active renal tubular epithelial cells (MDCK and LLC–PK$_1$) in culture: Prostaglandin biosynthesis and its regulation by peptide hormones and ionophore, *Prostaglandins* **21**:985–1001.

Heinlein, R. A., 1980, *The Number of the Beast,* Ballantine, New York, p. 508.

Herzlinger, D. A., Easton, T. G., and Ojakian, G. K., 1982, The MDCK epithelial cell line expresses a cell surface antigen of the kidney distal tubule, *J. Cell Biol.* **93**:269–277.

Hull, R. N., Cherry, W. R., and Weaver, W., 1976, The origin and characteristics of a pig kidney cell strain, LLC–PK, *In Vitro* **12**:670–677.

Ishizuka, I., Tadcno, K., Nagata, N., Niimura, Y., and Nagai, Y., 1978, Hormone-specific responses and biosynthesis of sulfolipids in cell lines derived from mammalian kidney, *Biochim. Biophys. Acta* **541**:467–482.

Johnson, J. P., Steele, R. E., Perkins, F. M., Wade, J. B., Preston, A. S., Green, S. W., and Handler, J. S., 1981, Epithelial organization and hormone sensitivity of toad urinary bladder cells in culture, *Am. J. Physiol.* **241**:F129–F138.

Katsuta, H., and Takaoka, T., 1973, Cultivation of cells in protein- and lipid-free synthetic media, in: *Methods in Cell Biology* (D. M. Prescott, ed.), Academic Press, New York, pp. 1–42.

Lang, M. A., Forrest, J. N., Preston, A. S., and Handler, J. S., 1982, Adenosine stimulates cAMP accumulation and sodium transport in epithelia formed by A6 (toad kidney) cells in culture, *Proc. Amer. Soc. Nephrol.,* p. 168A.

Leighton, J., Brada, Z., Estes, L. W., and Justh, G., 1969, Secretory activity and oncogenicity of a cell line (MDCK) derived from canine kidney, *Science* **163**:472–473.

Leighton, J., Estes, L. W., Mansukhani, S., and Brada, Z., 1970, A cell line derived from normal dog kidney (MDCK) exhibiting qualities of papillary adenocarcinoma and of renal tublar epithelium, *Cancer* **26**:1022–1028.

Lever, J. E., 1981, Regulation of dome formation in kidney epithelial cell cultures, *Ann. N.Y. Acad. Sci.* **372**:371–383.

Lever, J. E., 1982, Cell differentiation and dome formation in polarized epithelial cell monolayers, in: *Growth of Cells in Hormonally Defined Media,* Book A, Cold Spring Harbor Conferences on Cell Proliferation, Volume 9 (G. H. Sato, A. B. Pardee, and D. A. Sirbasku, eds.), Cold Spring Harbor Laboratory, pp. 541–554.

Levine, L., and Hassid, A., 1977, Epidermal growth factor stimulates prostaglandin synthesis by canine kidney (MDCK) cells, *Biochem. Biophys. Res. Commun.* **76**:1181–1187.

Levine, L., and Moskowitz, M. A., 1979, α- and β-Adrenergic stimulation of arachidonic acid metabolism in cells in culture, *Proc. Natl. Acad. Sci. USA* **76**: 6632–6636.

Lewis, M. G., and Spector, A. A., 1981, Differences in types of prostaglandins produced by two MDCK canine kidney cell sublines, *Prostaglandins* **21**:1025–1032.

Lewis, M. G., Kaduce, T. L., and Spector, A. A., 1981, Effect of essential polyunsaturated fatty acid modifications on prostaglandin production by MDCK canine kidney cells, *Prostaglandins* **22**:747–760.

Lifschitz, M. D., 1982, LLC–PK$_1$ cells derived from pig kidneys have a defect in cyclooxygenase, *J. Biol. Chem.* **257**:12611–12615.

Lin, M. C., Koh, S.–W. M., Dykman, D. D., Beckner, S. K., and Shih, T. Y., 1982, Loss and restoration of glucagon receptors and responsiveness in a transformed kidney cell line, *Exp. Cell Res.* **142**:181–189.

Ludens, J. H., Vaughn, D. A., Mawe, R. C., and Fanestil, D. D., 1978, Specific binding of deoxycorticosterone by canine kidney cells in culture, *J. Steroid Biochem.* **9**:17–21.

Madin, S. H., and Darby, N. B., 1958, Established kidney cell lines of normal adult bovine and ovine origin, *Proc. Soc. Exp. Biol. Med.* **98**:574–576.

Maleque, A., Endou, H., Koseki, C., and Sakai, F., 1980, Nephron heterogeneity: Gluconeogenesis from pyruvate in rabbit nephron, *FEBS Lett.* **116:**154–156.

McPherson, G. A., and Summers, R. J., 1982, A study of α_1-adrenoceptors in rat renal cortex: Comparison of [^{3}H]prazosin binding with the α_1-adrenoceptor modulating gluconeogenesis under physiological conditions, *Br. J. Pharmacol.* **77:**177–184.

McRoberts, J. A., Taub, M., and Saier, M. H., 1981, The Madin–Darby canine kidney (MDCK) cell line, in: *Functionally Differentiated Cell Lines* (G. Sato, ed.), Liss, New York, pp. 117–139.

Meier, K. E., and Insel, P. A., 1982, Clonal variation in the expression of catecholamine receptors in MDCK cells, *J. Cell Biol.* **95:**416a.

Meier, K. E., Snavely, M. D., and Insel, P. A., 1982, α- and β-adrenergic receptors in the MDCK renal epithelial cell line, *J. Cell Biochem.* Suppl. 6, p. 126.

Meier, K. E., Snavely, M. D., Brown, S. L., Brown, J. H., and Insel, P. A., 1983a, Alpha$_1$- and beta$_2$-adrenergic receptor expression in the Madin–Darby canine kidney epithelial cell line, *J. Cell Biol.* **97:**405–415.

Meier, K. E., Sternfeld, D. R., and Insel, P. A., 1983b, How different are alpha- and beta-adrenergic receptors? *Fed. Proc.* **42:**1875.

Meier, K. E., Sternfeld, D. R., and Insel, P. A., 1984, Alpha$_1$- and beta$_2$-adrenergic receptors co-expressed on cloned MDCK cells are distinct glycoproteins, *Biochem. Biophys. Res. Comm.* **118:**73–81.

Mills, J. W., Macknight, A. D. C., Dayer, J.-M., and Ausiello, D. A., 1979, Localization of [^{3}H]ouabain-sensitive Na$^+$ pump sites in cultured pig kidney cells, *Am. J. Physiol.* **2326:** C157–C162.

Misfeldt, D. S., and Sanders, M. J., 1981, Transepithelial transport in cell culture: D-Glucose transport by a pig kidney cell line (LLC–PK$_1$), *J. Membr. Biol.* **59:**13–18.

Misfeldt, D. S., Mamamoto, S. T., and Pitelka, D. R., 1976, Transepithelial transport in cell culture, *Proc. Natl. Acad. Sci. USA* **73:**1212–1216.

Moran, A., Handler, J. S., and Turner, R. J., 1982, Na$^+$-dependent hexose transport in vesicles from cultured renal epithelial cell line, *Am. J. Physiol.* **243:**C293–C298.

Morel, F., 1981, Sites of hormone action in the mammalian nephron. *Am. J. Physiol.* **240:**F159–F164.

Moss, N. G., 1982, Renal function and renal afferent and efferent nerve activity, *Am. J. Physiol.* **243:**F425–F433.

Mullin, J. M., Weibel, J., Diamond, L., and Kleinzeller, A., 1980, Sugar transport in the LLC–PK$_1$ renal epithelial cell line: Similarity to mammalian kidney and the influence of cell density, *J. Clin. Pathol.* **104:**375–389.

Mullin, J. M., Cha, C.-J. M., and Kleinzeller, A., 1982, Metabolism of L-lactate by LLC–PK$_1$ renal epithelia, *Am. J. Physiol.* **242:**C41–C45.

Munson, P. J., and Rodbard, D., 1980, LIGAND: A versatile computerized approach to characterization of ligand binding systems, *Anal. Biochem.* **107:**220–239.

Ohuchi, K., and Levine, L., 1978, Stimulation of prostaglandin synthesis by tumor-promoting phorbol-12,13-diesters in canine kidney (MDCK) cells: Cycloheximide inhibits the stimulated prostaglandin synthesis, deacylation of lipids, and morpholocial changes, *J. Biol. Chem.* **253:**4783–4790.

Perantoni, A., and Berman, J. J., 1979, Properties of Wilm's tumor line (TuWi) and pig kidney line (LLC–PK$_1$) typical of normal kidney tubular epithelium, *In Vitro* **15:**446–454.

Perkins, F. M., and Handler, J. S., 1981, Transport properties of toad kidney epithelia in culture, *Am. J. Physiol.* **241:**C154–C159.

Phillips, S. G., Lui, S.–L., and Phillips, D. M., 1982, Binding of epithelial cells to lectin-coated surfaces, *In Vitro* **18:**727–738.

Rabito, C. A., and Karish, M. V., 1982, Polarized amino acid transport by an epithelial cell line of renal origin (LLC–PK$_1$): The basolateral systems, *J. Biol. Chem.* **257**:6802–6808.

Rabito, C. A., and Karish, M. V., 1983, Polarized amine acid transport by an epithelial cell line of renal origin (LLC–PK$_1$): The apical systems, *J. Biol. Chem.* **258**:2543–2547.

Rabito, C. A., Tchao, R., Valentich, J., and Leighton, J., 1980, Effect of cell–substratum interaction on hemmicyst formation by MDCK cells, *In Vitro* **16**:461–468.

Rafferty, K. A., 1969, Mass culture of amphibian cells: Methods and observations concerning stability of cell type, in: *Biology of Amphibian Tumors* (M. Mizell, ed.), Springer, New York, pp. 52–81.

Richardson, J. C. W., Scalera, V., and Simmons, N. L., 1981, Identification of two strains of MDCK cells which resemble separate nephron tubule segments, *Biochim. Biophys. Acta* **673**:26–36.

Rindler, M. J., Chuman, L. M., Shaffer, L., and Saier, M. H., 1979, Retention of differentiated properties in an established dog kidney epithelial cell line (MDCK), *J. Cell Biol.* **81**:635–648.

Roy, C., and Ausiello, D. A., 1981, Characterization of (8-lysine) vasopressin binding sites on a pig kidney cell line (LLC–PK$_1$): Evidence for hormone-induced receptor transition, *J. Biol. Chem.* **256**:3415–3422.

Roy, C., Preston, A. S., and Handler, J. S., 1980, Insulin and serum increase the number of receptors for vasopressin in a kidney-derived line of cells grown in a defined medium, *Proc. Natl. Acad. Sci. USA* **77**:5979–5983.

Roy, C., Hall, D., Karish, M., and Ausiello, D. A., 1981, Relationship of (8-lysine)vasopressin receptor transition to receptor functional properties in a pig kidney cell line (LLC–PK$_1$), *J. Biol. Chem.* **256**:3423–3427.

Rugg, E. L., and Simmons, N. L., 1981, Catecholamine stimulation of adenylate cyclase in Madin-Darby canine kidney (MDCK) cells, *J. Physiol. London* **322**:26P–27P.

Saier, M. H., 1981, Growth and differentiated properties of a kidney epithelial cell line (MDCK), *Am. J. Physiol.* **240**:C106–C109.

Scatchard, G., 1949, The attraction of protein for small molecules and ions, *Ann. N.Y. Acad. Sci.* **51**:660–572.

Schmidt, U., and Guder, W. G., 1976, Sites of enzyme activity along the nephron, *Kidney Int.* **9**:233–242.

Schmitz, J. M., Graham, R. M., Sagalowsky, R., and Pettinger, W. J., 1981, Renal alpha-1 and alpha-2 adrenergic receptors: Biochemical and pharmacological correlations, *J. Pharmacol. Exp. Ther.* **219**:400–406.

Simmons, N. L., 1978, Hormone stimulation of net transepithelial Na transport in cell culture, *J. Physiol.* **276**:28P–29P.

Simmons, N. L., 1981, Ion transport in "tight" epithelial monolayers of MDCK cells, *J. Membr. Biol.* **59**:105–114.

Simmons, N. L., 1982, Cultured monolayers of MDCK cells: A novel system for the study of epithelial development and function, *Gen. Pharmacol.* **13**:287–291.

Snavely, M. D., and Insel, P. A., 1982, Characterization of alpha-adrenergic receptor subtypes in the rat renal cortex. Differential regulation of alpha$_1$- and alpha$_2$-adrenergic receptors by guanyl nucleotides and Na$^+$, *Mol. Pharmacol.* **22**:532–546.

Snavely, M. D., Motulsky, H. J., Moustafa, E., Mahan, L. C., and Insel, P. A., 1982, Beta-adrenergic receptor subtypes in the rat renal cortex: Selective regulation of beta$_1$-adrenergic receptors by pheochromocytoma, *Circ. Res.* **51**:504–513.

Stadel, J. M., DeLean, A., and Lefkowitz, R. J., 1982, Molecular mechanisms of coupling in hormone-receptor-adenylate cyclase systems, *Adv. Enzymol.* **53**:1–43.

Taub, M., 1982, Hormones control the growth and function of cultured kidney cells, in: *Growth of Cells in Hormonally Defined Media*, Book A, Cold Spring Harbor Conferences on Cell Proliferation, Volume 9 (G. H. Sato, A. B. Pardee, and D. A. Sirbasku, eds.), Cold Spring Harbor Laboratory, pp. 581–592.

Taub, M., and Saier, M. H., 1979, An established but differentiated kidney epithelial cell line (MDCK), *Methods Enzymol.* **38**:552–561.

Taub, M., Chuman, L., Saier, M. H., and Sato, G., 1979, Growth of Madin–Darby canine kidney epithelial cell (MDCK) line in hormone-supplemented, serum-free medium, *Proc. Natl. Acad. Sci. USA* **76**:3338–33422

Taub, M., Ü, B., Chuman, L., Rindler, M. J., Saier, M. H., and Sato, G., 1981, Alterations in growth requirements of kidney epithelial cells in defined medium associated with malignant transformation, *J. Supramol. Struct. Cell. Biochem.* **15**:63–72.

Thomas, S. R., Schultz, S. G., and Lever, J. E., 1982, Stimulation of dome formation in MDCK kidney epithelial cultures by inducers of differentiation: Dissociation from effects on trans-epithelial resistance and cyclic AMP levels, *J. Cell Physiol.* **113**:427–432.

Valentich, J. D., 1981, Morphological similarities between the dog kidney cell line MDCK and the mammalian cortical collecting tubule, *Ann. N.Y. Acad. Sci.* **372**:384–405.

Valentich, J. D., 1982, Basal-lamina assembly by the dog kidney epithelial cell line MDCK, in: *Growth of Cells in Hormonally Defined Media,* Book A, Cold Spring Harbor Conferences on Cell Proliferation, Volume 9 (G. H. Sato, A. B. Pardee, and D. A. Sirbasku, eds.), Cold Spring Harbor Laboratory, pp. 567–579.

Valentich, J. D., Tchao, R., and Leighton, J., 1979, Hemicyst formation stimulated by cyclic AMP in dog kidney cell line MDCK, *J. Cell. Physiol.* **100**:291–304.

Vassent, G., Roy, C., and Ausiello, D. A., 1981, Appendix: Methodology, *J. Biol. Chem.* **256**:3422.

Watlington, C. O., Perkins, F. M., Munson, P. J., and Handler, J. S., 1982, Aldosterone and corticosterone binding and effects on Na^+ transport in cultured kidney cells, *Am. J. Physiol.* **242**:F610–F619.

10

Monoclonal Antibodies to Integral Membrane Transport and Receptor Proteins

Novel Reagents for Protein Purification

J. CRAIG VENTER, URSINA SCHMIDT, BARBARA EDDY,
GIORGIO SEMENZA, and CLAIRE M. FRASER

1. INTRODUCTION

The isolation and molecular characterization of integral membrane proteins of major physiological significance, including receptors and transport proteins, are at best difficult tasks. For good results a multifaceted approach to elucidating the structure and function of membrane proteins is required. Useful approaches include affinity labeling and photoaffinity labeling (Ruoho *et al.*, 1984), radiation inactivation/target size analysis (Venter, 1983b; Venter *et al.*, 1983b; Lilly *et al.*, 1983), characterization of hydrodynamic properties (Davis, 1984), and immunological approaches using conventional antibodies, monoclonal antibodies, and antiidiotypic antibodies (Venter *et al.*, 1984 ; Fraser and Venter, 1980; Venter, 1983a; Venter and Fraser, 1983; Berzofsky, 1984; Berzofsky *et al.*, 1980, 1982; Tzartos and Linstrom, 1980; Gullick *et al.*, 1982; Yavin *et al.*,

J. CRAIG VENTER, URSINA SCHMIDT, BARBARA EDDY, and CLAIRE M. FRASER • Department of Molecular Immunology, Roswell Park Memorial Institute, Buffalo, New York 14263. GIORGIO SEMENZA • Laboratorium fur Biochemie der Eidgenossischen Technischen Hochschule ETH-Zentrum, CH-8092, Zurich, Switzerland. Work supported by grants AI-20863 and HL-31178 (to J. C. V.) from the National Institute of Health and AHA82-881 (to C. M. F.) from the American Heart Association.

1981; Beisiegel *et al.*, 1981; Greene *et al.*, 1980; Schreiber *et al.*, 1980; Wasserman *et al.*, 1982). This chapter will detail the monoclonal antibody approach that we have found to be successful in isolating and characterizing the structure of membrane transport proteins and hormone and neurotransmitter receptors.

2. MONOCLONAL ANTIBODY PRODUCTION

Details of monoclonal antibody production can be found in many excellent reviews (see Fraser and Lindstrom, 1983). Figure 1 depicts some of the essential features involved in this process and also the significance of the technique. When animals, e.g., BALB/c mice, are immunized with intact cells, isolated membranes, or purified proteins, an immune response develops in which B lymphocytes produce antibodies against many portions (epitopes) of the injected antigen(s). With repeated immunizations, a heterogeneous collection of antibodies

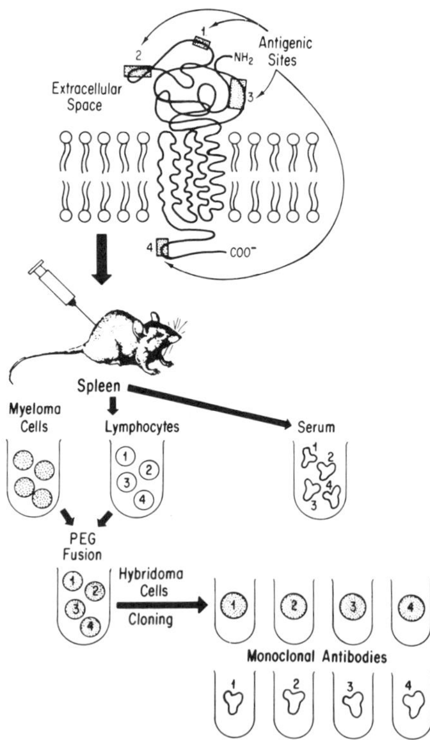

Figure 1. Antibody production to integral membrane proteins. Comparison of monoclonal antibodies to conventional serum antibodies (Venter *et al.*, 1984).

appears in the serum making antibody standardization an impossible task. Furthermore, it is often difficult to obtain serum antibodies to highly conserved proteins owing to the immunoregulatory environment *in vivo*.

The hybridoma technology developed by Kohler and Milstein (1975) has revolutionized the immunological approach to protein characterization. The nature of this technique is such that it is highly suited for the preparation of

Table I. Monoclonal Antibodies Produced to Functional Membrane Proteins

Monoclonal antibody	Antigen	Specificity
FV-101	Partly purified TEβR	β-Adrenergic receptor
FV-103		β-Adrenergic receptor
FV-104		β-Adrenergic receptor
M.6	Purified muscarinic receptor	Muscarinic cholinergic receptor
M.30		Muscarinic cholinergic receptor
M.42		Muscarinic cholinergic receptor
M.43		Muscarinic cholinergic receptor
M.44		Muscarinic cholinergic receptor
α1D8	Rat liver plasma membrane	α_1-Adrenergic receptor
α1F9		α_1-Adrenergic receptor
α3E3	Purified α_1-receptor	α_1-Adrenergic receptor
α3F3		α_1-Adrenergic receptor
α4C2		α_1-Adrenergic receptor
DA.18	Purified D_2 dopamine receptor	D_2 Dopamine receptor
DA.28		D_2 Dopamine receptor
RBC 4.5.5	Purified band 4.5	RBC–glucose carrier
RBC 4.5.21		RBC–glucose carrier
RBC 4.5.22		RBC–glucose carrier
RBC 4.5.23		RBC–glucose carrier
RBC 4.5.24		RBC–glucose carrier
RBC 4.5.25		RBC–glucose carrier
RBC 4.5.26		RBC–glucose carrier
RBC 4.5.27		RBC–glucose carrier
RBC 4.5.28	Purified band 4.5	RBC–glucose carrier
RBC 4.5.29		RBC–glucose carrier
RBC 4.5.30		RBC–glucose carrier
RBC 4.5.31		RBC–glucose carrier
RBC 4.5.32		RBC–glucose carrier
RBC 4.5.33		RBC–glucose carrier
RBC 4.5.34		RBC–glucose carrier
RBC 4.5.35		RBC–glucose carrier
RBC 4.5.36		RBC–glucose carrier
C 1.26	Guinea pig ileal membranes	Ca^{2+} channel
C 1.32		Ca^{2+} channel
C 1.47		Ca^{2+} channel

(*continued*)

chemically homogeneous antibodies to single components of impure antigen preparations (Venter, 1982). This laboratory has been successful in producing monoclonal antibodies to a number of integral membrane proteins (Table I) including α- and β-adrenergic receptors (Fraser and Venter, 1980; Venter *et al.*, 1984), muscarinic cholinergic receptors (Venter *et al.*, 1984), dopamine recep-

Table I. (*Continued*)

Monoclonal antibody	Antigen	Specificity
C 1.65		Ca^{2+} channel
C 1.69		Ca^{2+} channel
C 1.76		Ca^{2+} channel
C 1.99		Ca^{2+} channel
C 1.130		Ca^{2+} channel
C 1.136		Ca^{2+} channel
C 1.145		Ca^{2+} channel
C 1.151		Ca^{2+} channel
C 1.152		Ca^{2+} channel
C 1.156		Ca^{2+} channel
P 1.3	Type II pneumocyte membranes	Type II pneumocyte membranes
P 1.24		Type II pneumocyte membranes
P 1.40		Type II pneumocyte membranes
P 1.42		Type II pneumocyte membranes
P 1.44		Type II pneumocyte membranes
P 1.49		Type II pneumocyte membranes
P 1.56		Type II pneumocyte membranes
P 1.63		Type II pneumocyte membranes
P 1.83		Type II pneumocyte membranes
P 1.84		Type II pneumocyte membranes
P 1.88		Type II pneumocyte membranes
P 1.92		Type II pneumocyte membranes
P 1.97		Type II pneumocyte membranes
P 1.105		Type II pneumocyte membranes
P 1.106		Type II pneumocyte membranes
P 1.114		Type II pneumocyte membranes
P 1.117		Type II pneumocyte membranes
P 1.119		Type II pneumocyte membranes
P 1.124		Type II pneumocyte membranes
P 1.125		Type II pneumocyte membranes
P 1.128		Type II pneumocyte membranes
P 1.129		Type II pneumocyte membranes
P 1.139		Type II pneumocyte membranes
P 1.141		Type II pneumocyte membranes

(*continued*)

tors (Lilly *et al.*, 1983), various transport proteins including the sodium-dependent glucose transporter from brush border (Schmidt *et al.*, 1983a,b), the human erythrocyte glucose carrier, and the slow inward calcium channel, as well as a variety of cell-specific membrane proteins (Venter *et al.*, 1984; Fraser *et al.*, 1984). Monoclonal antibodies were produced as a means of purifying and characterizing each specific membrane protein or as a way of isolating specific cells by electronic cell sorting (Fraser *et al.*, 1984).

Table I.   (Continued)

Monoclonal antibody	Antigen	Specificity
P 1.152		Type II pneumocyte membranes
P 1.194		Type II pneumocyte membranes
IM.1	Brush border membranes	Brush border membranes
IM.2		Brush border membranes
IM.7		Brush border membranes
IM.8		Na^+-dependent glucose carrier
IM.9		Brush border membranes
IM.10		Brush border membranes
IM.11		Na^+-dependent glucose carrier
IM.18		Brush border membranes
IM.24		Na^+-dependent glucose carrier
IM.25		Brush border membranes
IM.30		Na^+-dependent glucose carrier
IM.31		Brush border membranes
IM.32		Brush border membranes
IM.34		Na^+-dependent glucose carrier
IM.35		Brush border membranes
IM.36		Brush border membranes
IM.38		Brush border membranes
IM.39		Brush border membranes
IM.41		Brush border membranes
IM.42		Na^+-dependent glucose carrier
IM.46		Brush border membranes
IM.55		Brush border membranes
IM.56		Brush border membranes
IM.57		Brush border membranes
IM.59		Brush border membranes
IM.60		Brush border membranes
IM.62		Brush border membranes
IM.63		Brush border membranes
IM.66		Na^+-dependent glucose carrier

This chapter will discuss the approaches we have found useful for immunization protocols and in screening for antibodies to functional proteins that exist in minute quantities in the cell membrane.

2.1. Monoclonal Antibodies to Transport Proteins

Ion and substrate transport proteins have been extremely difficult to isolate and characterize, and consequently this area of biochemistry has lapsed far behind the biochemistry of other important cellular proteins. The reasons for this are multifold. These macromolecules are integral membrane proteins and therefore require detergent for removal from the membrane and to remain soluble in aqueous solutions. As a result, the biological response characteristic of native ion or solute transport is lost. Reconstitution of solubilized biomolecules into lipid vesicles or native membranes is required to demonstrate that isolated proteins possess the characteristic biological functions (Klausner *et al.*, 1983). Second, owing to signal amplification from biological cascades such as calcium ion transport or cyclic AMP formation, some transport proteins and hormone receptors can exist on the cell surface in extremely low densities. For example, in cardiac muscle, β-adrenergic receptors and slow inward calcium channels exist at a density of one or less per square micron of membrane surface area (Venter *et al.*, 1983). Such low concentrations of cell protein have presented a major obstacle to detailed structural analysis of such molecules.

The use of monoclonal antibodies can overcome these and many other problems in protein isolation and characterization (Venter, 1982). The best starting point in attempting to produce monoclonal antibodies to functional membrane proteins is to have a clean, reproducible assay for the protein in question and to have a good source of material for use as antigen. Tissue culture cells provide many advantages over crude tissue extracts in that one can start with a relatively homogeneous cell population. Cells such as the cultured kidney epithelial cells (Chung *et al.*, 1982), described elsewhere in the volume, express many hormone receptors and transport functions (e.g., Taub and Saier, 1981; Rindler *et al.*, 1979; Rabito and Ausiello, 1980; Misfeldt *et al.*, 1976).

Brush border membranes of intestinal mucosal cells and proximal kidney tubules contain a Na^+/D-glucose cotransport system (Aronson and Sacktor, 1975; Hopfer *et al.*, 1973; Murer *et al.*, 1974; Tannenbaum *et al.*, 1977). This carrier can be identified with radioactive phlorizin, a competitive inhibitor of D-glucose transport (Diedrich, 1966), which binds with a relatively high affinity to membranes in the presence of Na^+ (Toggenburger *et al.*, 1978). The ability to measure Na^+-dependent glucose transport together with the assay of ^3H-phlorizin specific binding provided two important functional assays used in screening for monoclonal antibodies specific for the Na^+/D-glucose cotransporter (Schmidt *et al.*, 1983a,b).

2.1.1. Source of Material for Immunization

Immunization of BALB/c mice with intact cells or isolated plasma membranes produces a high yield of monoclonal antibodies to membrane proteins. In studies with isolated type II pneumocytes, immunization with intact cells or membranes produced hybridomas that secreted monoclonal antibodies to type II pneumocyte surface markers (Fraser *et al.*, 1984) in a high percentage of the clones derived from the cell fusion (Table II). Similar results have been obtained when intestinal brush border membranes are used as the antigen source (Fig. 2; Schmidt *et al.*, 1983a; Schmidt *et al.*, 1983b).

2.1.2. Screening Assays (First and Second Stage)

In screening for monoclonal antibodies to functional membrane proteins such as the Na^+-dependent glucose transporter, the first screening assay subsequent to cell fusion is usually either an enzyme-linked immunosorbent assay (ELISA) or solid-phase radioimmunoassay for detection of antibodies against any cell membrane antigen. Figure 2 illustrates the results of such a screening assay using spent culture medium from hybridomas obtained in a fusion of splenic lymphocytes from BALB/c mice immunized with intestinal brush border membranes. The cell fusion was performed 4 days subsequent to the second immunization. Hybrid cells were plated into 300 microtiter wells and spent medium removed for assay on the tenth day after cell fusion. The assay (solid phase) consisted of immobilization of brush border membranes on microtiter plates followed by incubation with spent culture medium from the hybridomas. The presence of monoclonal antibodies specific for brush border antigens was detected with $[^{125}I]$-antimouse immunoglobulin. As Fig. 2 illustrates, approximately one third of the wells contained hybridomas that secreted detectable anti-brush-border monoclonal antibodies.

The intent of this study was to obtain antibodies specific for the Na^+/D-glucose cotransporter. Preliminary identification of microtiter wells with anti-brush-border antibodies reduced the number of required functional screening assays by two thirds. However, it was felt that the number of positive wells identified in the first screening assay was still too large for rapid, functional screening. In order to further reduce the required functional assays and culture work, microtiter wells containing the 66 most positive hybridomas from the first screening assay were subcultured. Spent culture medium from the cells was reassayed using a solid-phase radioimmunoassay. These data demonstrated that after rescreening with $[^{125}I]$-antimouse IgG, approximately 50% of the cultures were significantly positive, suggesting that some of the hybridomas had stopped secreting monoclonal antibodies (Fig. 3). Two ampules of each positive hybridoma were frozen at this point.

Table II.   Screening of Monoclonal Antibodies Produced to Adult Rabbit Type II Pneumocyte Membranes

Clone no.	Freshly isolated type II pneumocytes	Type II pneumocytes cell culture	Rabbit lung fibroblasts cell culture	Human fibroblasts (foreskin): cell culture	Rat hepatocytes cell culture	Turkey erythrocyte membranes	Human erythrocyte membranes	Dog lung membranes	Rat brain membranes	Rat heart membranes
3	+	+	+	0	0	+	0	+	+	+
24	+	+	+	+	0	+	0	+	+	+
40	+	0	0	0	0	+	0	0	+	+
42	+	+	+	0	0	+	0	+	+	+
44	+	0	+	+	0	+	0	+	+	0
49	+	+	+	0	0	0	+	+	+	+
56	+	+	0	+	0	0	0	+	0	0
63	+	+	+	0	0	0	0	+	0	0
83	+	0	0	0	0	+	0	+	+	+
84	+	0	+	+	0	+	0	+	+	0
88	+	0	+	0	0	0	0	+	0	0
92	+	0	0	0	0	+	+	+	0	0
97	+	+	+	+	0	+	0	+	+	+
105	+	+	+	0	0	0	0	0	0	0
106	+	0	0	0	0	0	0	0	0	0
116	+	0	0	0	0	0	0	+	0	0
117	+	+	+	0	0	+	0	+	0	0
119	+	0	0	0	0	0	0	+	0	0
124	+	0	+	+	0	0	0	+	+	0
125	+	+	+	+	0	0	0	+	0	0
128	+	+	+	+	0	0	+	+	0	0
129	+	0	0	0	0	0	0	+	0	0
139	+	0	+	0	0	0	0	0	0	0
141	+	0	0	0	0	0	0	0	0	0
152	+	+	+	+	0	0	+	+	0	0
194	+	+	+	+	0	0	0	+	0	0

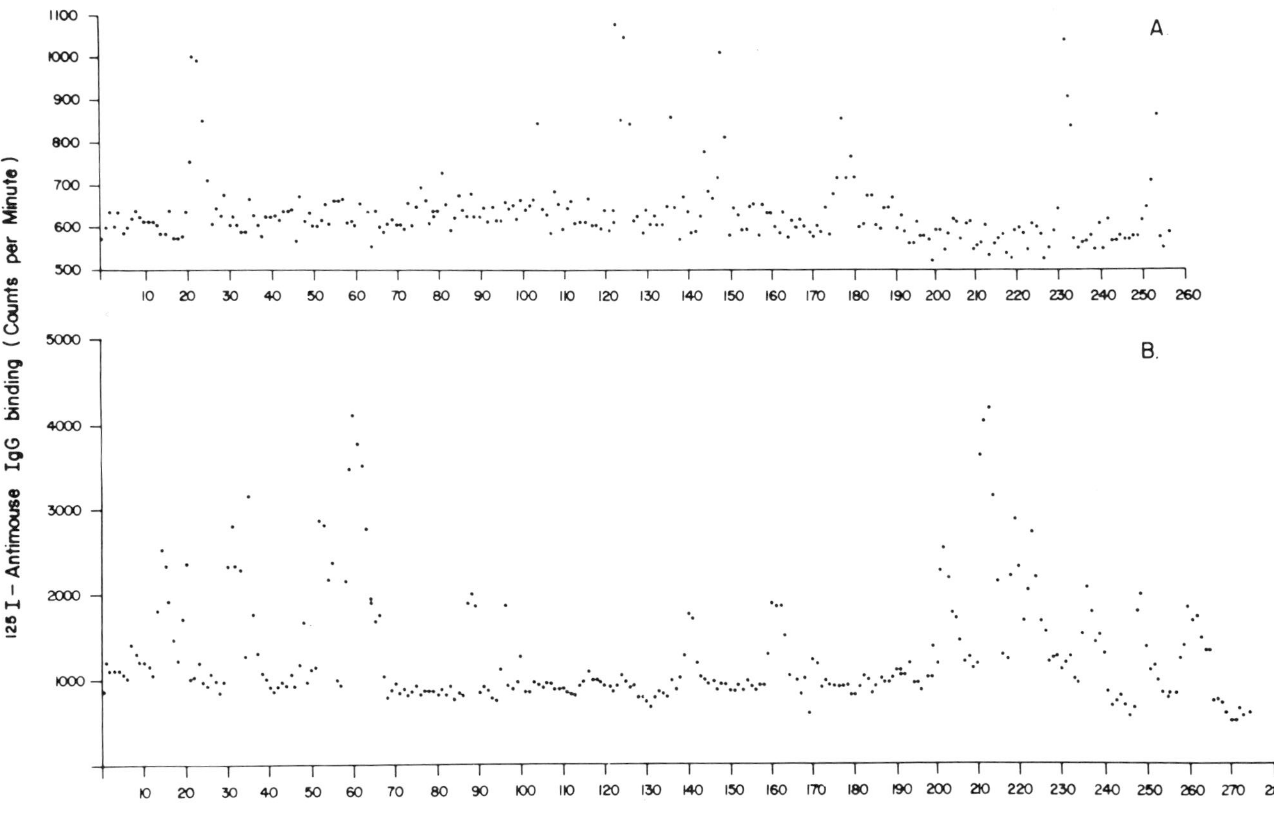

Figure 2. Monoclonal antibodies to intestinal brush border membrane antigens. Results of a solid-phase RIA used to screen hybridomas for production of monoclonal antibodies to brush border membranes. Brush border membranes were immobilized on PVC microtiter plates and incubated with samples of spent-culture medium. The association of putative monoclonal antibodies to brush border membrane antigens was detected by the subsequent binding of ^{125}I-antimouse IgG to membrane preparations. A and B represent the screening results from two separate cell fusions.

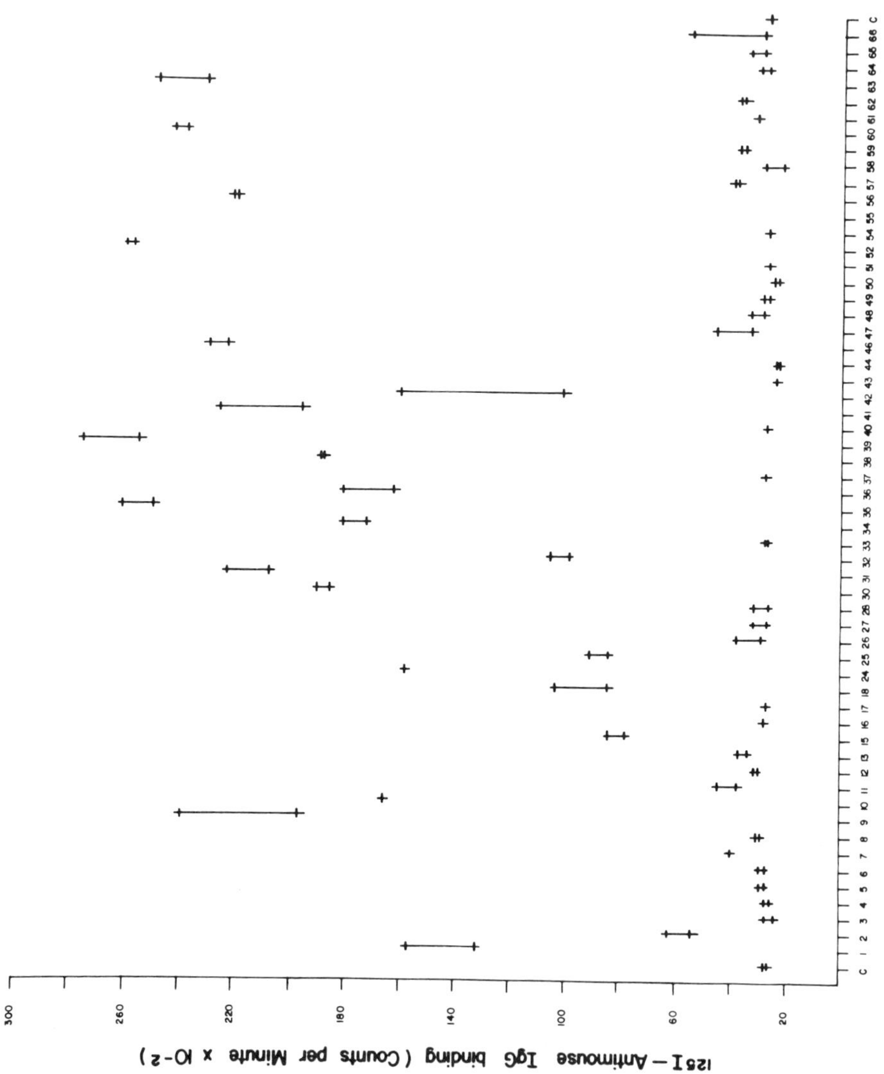

Figure 3. Second screening for monoclonal antibodies to brush border membranes subsequent to clonal expansion. See legend to Fig. 2 for experimental details.

The technique of initial screening followed by expansion and rescreening of the most positive clones reduced the number of functional assays required by approximately tenfold. This approach differs dramatically from that taken with purified antigens. For example, when splenic lymphocytes from BALB/c mice immunized with purified band 4.5 (the glucose carrier) of human red blood cells were used in a cell fusion and the resulting hybridomas were screened with a solid-phase radioimmunoassay against pure protein, positive hybridomas of the desired specificity could be identified in one step.

The approach taken with the brush border membranes clearly selects only for the most stable hybridomas but has a number of advantages when dealing with an unpurified protein that must be identified with functional assays. The initial screening assay is extremely rapid and simple, and the nature of the assay ensures the selection of relatively high-affinity antibodies. In addition, hybridomas determined to be positive in the second assay have already been shown to be stable to subculturing, cloning, and expansion. Although 50% attrition in the early stages is not uncommon, the surviving clones are usually very stable. It is an unusual event to lose clones after the second stage of screening.

2.1.3. Transport Protein Specific Assays

The use of functional assays as a primary means of identification of monoclonal antibodies in culture medium is one of the least reliable forms of antibody screening owing in part to the multitude of artifacts possible with binding and transport assays. Additional problems arise from the ion and solute content of culture medium together with the low levels of antibody usually present.

By performing primary and secondary screening assays as described previously, it becomes feasible to obtain milliliter quantities of spent culture medium, which permit antibody concentration, dialysis of culture medium, or antibody purification from the medium by ammonium sulfate precipitation and/or immunoaffinity purification of mouse immunoglobulins (Fraser and Lindstrom, 1983).

2.1.3a. [³H]-Phlorizin Binding. In the case of the Na^+/D-glucose cotransporter, monoclonal antibodies identified as positive in the second screening assay (Fig. 3) were tested for specificity for the Na^+/D-glucose cotransporter based on inhibition of ligand binding. [³H]-Phlorizin, a competitive antagonist of glucose transport, binds to brush border membranes in a manner consistent with binding to the glucose carrier. Spent culture medium from the hybridomas was tested at a number of dilutions for its ability to inhibit Na^+-dependent phlorizin binding to brush border vesicles (Fig. 4). Seven of the monoclonal antibodies significantly affected specific phlorizin binding. These hybridomas were cloned by limiting dilution and expanded. Monoclonal antibodies were purified by

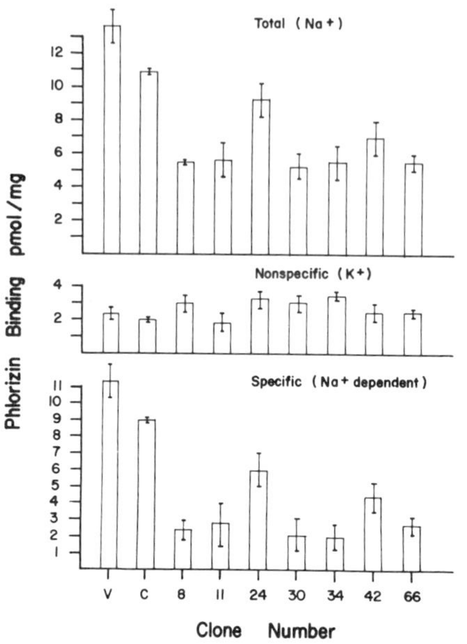

Figure 4. Inhibition of phlorizin binding to the Na^+-dependent glucose carrier of brush border by monoclonal antibodies. Na^+-dependent [³H]phlorizin binding to brush border vesicles was assayed in the presence and absence of anti-brush-border monoclonal antibodies. Monoclonal antibody from clone no. 8 demonstrated significant inhibition of Na^+-dependent binding, suggesting possible specificity for the glucose carrier.

immunoaffinity chromatography and rescreened with the phlorizin-binding assay. Three antibodies were selected for extensive characterization (Schmidt *et al.*, 1983a, b).

2.1.3b. Na^+-Dependent Glucose Transport.

The final and perhaps most important functional assay for identification of monoclonal antibodies to the Na^+/D-glucose cotransporter was measurement of the monoclonal antibody effect on Na^+-dependent glucose transport. D-Glucose uptake into brush border vesicles was measured as equilibrium exchange. Only immunoaffinity-purified antibodies were used in these studies as culture medium and ascites fluid had nonspecific deleterious effects on transport. Vesicles preincubated with antibody were equilibrated with 1 mM D-glucose, 100 mM NaCl or KCl, 300 mM mannitol, 10 mM Hepes–Tris buffer, pH 7.0, for 1 hr at 4°C. Uptake was initiated by mixing 10 µl of membrane vesicles with 10 µl of equilibration buffer containing traces of radiolabeled D-glucose. The reaction was stopped in an automated apparatus after 10 sec by the addition of 2.5 ml of a stop solution at 4°C. Mouse IgG at identical protein concentrations served as the control. Figure 5 illustrates monoclonal-antibody-specific inhibition of glucose transport (Schmidt *et al.*, 1983a,b).

All the evidence to date is consistent with monoclonal antibody specificity for the Na^+/D-glucose cotransporter of brush border membranes. Preliminary

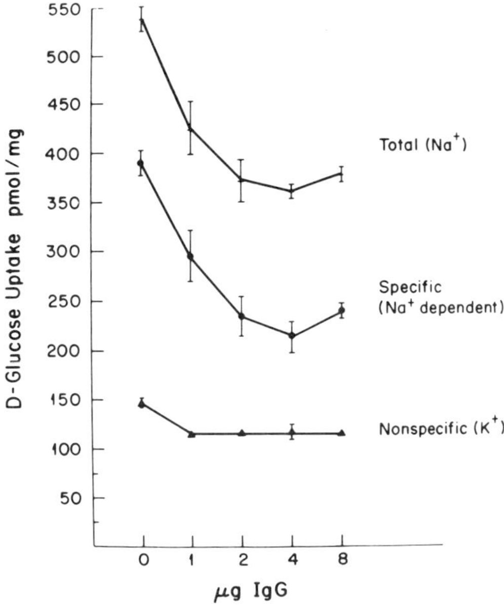

Figure 5. Monoclonal antibody inhibition of specific (Na+-dependent) glucose transport in brush border vesicles. Specific Na+-dependent D-glucose uptake was measured under tracer equilibrium conditions as the difference between total (Na+) and nonspecific (K+) D-glucose uptake. The inhibition of this specific Na+-dependent D-glucose uptake was assayed with different amounts of an immunoaffinity-purified IgG fraction of ascites fluid, IM-24.

data indicate that at least one of these antibodies cross-reacts with the Na+/D-glucose cotransporter of cultured kidney cells (Schmidt *et al.*, unpublished observation). One of these antibodies has been extensively studied and utilized as an immunoaffinity adsorbent to purify the glucose transporter of brush border membranes (Schmidt *et al.*, 1983a,b).

2.2. *Monoclonal Antibodies to the Slow Inward* Ca^{2+} *Channel*

We have taken a similar approach to that utilized for the Na+/D-glucose cotransporter in making monoclonal antibodies to the slow inward calcium channel of smooth muscle (Robinson *et al.*, in preparation). The slow inward calcium channel plays an important role in the regulation of both cardiac and smooth muscle contraction (Janis and Triggle, 1983). With the application of radiation inactivation and covalent affinity labeling, we have demonstrated that the intact channel in the membrane has a molecular weight of 278,000 and a regulatory subunit with a molecular weight of 45,000 (Venter *et al.*, 1983).

The calcium channel in membranes can be identified by radioligand binding using [3H]-nitrendipine (Bolger *et al.*, 1982, 1983). Therefore, as with the glucose carrier, a functional screening assay exists that can allow for detection of monoclonal antibodies with specificity for the slow inward calcium channel.

Table III. Molecular Properties of Neurotransmitter Receptors[a]

Receptor	α_1-Adrenergic	α_2-Adrenergic	β_1-Adrenergic	β_2-Adrenergic	D_2-Dopaminergic	Muscarinic
Hydrodynamic Mr Stokes Radus	83 kda Rat liver 6.0 nm Rat liver		65 kda Dog heart 4.2 nm Dog heart 4.8 nm Rat heart 4.5 nm Turkey RBC	90 kda Dog lung/liver 5.8 nm Dog lung/liver Calf lung Frog RBC	123 kda Dog brain	86 kda Rat brain
Isoelectric Point (pI)	3.9 Rat liver		5.5 Turkey RBC	4.2 Dog lung	4.0 Dog striatum	5.9 Rat brain Rat heart, *Drosophila* head
SDS–PAGE Mr	85 kda Rat liver		65–70 kda Dog heart, tur-key RBC	58 kda Dog lung		80 kda Human, rat, dog, guinea pig brain; dog heart, rat heart, *Drosphila* head
Radiation Inactivation Target size Analysis	160 kda Rat liver	160 kda Human platelets	90 kda Turkey RBC 120 kda Dog heart	109 kda Dog lung	123 kda Dog and human brain	80 kda Rat, human brain Dog heart, guinea pig ileum

[a]kda, Kilo dalton; RBC, red blood cells.

Table IV.   Uses of Monoclonal Antibodies to Membrane Antigens

Immunoaffinity purification of receptor and transport proteins
Phylogenetic and cross-reactivity studies
Radioimmune quantitation assays
Cell and tissue localization studies at both the light and EM level (immunocytochemical studies)
Clinical assays
Therapeutic agents

We immunized BALB/c mice with highly purified membranes isolated from the longitudinal smooth muscle of the guinea pig ileum. These membranes contain approximately 1 pmol calcium channel per milligram of membrane protein (Venter *et al.*, 1983; Bolger *et al.*, 1982, 1983). Cell hybridization was performed with SP/2 cells as previously described (Fraser and Venter, 1980; Fraser, 1984; Fraser and Lindstrom, 1984). The wells containing hybridomas (300/300 plated) were screened for antibody binding to smooth-muscle membranes in a solid phase assay as described previously for the glucose carrier. The positive hybridomas from this assay were further screened against [3H]-nitrendipine specific binding. Ten hybridomas were selected that had high membrane binding and a high degree of inhibitory activity in the calcium channel specific assay. The cells in these wells were expanded, cloned, and rescreened with both solid-phase radioimmunoassays and functional assays. These antibodies are presently being utilized to purify, characterize, and localize the slow inward calcium channel in various tissues (Robinson *et al.*, in preparation).

2.3. Monoclonal Antibodies to Neurotransmitter Receptors

The molecular propreties of a number of neurotransmitter receptors under study in this laboratory are summarized in Table III. Monoclonal antibodies made against some of these receptors have had a major impact on the isolation and molecular characterization of the receptors (Venter, 1983a; Fraser and Venter, 1980; Venter *et al.*, 1981, Venter *et al.*, 1984; Fraser, 1984).

The approach to producing monoclonal antibodies to these important regulatory molecules has relied on functional assays as described previously in addition to specific immunoprecipitation assays of the receptor proteins involved. An example of a receptor immunoprecipitation assay is illustrated in Fig. 6, which shows the immunoprecipitation of affinity-labeled, partly purified muscarinic acetylcholine receptor from rat brain by a series of antireceptor monoclonal antibodies. This assay demonstrates that the loss of receptor from the supernatant corresponds directly with the appearance of the receptor in the immunoprecipitate (Venter *et al.*, 1984).

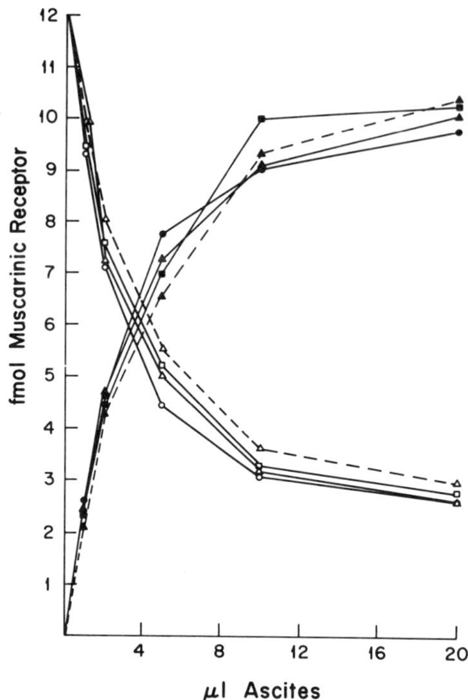

Figure 6. Immunoprecipitation of muscarinic cholinergic receptors with monoclonal antibodies. The open symbols represent muscarinic receptors remaining in the supernate after immunoprecipitation with monoclonal antibodies whereas the closed symbols represent the corresponding appearance of receptors in the pellet.

A summary of the uses of antireceptor and transport protein monoclonal antibodies is given in Table IV.

3. IMMUNOAFFINITY CHROMATOGRAPHY PURIFICATION OF MEMBRANE PROTEINS WITH MONOCLONAL ANTIBODIES

One of the uses of monoclonal antibodies produced to functional proteins is to purify and characterize receptors and transporters where conventional approaches have not proven useful. We will describe the affinity purification of two proteins, the β-adrenergic receptor and the Na/D-glucose cotransporter.

3.1. Immunoaffinity Purification of β-Adrenergic Receptors

In 1980 we described the production of monoclonal antibodies to the turkey erythrocyte β-adrenergic receptor (Fraser and Venter, 1980). One of the monoclonal antibodies from this study (FV-104) was found to be specific for the ligand binding site of the receptor (Fraser and Venter, 1980, 1981). The characteristics

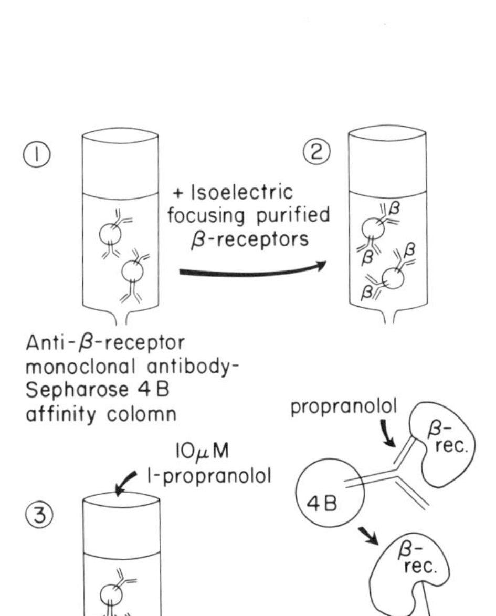

Figure 7. Immunoaffinity column preparation. Ascites fluid is collected from tumor-bearing mice (1), and immunoglobulins are purified from ascites fluid by isoelectric focusing or ammonium sulfate precipitation (2). Purified immunoglobulins are coupled to a suitable matrix such as CNBr-activated Sepharose 4B (3), and immunoaffinity columns of the desired size are prepared for use in protein purification (4).

Figure 8. Ligand specific elution of purified proteins from immunoaffinity columns. Detergent extracts of membrane proteins are applied to immunoaffinity columns (1, 2) which retain the desired protein (in this case, the β-adrenergic receptor) via antibody–antigen binding. Adsorbed protein is eluted from immunoaffinity columns with physiological concentrations of ligand, such as *l*-propranolol for the β-adrenergic receptor (3). Ligand-specific elution of proteins can be accomplished if monoclonal antibodies are directed to epitopes within the active site of a protein molecule.

of antibody binding to the β-receptor were such that adrenergic antagonists and antibody were mutually antagonistic. The affinity of antibody 104 for the turkey erythrocyte β receptor was only moderate (equilibrium dissociation constant antigen $K_d = 1 \times 10^{-7}$ M). Therefore, this antibody seemed like a perfect candidate for the affinity purification of the β receptor. Figures 7, 8, and 9 illustrate the receptor purification scheme. The key to receptor purification is the unique antibody specificity. Owing to the competitive nature of the receptor interaction between the antibody and adrenergic antagonists, the β-receptor blocking agent propranolol was utilized to desorb the receptor from immunoaffinity column (Fig. 9). The molecular properties of the isolated receptor are summarized in Table III.

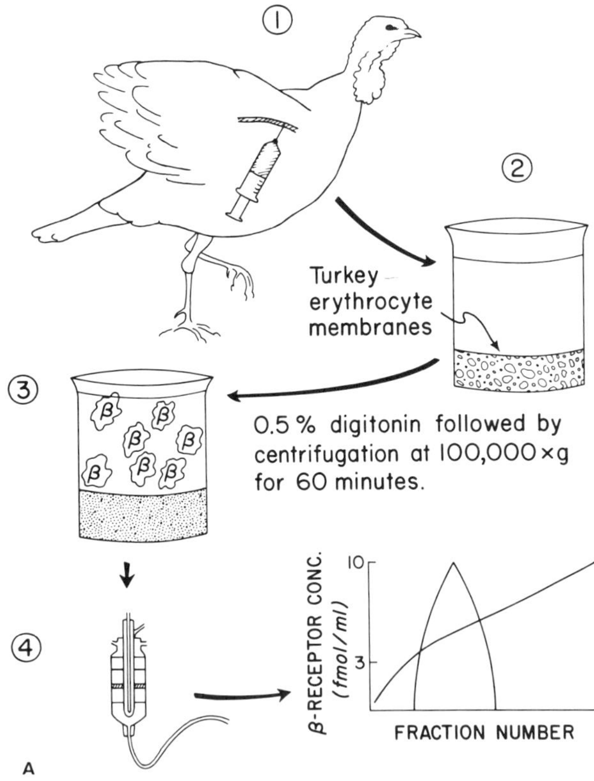

Figure 9. Protocol for purification of β-adrenergic receptors from turkey erythrocytes. (A) Freshly isolated turkey erythrocytes (1) are used to prepare membranes containing β-adrenergic receptors (2). Receptors are solubilized from membranes using 0.5% digitonin (3) and partly purified by preparative isoelectric focusing (4). (B) Isoelectric-focusing-purified receptors are applied to mono-

The lung β_2-adrenergic receptor subunit was also purified using the same immunoaffinity column (Fraser and Venter, 1982) (Table III).

3.2. Immunoaffinity Purification of the Na^+/D-Glucose Cotransporter

Purified monoclonal antibody specific for the phlorizin and D-glucose binding site of the Na^+/D-glucose cotransporter of brush border membranes (see previous discussion) was coupled to Sepharose 4B to create an immunoaffinity column (Schmidt *et al.*, 1983a&b). Digitonin-solubilized glucose carrier was applied to the immunoaffinity columns. Subsequent to several buffer washes, low concentrations of phlorizin (μM) or D-glucose (mM) were used to specifical-

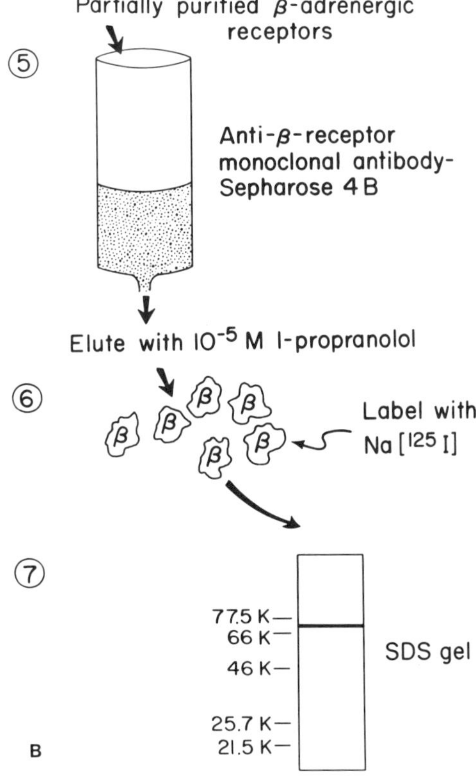

clonal antibody immunoaffinity columns (5), and purified receptors are eluted with 10 μM *l*-propranolol, a β-receptor antagonist (6). The propranolol eluate is analyzed on SDS–polyacrylamide gels (7). A single protein, $M_r = 70,000$, is obtained, demonstrating the great utility of monoclonal antibodies in the rapid purification of membrane proteins present in extremely low concentrations in the native membrane.

ly elute the glucose carrier (Schmidt *et al.*, 1983a,b). SDS–PAGE analysis of the specific eluates (Fig. 10) demonstrated that a single protein Mr = 72,000 daltons was purified (Schmidt *et al.*, 1983a,b). This molecular weight is consistent with that obtained by other methods, e.g., photoaffinity labeling for this same carrier (Schmidt *et al.*, 1983a,b).

4. CONCLUSION AND FUTURE APPROACHES

Monoclonal antibodies have numerous advantages for isolating and characterizing membrane proteins, particularly those which exist in minute quantities such as some receptors and transporters.

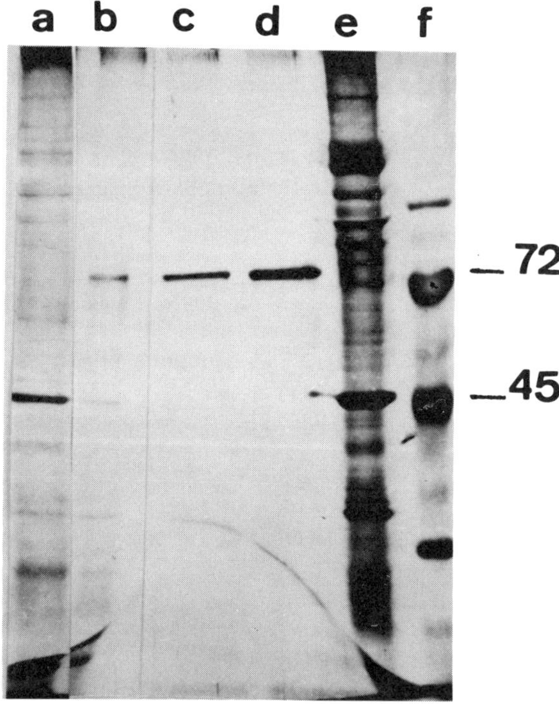

Figure 10. SDS–polyacrylamide gel electrophoresis of deoxycholate-extracted brush border membranes and immunoaffinity-purified Na$^+$/D-glucose cotransporter. Brush border membrane vesicles (10 mg protein) were solubilized with 0.5% digitonin, 0.2 mM EGTA in phosphate-buffered saline (PBS) for 5 min at 22°C, were centrifuged for 20 min at 40,000 g, and the supernatant was immediately applied to an immunoaffinity column followed by extensive washing with PBS. (a) Pellet from the deoxycholate extraction; (b) digitonin extract therefrom; (c) phlorizin (15 μM) eluate of absorbed protein; (d) second phlorizin (100 μM) eluate; (e) initial, intact brush border membrane vesicles; (f) protein standards, from the top: muscle glycogen phosphorylase *b*, 94 kda; bovine serum albumin, 67 kda; chicken ovalbumin, 43 kda; carbonic anhydrase, 30.1 kda; trypsin inhibitor, 20.1 kda. Gels were silver stained.

This chapter discusses some approaches that have resulted in monoclonal antibodies specific for the binding sites of functional proteins. The competitive nature of antibody and ligand binding creates an ideal situation for the immunoaffinity purification of the proteins of interest. Application of these approaches has provided some of the first molecular information on receptors and transport systems.

REFERENCES

Aronson, P. A., and Sacktor, B., 1975, The Na^+ gradient-dependent transport of D-glucose in renal brush border membranes, *J. Biol. Chem.* **250**:6032–6039.

Beisiegel, U., Schneider, W. J., Golstein, J. L., Anderson, R. G. W., and Brown, M. S., 1981, Monoclonal antibodies to the low density lipoprotein receptor as probes for study of receptor-mediated endocytosis and the genetics of familial hypercholesterolemia, *J. Biol. Chem.* **256**:11923–11931.

Berzofsky, J. A., 1984, Monoclonal antibodies as probes of antigenic structure, in: *Receptor Biochemistry and Methodology,* Volume IV (J. C. Venter, C. M. Fraser, J. M. Lindstrom, eds.), Liss, New York, in press.

Berzofsky, J. A., Hicks, G., Fedorko, J., and Minna, J., 1980, Properties of monoclonal antibodies specific for determinants of a protein antigen, myoglobin, *J. Biol. Chem.* **255**:11188–11191.

Berzofsky, J. A., Buckenmeyer, G. K., Hicks, G., Gurd, F. R. N., Feldmann, R. J., and Minna, J., 1982, Topographic antigenic determinants recognized by monoclonal antibodies to sperm whale myoglobin, *J. Biol. Chem.* **257**:3189–3198.

Bolger, G. T., Gengo, P. T., Luchowski, E. M., Siegel, H., Triggle, D. J., and Janis, R. A., 1982, High affinity binding of a Ca^{++} channel antagonist to smooth and cardiac muscle, *Biochem. Biophys. Res. Commun.* **104**:1604–1609.

Bolger, G. T., Gengo, P. T., Kockowski, R., Luchowski, E. M., Siegel, H., Janis, R. A., Triggle, A. M., and Triggle, D. J., 1983, Characterization of binding of the Ca^{++} channel antagonist [^{3}H]Nitrendipine, to guinea-pig Ileal smooth muscle, *J. Pharmacol. Exp. Ther.* **225**:291–309.

Chung, S. D., Alan, N., Livingston, D., Hiller, S., and Taub, M., 1982, Characterization of primary rabbit kidney cultures that express proximal tubule functions in a hormonally defined medium, *J. Cell Biol.* **95**:118–126.

Davis, A., 1984, Determination of the hydrodynamic properties of detergent solubilized proteins, in: *Receptor Biochemistry and Methodology,* Volume III (J. C. Venter and L. C. Harrison, eds.), Liss, New York, pp. 161–178.

Diedrich, D. F., 1966, Competitive inhibition of intestinal glucose transport by phlorizin analogs, *Arch. Biochem. Biophys.* **117**:248–256.

Fraser, C. M., 1984, Monoclonal antibodies to β-adrenergic receptors and receptor structure in: *Receptors and Recognition* (P. Cuatrecasas and M. F. Greaves, eds.), Chapman and Hall, London, in press.

Fraser, C. M., and Lindstrom, J. M., 1984, The production and use of monoclonal antibodies in receptor characterization and purification, in: *Receptor Biochemistry and Methology,* Volume III (J. C. Venter, and L. C. Harrison, eds.), Liss, New York, pp. 1–30.

Fraser, C. M., and Venter, J. C., 1980, Monoclonal antibodies to β-adrenergic receptors: Use in purification and molecular characterization of β-receptors, *Proc. Natl. Acad. Sci. USA* **77**:7034–7038.

Fraser, C. M., and Venter, J. C., 1982, The size of the mammalian lung β_2-adrenergic receptor as determined by target size analysis and immunoaffinity chromatography, *Biochem. Biophys. Res. Comm.* **109**:21–29.

Fraser, C. M., Venter, J. C., Finkelstein, J. N., and Shapiro, D. L., 1984, Monoclonal antibodies to surface antigens of rabbit type II pneumocytes, *Am. Rev. Resp. Dis.*, in press.

Greene, G. L., Fitch, F. W., and Jensen, E. V., 1980, Monoclonal antibodies to estrophilin: Probes for the study of estrogen receptors, *Proc. Natl. Acad. Sci. USA* **77**:157–161.

Gullick, W. J., Tzartos, S., and Lindstrom, J. M., 1982, Monoclonal antibodies as probes for acetylcholine receptor structure. I. Peptide mapping, *Biochemistry* **20**:2173–2180.

Hopfer, U., Nelson, K., Perroto, J., and Isselbacher, K. J., 1973, Glucose transport in isolated brush border membrane from rat small intestine, *J. Biol. Chem.* **248**:25–32.

Janis, R. A., and Triggle, D. J., 1983, New developments in Ca^{++} channel antagonist, *J. Med. Chem.* **26**:775–785.

Klausner, R. D., van Renswoude, J., Blumenthel, R., and Riunay, B., 1984, Reconstitution of membrane receptors, in: *Receptor Biochemistry and Methodology,* Volume III (J. C. Venter and L. C. Harrison, eds.), Liss, pp. 209–240.

Kohler, G., and Milstein, C., 1975, Continuous cultures of fused cells secreting antibody of pre-defined specificity, *Nature* **256**:495–497.

Lilly, L., Fraser, C. M., Jung, C. Y., Seeman, P., and Venter, J. C., 1983, Molecular size of the canine and human brain D2 dopamine receptor as determined by radiation inactivation, *Mol. Pharm.* **24**:10–14.

Misfeldt, D. S., Hammamoto, S. T., and Pitelka, D. R., 1976, Transepithelial transport in cell culture, *Proc. Natl. Acad. Sci. USA* **73**:1212–1216.

Murer, H., Hopfer, U., Kinne-Saffron, E., and Kinne, R., 1974, Glucose transport in isolated brush-border and lateral–basal plasma–membrane vesicles from intestinal epithelial cells, *Biochim. Biophys. Acta* **345**:170–179.

Rabito, C. A., and Ausiello, D. A., 1980, Na^+-dependent sugar transport in a cultured epithelial cell line from pig kidney, *J. Membrane Biol.* **54**:31–38.

Rindler, M. J., Taub, M., and Saier, M. H., Jr., 1979, Uptake of $^{22}Na^+$ by cultured dog kidney cells (MDCK), *J. Biol. Chem.* **254**:11935–11939.

Ruoho, A. E., Rashidbaigi, A., and Roeder, P. E., 1984, Approaches to the identification of receptors using affinity labels, in: *Receptor Biochemistry and Methodology,* Vol. I (J. C. Venter and L. C. Harrison, eds.), Liss, Inc., New York, pp. 119–160.

Schmidt, U. M., Eddy, B., Fraser, C. M., Semenza, G., and Venter, J. C., 1983a, Monoclonal antibodies and immunoaffinity purification of the Na^+,D-glucose cotransporter of intestinal brush border, submitted for publication.

Schmidt, U. M., Eddy, B., Fraser, C. M., Venter, J. C., and Semenza, G., 1983b, Purification of the Na^+,D-glucose cotransporter of rabbit intestinal brush border membranes using monoclonal antibodies, *FEBS Letters* **161**:279–283.

Schreiber, A. B., Couraud, P. O., Andre, C., Vray, B., and Strosberg, A. D., 1980, Anti-alprenolol anti-idiotypic antibodies bind to β-adrenergic receptors and modulate catecholamine-sensitive adenylate cyclase, *Proc. Natl. Acad. Sci. USA* **77**:7385–7389.

Tannenbaum, C., Toggenburger, G., Kessler, M., Rothstein, A., and Semenza, G., 1977, High affinity phlorizin binding to brush border membranes from small intestine: Identity with (a part of) the glucose transport system. Dependence on the Na^+-gradient, partial purification, *J. Supramol. Struct.* **6**:519–533.

Taub, M., and Saier, M. H., Jr., 1981, Amiloride-resistant Madin–Darby canine kidney (MDCK) cells exhibit decreased calcium transport, *J. Cell. Phys.* **106**:191–199.

Toggenburger, G., Kessler, M., Rothstein, A., Semenza, G., and Tannenbaum, G., 1978, Similarity in effects of Na^+ gradients and membrane potentials on D-glucose transport by, and phlorizin binding to, vesicles derived from brush borders of rabbit intestinal mucosal cells, *J. Membrane Biol.* **40**: 269–290.

Tzartos, S., and J. M. Lindstrom, 1980, Monoclonal antibodies used to probe acetylcholine receptor structure: Localization of the main immunogenic region and detection of similarities between subunits, *Proc. Natl. Acad. Sci. USA* **77:** 755–759.

Venter, J. C., 1982, Monoclonal antibodies and autoantibodies in the isolation and characterization of neurotransmitter receptors: The future of receptor research, *J. Mol. Cell. Cardiol.* **14:**687–693.

Venter, J. C., 1983a, Monoclonal and anti-idiotypic antibodies and the elucidation of receptor structure, *Surv. Immunol. Res.* **2:**302–305.

Venter, J. C., 1983b, Muscarinic cholinergic receptor structure I. Receptor size, membrane orientation and absence of major phylogenetic structural diversity, *J. Biol. Chem.* **258:**4842–4848.

Venter, J. C., and Fraser, C. M., 1981, The development of monoclonal antibodies to β-adrenergic receptors and their use in receptor purification and characterization, in: *Monoclonal Antibodies in Endocrine Research* (G. Eisenbarth and R. Fellows, eds.), Raven Press, New York, pp. 119–134.

Venter, J. C., and Fraser, C. M., 1983, β-adrenergic receptor isolation and characterization with immobilized drugs and monoclonal antibodies, *Fed. Proc.* **42:**273–278.

Venter, J. C., Eddy, B., Hall, L. M., and Fraser, C. M., 1984, Monoclonal antibodies detect the conservation of muscarinic cholinergic receptor structure from Drosophila to human brain and possible structural homology with α-adrenergic receptors, *Proc. Nat. Acad. Sci. USA* **80:**272–276.

Venter, J. C., Eddy, B., Schmidt, U., and Fraser, C. M., 1984, Production of monoclonal antibodies to integral membrane transport and receptor proteins and their use in structural elucidation, in: *Idiotypic Manipulations in Biological Systems* (H. Kohler, P. A., Cazenave, and J. Urbain, eds.), Academic Press, New York, pp. 273–301.

Venter, J. C., Fraser, C. M., Schaber, J. S., Jung, C. Y., Bolger, G., and Triggle, D. J., 1983, Molecular properties of the slow inward calcium channel: Molecular weight determinations by radiation inactivation and covalent affinity labelling, *J. Biol. Chem.* **258:**9344–9348.

Venter, J. C., Fraser, C. M., Soiefer, A., Jeffrey, D. R., Strauss, W. L., Charlton, R. R., and Greguski, R., 1981, Autoantibodies and monoclonal antibodies to β-adrenergic receptors: Their use in receptor purification and characterization, in: *Advances in Cyclic Nucleotide Research,* Vol. 14 (J. Dumont, P. Greengard, and G. A. Robison, eds.), Raven Press, New York, pp. 135–143.

Wasserman, N. H., Penn, A. S., Freimuth, P. I., Treptow, N., Wentel, S., Cleveland, W. L., and Erlanger, B. F., 1982, Anti-idiotypic route to anti-acetylcholine receptor antibodies and experimental myasthenia gravis, *Proc. Natl. Acad. Sci. USA* **79:**4810–4814.

Yavin, E., Yavin, Z., Schneider, M., and Kohn, L. D., 1981, Monoclonal antibodies to the thyrotropin receptor: Implications for receptor structure and the action of autoantibodies in Graves disease, *Proc. Natl. Acad. Sci. USA* **78:**3180–3184.

V

EPITHELIAL CELL CULTURES AND HORMONALLY DEFINED MEDIUM

A number of epithelial cell culture systems are available for study. Kidney epithelial cell lines such as MDCK and LLC–PK$_1$ have been established from normal renal tissue. The cell lines are of particular interest to study with regard to their differentiated transport functions. In addition, kidney epithelial cell lines, such as H-301, have been developed from renal tumors. These tumor cell lines are of interest to study from the standpoint of their growth regulation. H-301 cells form estrogen-dependent tumors *in vivo*. Thus tissue culture techniques may be used to study the mechanism of tumor induction by estrogens. The differentiated functions of H-301 cells are also of interest. Such estrogen-dependent tumor cells may originate from the proximal tubule. The alterations that have occurred in hormone responsiveness, metabolism, and transport in such cells is of interest to study. Such alterations may be relevant to the problem of malignancy. In these regards primary cultures of normal epithelial cells may prove to be important in the study of the transformation process.

Hormonally defined serum-free medium is an important new technique for the study of transporting epithelial cells *in vitro*. The availability of hormonally defined media permits an examination of the mechanism by which the growth and differentiated transport functions of cultured epithelial cells are regulated by hormones. The hormonally defined serum-free media also permit additional types of transporting epithelial cells to be studied *in vitro*, by facilitating the growth of the appropriate cells in primary cultures. In this manner additional types of transport systems can be examined *in vitro*.

11

Estrogen-Dependent Kidney Tumors

JOACHIM G. LIEHR and DAVID A. SIRBASKU

1. INTRODUCTION

Kidney tumors in laboratory animals, in particular, hormone-dependent tumors, were subjected to detailed investigations in the hope of obtaining insight concerning genesis and treatment of human kidney cancer. Shared characteristics between renal adenocarcinoma in human beings (reviewed elsewhere by Kantor, 1977, and Clark and Anderson, 1976) and in laboratory animals led to suggestions (Bloom *et al.*, 1967; Bloom and Wallace, 1964) that a hormonal influence on renal carcinoma in human beings may exist, although such relationships were never established. Also, kidney development from the embryonic urogenital ridge suggested an influence of hormones in kidney function and neoplastic transformation.

In most species of laboratory animals, the incidence of spontaneous kidney tumors was found to be quite low. Spontaneous renal cortical adenomas or carcinomas were rare in rats, mice, guinea pigs, and hamsters (Murphy *et al.*, 1967; Grabstald, 1973). Several cases of spontaneous kidney tumors in Syrian hamsters were described by Kirkman (1974) but were very rare. Kirkman's

JOACHIM G. LIEHR • Department of Pharmacology, The University of Texas Medical School at Houston, Houston, Texas 77225. DAVID A. SIRBASKU • Department of Biochemistry and Molecular Biology, The University of Texas Medical School at Houston, Houston, Texas 77225. Financial support was provided by a National Institutes of Health National Cancer Institute Grant (CA27539 to J. G. L.). D. A. S. is the recipient of an American Cancer Society Faculty Research Award (FRA-212), a National Institutes of Health National Cancer Institute Grant (CA26617), and an American Cancer Society Grant (BC-255).

report listed four spontaneous renal tumors found in a colony of approximately 4000 hamsters.

Chemically induced renal lesions have been known for some time. In 1939 Sempronj and Morelli induced proliferative changes resembling adenocarcinomas in rat kidney using β-anthraquinoline. Alkylating agents are known to induce renal sarcomas (Grabstald, 1973). Cycasine and also dimethylnitrosamine may induce clear-cell renal tumors in the rat (Murphy *et al.*, 1967; Williams and Murphy, 1971). Also, heavy metals including lead acetate were used to induce experimental kidney tumors in rat (Murphy *et al.*, 1967). Other kidney-neoplasm-inducing agents are described by Grabstald (1973). In Syrian hamsters, a renal adenoma was mentioned by Tomatis *et al.* (1964) occurring in a group of hamsters fed with dimethylnitrosamine. Toth *et al.* (Toth, 1971; Toth *et al.*, 1961) report a clear-cell renal adenoma and two carcinomas in a large group of urethane-treated hamsters. Renal angiosarcomas have been reported in hamsters after administration of 7,12-dimethylbenzanthracene (Lee *et al.*, 1963) or administration of urethane associated with X rays.

In the following review, only the hormone-dependent tumors of the kidney and cell lines derived from hormone-dependent kidney tumors will be discussed. Tumorigenesis, the influence of hormones on tumor initiation and tumor growth, and the receptor states in kidney and tumor tissue will be described. Mechanisms of tumor induction and of growth control by hormones will be discussed.

2. PRIMARY HORMONE-DEPENDENT TUMORS

2.1. Hyperplasia of Bowman's Capsule

Hyperplasia of the parietal layers of Bowman's capsule was described by Crabtree (1941) in two different inbred mice strains. The lesion was found in normal male animals but did not develop in males castrated at 5 wk of age. This hyperplasia was found to be testosterone dependent since injections of testosterone propionate into castrated animals restored the lesions. These alterations have not been studied further.

2.2. Induction of Estrogen-Dependent Tumors of the Hamster

The most commonly studied primary kidney tumor is the estrogen-induced and -dependent clear-cell carcinoma of the Syrian hamster (Fig. 1). A kidney tumor in an estrogen-treated Syrian hamster was observed first by Vasquez–Lopez (1944), but it was considered to be a metastasis from an estrogen-induced pituitary tumor. These kidney neoplasms were later proposed by Kirkman (1959a) to be identical with the primary renal clear-cell carcinomas, which he

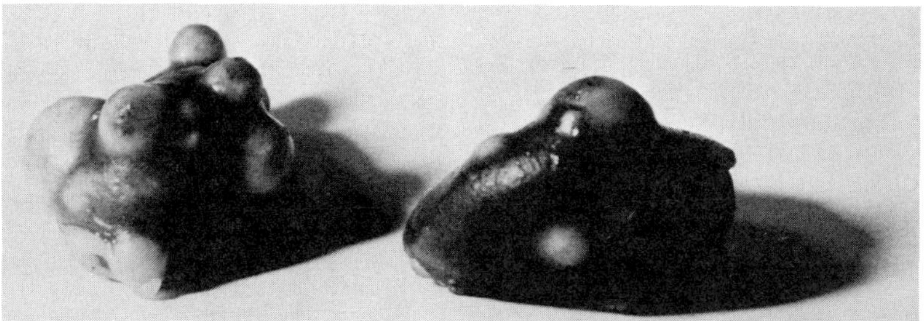

Figure 1. Kidneys excised from a male Syrian hamster treated with hexestrol for 7 mo.

had previously studied in great detail (Kirkman, 1951, 1957, 1959a,b). These kidney neoplasms were also investigated by Horning (1954, 1956a,b) and Ward (McGregor *et al.,* 1960; Ward *et al.,* 1964). As described by Kirkman (1959a), tumors were induced in male Syrian hamsters approximately 6–8 mo. after administration of estradiol or diethylstilbestrol (DES). The induction period was about the same for natural estradiol and the synthetic estrogen DES. After this time tumors usually were observed in 80–100% of the estrogen-treated animal population. Typically, estrogens were administered as 20-mg subcutaneous implants consisting of pure estrogen (Kirkman, 1959a) or 25-mg subcutaneous implants consisting of 90% estrogen and 10% cholesterol (Sirbasku and Kirkland, 1976). Depending on the purity of the estrogen used, the degree of packing in the pellets, and other physical characteristics of the dosage form administered, absorption from these pellets ranged from approximately 80 μg/day (Kirkman, 1959a) to 630 μg/day (Ward *et al.,* 1964). Fresh pellets were reimplanted every 90–150 days to ensure constant absorption. A minimum dosage necessary to induce tumors has not been precisely determined. However, Kirkman (1959a) observed that a single subpannicularly implanted 20-mg pellet of DES:cholesterol (1:4, w/w) failed to result in renal tumors (such a pellet contained only 4 mg of estrogen).

The neoplasms that arise in estrogen-treated hamster kidneys have been thought to originate in tubular epithelium (Kirkman and Robbins, 1959; Llombart–Bosch and Peydro, 1975). Histological examination (Fig. 2A) of the induced primary tumor suggests tubule formation and also provides clear evidence of the epithelial or adenocarcinoma character of the induced tumors. This histological pattern persists even during multiple transplant generations, indicating the stable morphological properties of this tumor.

2.3. Role of Hormones in Hamster Kidney Tumor

The tumorigenic effects of estrogens have several unique characteristics. First, estrogens alone, without contribution of other known carcinogens or cocar-

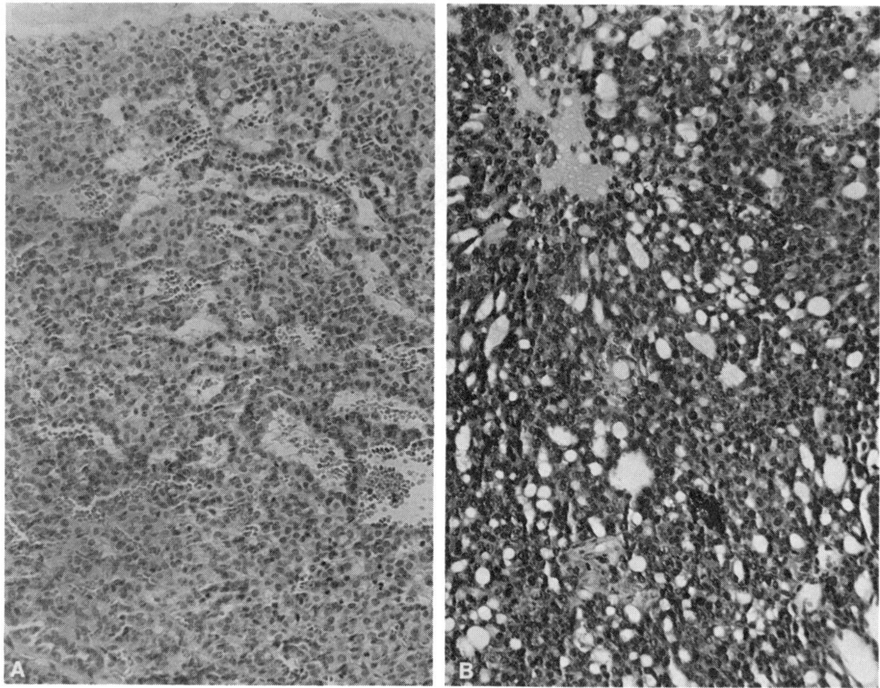

Figure 2. (A) Histology of a primary estradiol-induced renal tumor in Syrian hamster. (B) Histology of an estrogen-dependent tumor arising from H-301 cells inoculated into an estradiol treated male Syrian hamster (X100).

cinogens, are capable of initiating and promoting tumor growth. These dual effects of estrogens are unusual as compared to classical chemical carcinogen action (Farber, 1982). Second, sustained estrogen treatment is needed. This observation suggests that the carcinogenic process by estrogen requires some long duration of exposure, unlike the carcinogenic action of a known tumor-initiating agent (Farber, 1982) which causes malignancies after a single exposure.

A variety of estrogens have been used successfully to induce renal clear-cell carcinoma in male Syrian hamsters (Table I). Estradiol metabolites such as estriol and estrone were found to cause kidney tumors in males and orchiectomized male hamsters with an estimated mean latent period of 13–14 mo (Kirkman, 1959a). Synthetic estrogens other than DES, such as ethynyl estradiol

*Table I. Induction of Renal Clear-Cell Carcinoma of Male Syrian Hamsters
by Various Estrogens*

Estrogen	Tumor incidence (%)	Estimated latent period (months)[a]	Reference
Estradiol	100	6–8	Kirkman, 1959a
2,4-Dideuterioestradiol	100	6–8	Liehr, 1983
Estriol	55	13–20	Kirkman, 1959a
Estrone	88	12–18	Kirkman, 1959a
2-Hydroxyestradiol	0	6	Liehr and Sirbasku unpublished results
4-Hydroxyestradiol	80	6	Liehr and Sirbasku unpublished results
2-Methoxyestradiol	0	6	Liehr and Sirbasku unpublished results
Equilin	75	8.7	Li *et al.*, 1983
d-Equilenin	0	8.7	Li *et al.*, 1983
α-Dienestrol	88	8.7	Li *et al.*, 1983
DES 3,4-oxide	86	8.7	Li *et al.*, 1983
DES	100	6–8	Kirkman, 1959a
Hexestrol	92	6	Liehr *et al.*, 1984 and Li *et al.*, 1983
3′,3″,5′,5″-Tetradeuteriohexestrol	100	6	Liehr, *et al.*, 1984
Ethynylestradiol	29	10–13	Kirkman, 1959a
Fenocylin	8	12–13	Kirkman, 1959a
2-Fluoroestradiol	0	6–9	Liehr, 1983
4-Fluoroestradiol	40	6–9	Liehr, 1983
3′,3″,5′,5″-tetrafluorodiethylstilbestrol	100	9	Liehr *et al.*, 1983b

[a]Latent periods as reported. Large differences in the estimated latent periods may be explained by different experimental conditions in the various laboratories.

(estimated latent period: 10–13 mo), 7-methylbisdehydroisynolic acid (Fenocylin, estimated latent period: 12–14 mo) (Kirkman, 1959a), and hexestrol (estimated latent period: 6 mo; Liehr *et al.*, 1984; Li *et al.*, 1983), may be used as implants. The role of fluorinated estrogens (Liehr, 1983) will be discussed in Section 2.8.

DES-induced renal carcinogenesis in male Syrian hamsters may be prevented by coadministration of hormones, antiestrogens, or vitamins (Table II). As can be seen in Table II, both enclomiphene and nafoxidine were highly effective in preventing estrogen-induced renal neoplasms in the Syrian hamster (Antonio *et al.*, 1974; Li *et al.*, 1980). The ability to suppress estrogen carcinogenesis in hamsters was correlated (Li *et al.*, 1980) with the ability of

Table II.   Prevention of Estrogen-Induced Carcinogenesis in Male Syrian Hamsters by Coadministration of Hormones, Antiestrogens, or Vitamins

Estrogen administered	Compound coadministered	Tumor incidence (%)	Latent period (months)	Reference
DES	Testosterone	0	11–20	Kirkman, 1959a
	Testosterone propionate	3	10–20	Kirkman, 1959a
	Deoxycortiesterone acetate	9	9–15	Kirkman, 1959a
	Progesterone	0	11–16	Kirkman, 1959a
DES	—	62–75	7–9	Li *et al.*, 1980
	Nafoxidine	0	7–9	
DES	—	100	9.6	Li *et al.*, 1980
	Enclomiphene	0	9.6	Li *et al.*, 1980
	1-(*p*-2-Diethylaminoethoxy-phenyl)-1-phenyl-2-p-methoxy-phenyl ethanol	100	9.6	Li *et al.*, 1980 Li *et al.*, 1980
DES	—	85	6	Liehr and
	L-Ascorbic acid	50	6	Wheeler, 1983
Estradiol	—	70	6.5	
	L-Ascorbic acid	30	6.5	

antiestrogens to inhibit estradiol binding to the estrogen receptor in the un-transformed, DES-primed hamster kidney. Furthermore, an antiestrogen, 1-2-[*p*-(3,4- dihydro-6-methoxy-2-phenyl-1-naphthyl)phenoxy]ethyl pyrrolidine hydrochloride, markedly inhibited the growth of transplanted estrogen-induced renal tumor tissue in hamsters (Bloom *et al.*, 1967).

Kirkman (1959a) studied the influence of various hormones on the estrogen-induced renal carcinoma in male Syrian hamsters (Table II). He found that testosterone propionate, progesterone, and deoxycorticosterone acetate, in addition to administered estrogens, will prevent induction of renal tumors in male hamsters. Cortisone appeared to increase the incidence of primary DES-induced renal tumors. The influence of ACTH on tumor formation also was tested. Four male DES-treated hamsters were injected daily with ACTH. Although the period of survival was short for all four hamsters, one acquired tumors bilaterally.

Interestingly, testosterone propionate, progesterone, and deoxycorticosterone acetate were highly effective in inhibiting tumor induction (Table II) by DES but did not prevent growth of transplanted tumor tissue in DES-treated host animals. Tumor transplant growth stimulated by DES alone was measured to be 126 mg of tumor tissue per day. Coadministration of progesterone or deoxycorticosterone acetate slowed DES-stimulated tumor transplant growth to 16 mg/day or 68 mg/day, respectively. Coadministration of androgen appeared to stimulate transplant growth (287 mg/day).

Renal traumatization also facilitated the development of kidney neoplasms (Kirkman, 1959a). Needle punctures of one kidney followed by administration of DES for 200 days led to the appearance of tumors along each scar. Also, Horning (1954) found large, rapidly growing tumors developing from residual pieces of kidney tissue left during unilateral nephrectomies in male hamsters treated subsequently with DES. The role of traumatization in estrogen-induced carcinogenesis is not clear. However, hormones or hormonelike substances induced by wounding may possibly be involved in trauma-facilitated renal tumor induction (an effect of DES treatment) in analogy to the involvement of traumatization in skin tumorigenesis (Marks *et al.*, 1982).

2.4. Sex Differences in the Hamster Kidney Tumor

The inhibiting role of other hormones such as progesterone in tumor initiation, mentioned previously, suggests a sex difference with regard to estrogen induction of hamster kidney tumors. Indeed, the early explorations (Kirkman, 1957, 1959a; Kirkman and Bacon, 1950, 1952a,b; Horning, 1954, 1956a,b; Horning and Whittick, 1957; Bloom *et al.*, 1963a,b) of the DES- or estradiol-induced hamster kidney neoplasm were all carried out using male animals. Only occasionally were renal tumors reported to have occurred after gonadectomy (Kirkman, 1951), in pregnant females and in females treated with estrogen before the first ovulation (Kirkman and Bacon, 1952b). The striking sex differences in the incidence of estrogen-induced renal tumors was detailed by Kirkman (Kirkman and Wurster, 1957; Kirkman, 1959b). He found that inbred mature female hamsters did not develop renal carcinomas as a result of estrogen treatment. Kirkman (1959b) suspected that normal progesterone secretion in females was responsible for the failure of these animals to acquire tumors. Indeed coadministration of progesterone inhibited kidney tumor induction in males. Therefore, Kirkman attempted to induce renal carcinomas in female hamsters by starting estrogen treatment under conditions of little or no progesterone secretion. With all the different methods used, the tumor incidence rates were very high. Renal clear-cell carcinoma in female hamsters was induced by estrogen, when administration of hormones was begun at times of lowest progesterone secretion, i.e., late in day 4 of the cycle, or if treatment was initiated before the animals reached reproductive maturity. Furthermore, tumors were induced in ovariectomized females or in animals whose pituitary glands had been irreversibly "masculinized" by treatment with testosterone propionate at birth.

2.5. Hormone Dependence of Tumor Growth

Renal clear-cell carcinoma was not only induced by estrogen but was also dependent on estrogen for growth. The tumor continued to grow as long as the supply of estrogen was renewed regularly in experimental animals. If the source of estrogen was removed, tumors underwent a marked regression (Kirkman, 1959a). Eventually, the microscopic tumors transformed into blisterlike vesicles filled with a colorless fluid and disappeared by liquefaction. After a regression period of 200 days, stilbestrol was resupplied to the hamsters. Tumor nodules reappeared within 2 mo. or less of this second estrogen treatment and resumed their growth (Kirkman, 1959a).

The estrogen dependence of the renal clear-cell carcinoma was further demonstrated by its failure to grow when transplanted in untreated or gonadectomized hosts (Kirkman, 1959a). In DES-treated hosts, transplants grew through 34 serial passages. After the thirtieth transfer, latent periods were shorter (15–17 days) than in the initial passages. The mean average growth rate in DES-treated hosts was 100 mg/day. Tumor transplants were carried serially as estrogen-dependent subpannicular growth in intact or in gonadectomized hamsters of either sex. Cheek pouch transplants also were made and were viable. However, the success rates with such cheek pouch transplants were not as high as with subpannicular transplants.

2.6. Hamster Kidney Tumor Transplants

Transplants of the estrogen-induced hamster kidney tumor had been used for a number of investigations. Uptake of estrogen by tumor tissue was studied by Steggles and King (1967, 1978). They found that primary renal tumors took up more estradiol than transplanted tumor tissue did and attributed this difference to a change in receptor activity in the transplanted material. Bloom *et al.* (1967) inhibited growth of transplanted tumor tissue *in vivo* by coadministering an antiestrogen. The inhibitory effect of 1-2[*p*(3,4-dihydroxy-6-methoxy-2-phenyl-1-naphthyl)phenoxy]ethyl pyrrolidine hydrochloride was abolished by simultaneous administration of estradiol.

Kirkman (1974) transplanted the estrogen-dependent renal carcinoma into androgen-treated hosts. Estrogen dependency of the neoplasm was usually lost between the twelfth and twentieth serial passages. These ''autonomous'' derivations of hormone-dependent renal tumors retained some degree of hormone responsiveness. They were never as completely autonomous as the renal carcinoma of spontaneous origin.

2.7. Metastasis of Hamster Renal Carcinoma

Kirkman and Bacon (1952a) in one of their early publications on this tumor observed metastases of the primary renal carcinoma. Metastasis occurred in animals with well-developed estrogen-induced kidney tumors. Indeed, 45% of animals treated with estrogen for 250 days or longer showed metastases. Metastases were not found outside the abdominal cavity, although tumor cells had been shown to be capable of metastasis through the blood and lymph streams.

2.8. Mechanistic Studies of Estrogen-Induced Carcinogenesis

The mechanism by which estrogen acts as a carcinogen in Syrian hamster remains unknown. The earliest proposal concerning the carcinogenic action of estrogen in this particular tumor model came from Horning (1954), who suggested that estrogen-metabolizing enzymes in hamster liver may not be as active as those in livers of other rodents. This proposal implicated estradiol, not a reactive metabolic intermediate, in tumorigenesis. Kirkman and Bacon (1952b) studied this possibility by using various estrogens and estradiol metabolites to induce tumors in Syrian hamsters. Their studies initially appeared to implicate an estrogen metabolite as the ultimate carcinogenic agent. However, in a subsequent investigation implantation of estradiol pellets in the spleen for as long as 300 days failed to induce kidney tumors in male Syrian hamsters (Kirkman, 1959a). This experimental result indicated to Kirkman (1959a) that estrogen itself, rather than a reactive metabolite, was responsible for tumor induction.

In a more recent review Jacob Furth (1982) discussed a hypothesis of hormonal carcinogenesis without extrinsic carcinogens. He suggested that many tumors could be induced without extrinsic carcinogenic agents, if specific growth stimulation of cells was sustained for long time periods. Such a condition could be obtained either by removal of a cell's homeostatic growth inhibitor or by sustained application of its physiological growth stimulant. Such types of tumors were specifically enriched with the hyperstimulated cell type. According to Furth, their transportability was, at first, dependent on the inducer. Later, such neoplasms develop into fully hormone-independent tumors. Although Furth elaborated this mechanistic explanation mainly with regard to pituitary and breast tumors, the mechanistic outline contained features applicable to the estrogen-induced renal clear-cell carcinoma of Syrian hamsters.

A different approach to the mechanism of tumor induction by estrogens in Syrian hamsters was chosen by other authors (Metzler and McLachlan, 1978; Liehr et al., 1982; Lin et al., 1982). Since the renal clear-cell carcinoma is both estrogen induced and estrogen dependent, these authors proposed that estrogens may act as complete carcinogens, i.e., as initiators and as promoters. In the laboratories of Metzler and McLachlan (1978) and Liehr (Liehr et al., 1982) the tumor-initiating activity of estrogens was investigated. At the same time Villee (Lin et al.,1982) and Sirbasku (Section 3) studied the mechanism of tumor promotion and growth. Tumor initiation by estrogens in Syrian hamster was proposed (McLachlan et al., 1978; Liehr et al., 1982) to proceed via metabolic activation of the administered estrogen to a carcinogenic metabolite. This active metabolite was thought either (1) to interact with DNA, and as a consequence cause DNA damage, or (2) to interact with other important cellular target macromolecules. In the search for the carcinogenic intermediate, Metzler studied the metabolism of DES in Syrian hamsters (Gottschlich and Metzler, 1980) and in other species *in vivo* (Metzler, 1981). He suggested the formation of a semiquinone or quinone intermediate, which could not be identified, however, owing to its rapid conversion to β-dienestrol.

These results confirmed earlier experiments of cytochrome P-450 or peroxidase-mediated binding of DES to DNA (Liehr *et al.,* 1982; Metzler and McLachlan, 1978) or protein (Metzler and McLachlan, 1978) *in vitro*. In these enzyme-mediated binding experiments DES quinone or semiquinone was postulated as a reactive intermediate.

DES quinone, one of the postulated reactive intermediates of DES, was synthesized and studied by Liehr (Liehr *et al.,* 1983c) (Fig. 3). It was observed to bind to calf thymus DNA at levels significantly greater than those obtained with DES *in vitro*. Lutz (1979) studied this binding of radioactive DES to liver and kidney DNA in rats and hamsters *in vivo*. He found elevated levels of DES binding in hamster liver and kidney when compared to rat. The nature of the binding found *in vivo* or *in vitro* has not been elucidated.

Another mechanistic possibility, discussed by Tsutsui *et al.* (1983, 1984), is the induction of unscheduled DNA synthesis and also of aneuploidy in Syrian hamster embryo cells. Parallel dose-response curves for cell transformation and aneuploidy induction by DES were found when the synchronized cultures were treated during the mitotic phase of the cell cycle. These studies supported a role of gene rearrangement in DES carcinogenesis.

A different approach to studying the mechanism of tumor induction in Syrian hamsters by estrogens was taken by Li and Li (1981) and Liehr (1983). Both groups suspected that a carcinogenic metabolite (a quinone/semiquinone formed from DES or catechol estrogens) was the carcinogenic intermediate. Thus, prevention of tumor formation by estrogens was attempted by blocking the formation of the suspected carcinogenic intermediate. Li and Li (1981) reduced tumor formation by DES in Syrian hamsters by coadministration of inhibitors of cytochrome P-450. Liehr and Wheeler (1983) (Table II) lowered the tumor frequency rate of DES or estradiol in male Syrian hamsters by the continuous

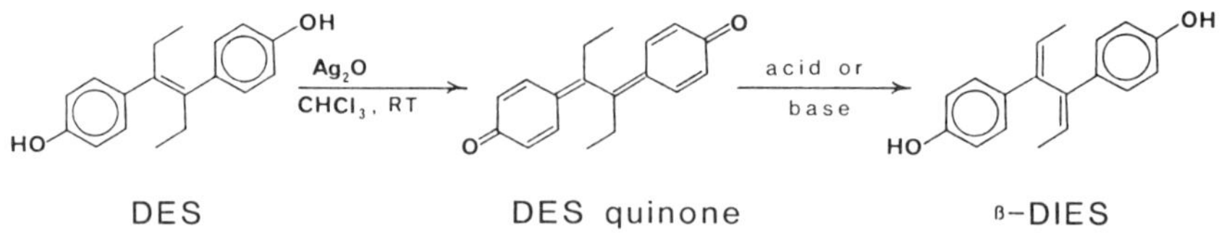

Figure 3. Synthesis and decomposition of DES quinone, a potential intermediate in DES metabolism.

feeding of the intracellular reducing agent L-ascorbic acid (vitamin C). Liehr *et al.* (1983a) had previously shown that L-ascorbic acid indeed reduced synthetic DES quinone to *cis-* and *trans*-DES in the test tube. Furthermore, the tumorigenicity of 2-fluorestradiol was also tested. This potent estrogen has a markedly lowered conversion to catechol (Krey *et al.*, 1983), presumably owing to the substitution of a fluorine in the 2-position. The studies indicated that 2-fluorestradiol was not carcinogenic in male Syrian hamsters (Liehr, 1983). These experiments implicated quinone or semiquinone formation from catechol estrogens or from DES as crucial steps in kidney tumorigenesis by estrogens in Syrian hamster (Liehr *et al.*, 1982).

Villee (Lin *et al.*, 1982) studied the tumor-promoting role of estrogens in renal clear-cell carcinoma of Syrian hamster. The growth-promoting property of estrogen in a variety of tumors has been known for some time. Estradiol had been shown to induce ornithine decarboxylase in Syrian hamster kidney, just as do other tumor-promoting agents of the phorbol ester type (Nawata *et al.*, 1981). Villee suspected that pituitary factors induced by estrogen were involved in kidney carcinogenesis as either promoters or cocarcinogens. Castrated hypophysectomized hamsters (and normal males as controls) were supplied with DES implants. After 12 to 15 mo none of the hypophysectomized hamsters had developed kidney tumors whereas almost all of the controls did. Hypophysectomized castrated males, implanted with DES pellets, were injected daily with 1 μg each of follicle-stimulating hormone and luteinizing hormone or with 1 μg each of follicle-stimulating hormone, luteinizing hormone, and prolactin. None of these hamsters developed kidney tumors. Furthermore, tumor tissue transplanted under the kidney capsule of hypophysectomized, DES-treated hamsters did not grow as it did in normal estrogen-treated males. These studies suggested that some pituitary factor other than prolactin, luteinizing hormone, or follicle-stimulating hormone played a necessary part in the development of renal clear-cell carcinoma induced by estrogen in Syrian hamsters.

Sirbasku and Kirkland (1976) studied the role of estrogen-induced growth factors in tumor growth and, in particular, in the growth control of hamster kidney tumors. Much of this work was done using a cell line derived from the primary estrogen-induced renal clear-cell carcinoma. These studies are described in detail in the next section.

3. HORMONE-DEPENDENT KIDNEY TUMOR CELL LINES

The only hormone-dependent kidney tumor cell line developed is the H-301 line (Sirbasku and Kirkland, 1976) derived from the primary estrogen-induced and dependent renal clear-cell carcinoma of male Syrian hamsters. In this section the development and the characteristics of this cell line will be discussed. Its properties will be compared with those of the primary renal clear-cell carcinoma.

3.1. Development of H-301 Tumor Cell Line

A cell line was derived from the estrogen-dependent renal clear-cell carcinoma in male Syrian hamsters by Sirbasku and Kirkland (1976). This cell line was designated the H-301 cell line. Tumor cells were selected from mixed populations in the following way: primary tumors were grown in male DES-treated hamsters over a period of 6 mo. and then were transplanted into DES-treated males. After a 6- to 8-mo. latent period, transplanted tumors were excised and grown in monolayer cultures by the collagenase procedure. Kidney tumor cells were selected by a procedure of alternate culture and animal passage. Cells obtained after repeated cycles of animal passage enrichment had an *in vivo* latent period of tumor growth of 1 mo. (approximately $1-3 \times 10^6$ cells were inoculated per animal in these studies). After several animal passages, epithelial tumor cell colonies became visible in culture, indicating an increased tumor growth rate.

Contamination of the cultures by fibroblastic growth was reduced by alternate passage of the mixed cell populations on tissue culture dishes and Petri dishes. Tumor cells grew well on both types of dishes, but hamster kidney fibroblasts grew only on tissue culture dishes. Tissue culture dishes have a wettable, treated plastic surface which promotes attachment and growth of mammalian cells. By a procedure of alternate growth in tissue culture in Petri dishes a kidney tumor cell population was selected that was free of fibroblastic contamination. These cells have been designated the H-301 line by Sirbasku and Kirkland (1976).

3.2. Morphology and Histology of H-301 Cells

Figure 2B illustrates the histology of estrogen-dependent tumors that had formed in hamsters after their inoculation with H-301 cells. Their histology indicates that these tumors are composed of epithelial (adenocarcinoma) cell types and that they bear some degree of resemblance to the histological pattern of the parent induced tumors. However, it should be noted that the cells from the cultured line do not tend to form the distinct tubule like structures seen with primary induced tumors.

3.3. Estrogen Dependence of H-301 Cell Growth

In view of the demonstrated estrogen dependence of growth of the primary hamster kidney tumor, the growth properties of the newly developed H-301 cell line were studied and compared with those of the original tumor. Inoculation of male or female hamsters, normal or castrated, with H-301 cells did not result in tumor formation (Sirbasku and Kirkland, 1976). Tumors formed, however, upon inoculation of H-301 cells into estradiol- or DES-treated hamsters (Fig. 4). Tumors, which were excised 34 days after inoculation of animals with 1.0×10^6 H-301 cells, each had attained a mass of approximately 4–7 g independent of the sex of the hamsters. The estrogen dependence of H-301 cells *in vivo* was also clearly demonstrated by the change that was observed in tumor mass when estrogen was removed from the animals. Tumors were grown in estrogen-treated hamsters inoculated with H-301 cells. After 46 days the estrogen pellets were removed from one half of the animal population. Within 10 days after removal of the estrogen treatment, tumor sizes had shrunk to approximately 10% of the size measured immediately before removal of the estrogen pellets. In the control group, a continued supply of estrogen led to continued growth of tumors over the same time period (Leland *et al.*, 1981).

Other steroid hormones have a unique influence on estrogen-dependent primary kidney tumor induction and on growth of transplants (Kirkman, 1959a). Thus the growth of H-301 tumor cells *in vivo* was investigated as a function of steroid hormone supply (Sirbasku and Kirkland, 1976). After inoculation of

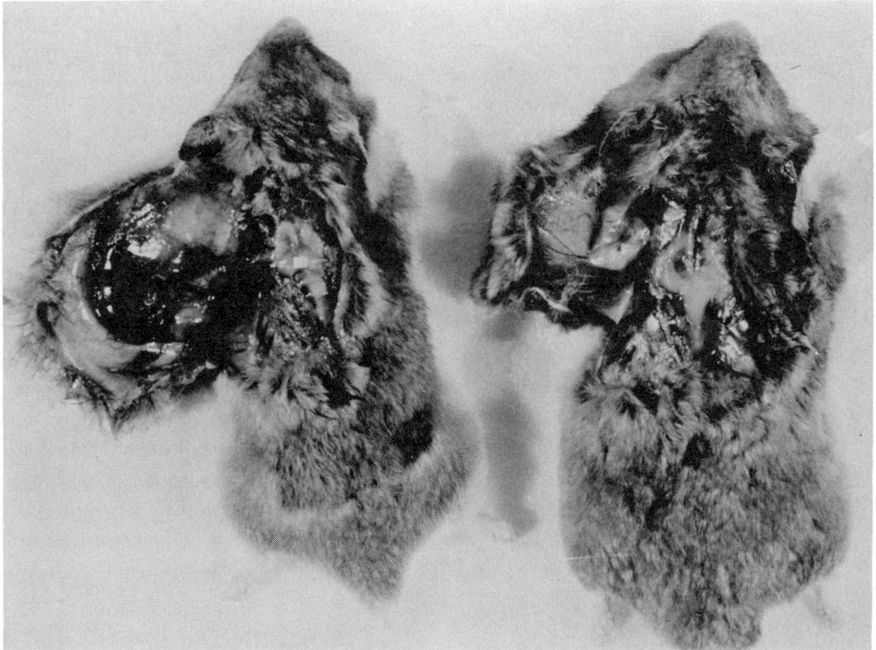

Figure 4. One $\times$ 10⁶ H-301 cells inoculated into an estradiol-treated male Syrian hamster (left) and into an untreated male Syrian hamster (right). No large tumor on the left side of the right animal.

H-301 cells into estradiol-treated male Syrian hamsters, tumor growth was inhibited, but not stopped, by coadministration of progesterone, hydrocortisone, or testosterone. Deoxycorticosterone acetate, which inhibited DES-supported growth of primary tumor explants, did not influence the estradiol-dependent growth of H-301 cells. Each of the nonestrogenic steroids (testosterone, progesterone, hydrocortisone, and deoxycorticosterone acetate) administered alone to male Syrian hamsters (i.e., without concomitant estrogen treatment) completely stopped growth of H-301 cells *in vivo*. Testosterone alone, however, supported limited tumor formation by H-301 cells in female animals.

The influence of steroid hormones on the growth of H-301 cells *in vivo* resembled the effects of steroid on the growth of primary tumor transplants (Kirkman, 1959a). Both tumor types were estrogen dependent. Administration of progesterone in addition to estrogen inhibited growth of both types of tumors. Tumor transplant growth may be stimulated by the addition of testosterone propionate to DES, whereas such combined hormone treatment was inhibitory to proliferation of H-301 cells in males. Deoxycorticosterone acetate plus estrogen did not inhibit H-301 cell growth but inhibited growth of kidney tumor transplants.

In an attempt to understand the mechanism of estrogen-stimulated H-301 tumor cell growth support (i.e., H-301 cell growth support), a series of synthetic estrogens, and also a large number of estrogen metabolites, were tested with regard to H-301 cell growth *in vivo*. With the exception of 2-methoxyestradiol, all hormone metabolites did support growth of H-301 cells to varying degrees (Liehr and Sirbasku, unpublished results). As judged by the tumor masses formed from H-301 cells inoculated into hamsters, the tumor growth support of the various estradiol metabolites appeared to correlate with their ability to bind estrogen receptors (Table III). In contrast to the correlation between estrogen receptor binding and H-301 cell growth support, Robison *et al.* (1983) observed that *o,p'*-DDT, an insecticide with some estrogenic properties, did not support H-301 tumor growth. *o,p'*-DDT was shown to have estrogen properties. In particular, *o,p'*-DDT was demonstrated to compete with estradiol for binding to cytoplasmic receptors from H-301 tumor cells or from uterine cells. The failure of *o,p'*-DDT to maintain H-301 estrogen-dependent tumor growth in Syrian hamsters was explained by its inability to cross the blood–brain barrier and as a consequence by its failure to elicit the synthesis of pituitary factors (Sirbasku and Benson, 1979) that are responsible for tumor growth. Several fluorinated estrogens were also tested (Fig. 5) for their support of H-301 tumor cell growth. The studies were initiated because 2-fluoroestradiol also did not induce primary renal carcinoma in male Syrian hamsters despite its estrogenic properties. All modified estrogens tested were found to cause the growth of tumors in hamsters inoculated with H-301 cells. Tumor masses correlated with the receptor binding of the modified estrogens. Thus we concluded that 2-fluoroestradiol prevented

Table III. Growth of H-301 Cells Supported by Estrogens

Estrogen implanted	No. with tumors / No. of animals	Tumor mass[a]	Reference	Receptor binding[b]	Reference
—	0/10	—	Sirbasku and Kirkland, 1976		
Estradiol	8/8	6.7±2.8	Liehr and Sirbasku, unpublished results	100	Martucci and Fishman, 1976
2-Hydroxyestradiol	4/4	4		23	Martucci and Fishman, 1976
4-Hydroxyestradiol	5/5	10		43	Martucci and Fishman, 1976
2-Methoxyestradiol	0/5	—	Liehr and Sirbasku, unpublished results	0.05	Martucci and Fishman, 1976
Estrone	8/8	82	Sirbasku and Kirkland, 1976	11	Martucci and Fishman, 1976
Estriol	8/8	64		10	Martucci and Fishman, 1976
Diethylstilbestrol	13/13	4.3±2.0	Sirbasku and Kirkland, 1976		
3′,3″,5′,5″-Tetrafluoro-diethylstilbestrol	6/6	91	Liehr et al., 1983b	n.d.	
2-Fluoroestradiol	4/4	21	Liehr and Sirbasku, unpublished results	86	Heiman et al., 1980
4-Fluoroestradiol	4/4	76		128	

[a]Since tumor-growth-promoting properties were measured in different experiments under varying conditions, tumor growth is expressed as a percentage of that of estradiol or diethylstilbestrol which has always been tested as positive control. Only tumor masses obtained by estradiol or diethylstilbestrol stimulation are expressed in grams.
[b]Relative binding affinities to rat uterine cytosol estrogen receptor.

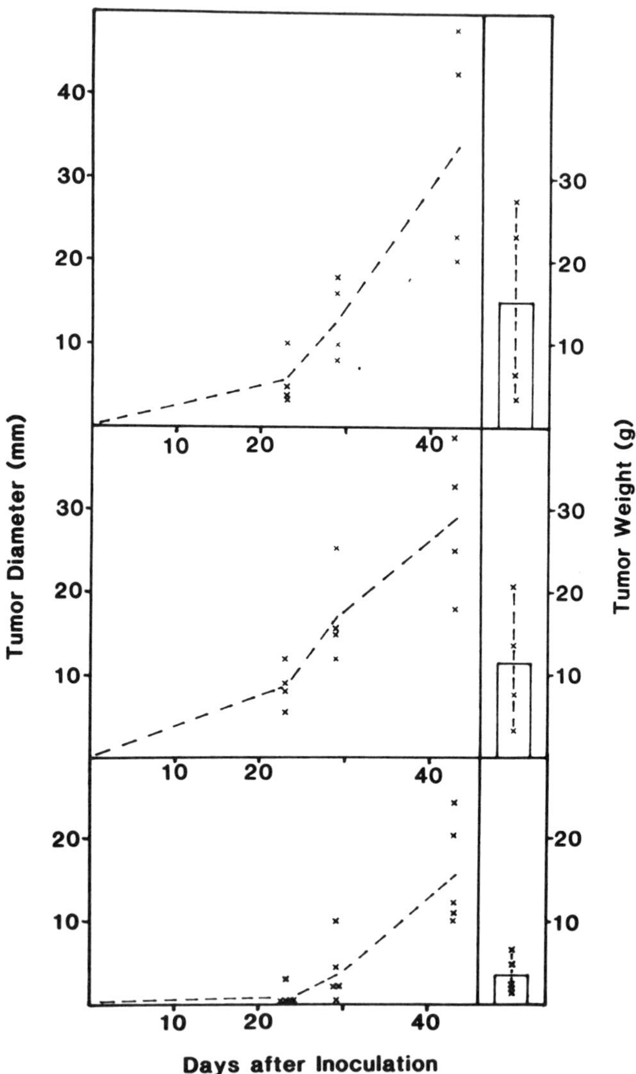

Figure 5. Growth of H-301 tumor cells in male Syrian hamster treated with 4-fluoroestradiol (upper panel), estradiol (middle panel), and 2-fluoroestradiol (lower panel). Tumor sizes were measured 23, 29, and 43 days after inoculation of hamsters with 4.6×10^6 cells/animal. After the last measurement hamsters were killed, the tumors were excised cleanly and weighed (right panels) to correlate weights with diameters.

initiation of primary renal clear-cell carcinoma in Syrian hamsters but did support tumor growth. The ability of 2-fluoroestradiol to support tumor growth was attributed to its estrogenicity (Liehr and Sirbasku, unpublished results).

The effect of vitamin C on estradiol-dependent H-301 cell growth *in vivo* was studied (Fig. 6). L-Ascorbic acid was chosen for study because it inhibits primary tumor formation when administered to estrogen-treated male hamsters

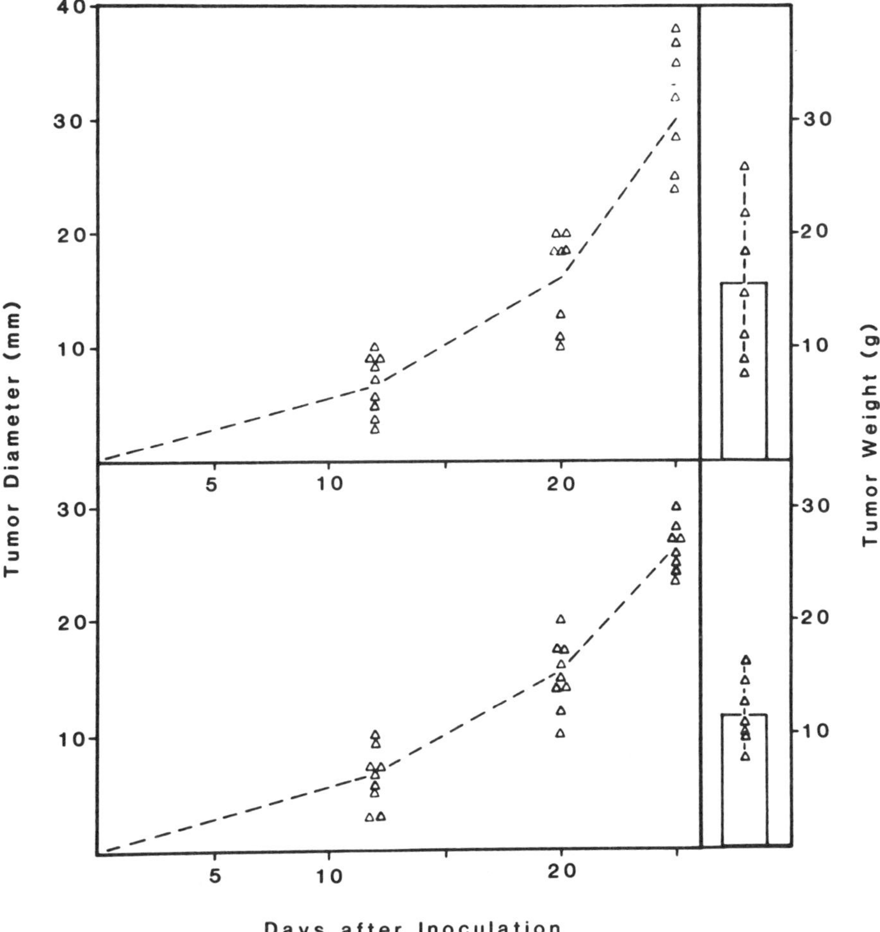

Figure 6. Estradiol-dependent growth of H-301 tumor cells in male Syrian hamsters fed 10% L-ascorbic acid solution *ad libitum* (lower panel) and control (upper panel). Tumor sizes were measured on the twelfth, twentieth, and twenty-fifth day after inoculation of each animal with 2.9×10^6 H-301 cells. After the last measurement hamsters were killed and the tumors were excised cleanly and weighed (right panels).

(Table II; Liehr and Wheeler, 1983). Vitamin C was provided in the drinking water to estrogen-treated hamsters simultaneously with their inoculation with 2.9×10^6 H-301 cells. Tumor size was measured on various days after inoculation. As seen in Fig. 6, vitamin C supplementation to the diet of these animals did not significantly influence the size of tumors. When the animals were killed on day 25 after injection of H-301 cells, the weights of excised tumors from vitamin-C-treated and untreated animals did not differ statistically significantly. Thus the prevention of estrogen carcinogenesis in Syrian hamsters by coadministration of vitamin C may result from inhibition of the tumor-initiating step rather than from inhibition of tumor growth. L-Ascorbic acid was suspected to act by reducing the ultimately carcinogenic metabolite; thus the concentration of this metabolite was lowered within the animals to levels that were less carcinogenic. The role of pituitary hormones on estrogen carcinogenesis in Syrian hamsters has been examined by Lin *et al.* (1982). These investigators treated hypophysectomized male hamsters with estrogen. Primary tumors failed to develop in these animals even after prolonged exposure times. Thus these authors concluded that pituitary factors played a permissive role in estrogen induction of cancer. In order to analyze the role of pituitary hormones further, the effect of hypophysectomy on tumor formation by H-301 cells was studied. The H-301 cells were inoculated into estrogen-treated females (Sirbasku and Kirkland, 1976) as well as male hamsters after hypophysectomy (Fig. 7; Liehr and Sirbasku, unpublished results). In both sexes tumors did develop in estrogen-treated hypophysectomized animals. In both studies, tumor masses were approximately one half of those found in intact animals. Thus, pituitary hormones aided H-301 tumor growth but did not play the strict permissive role found in initiation of primary renal clear-cell carcinoma.

3.4. Mechanistic Studies of Estrogen Dependence of H-301 Growth

Other experiments were carried out on the role of estrogens in estrogen-dependent growth of H-301 cells. First, the uptake of estrogen and metabolites by H-301 cells was measured. Second, H-301 cell proliferation in culture was measured. Extracts of various organs derived from estrogen-treated hamsters were added to the culture media, and their growth-stimulatory effects were measured.

The uptake of ^{3}H-labeled estrogens was measured in male or female hamsters with large tumors derived from H-301 cells (Liehr and Harden, 1981; Liehr and Sirbasku, unpublished results). In order to stimulate tumor growth, the animals were pelleted with DES. The estrogen source was removed from the animals 2 days prior to injection of tritiated estrogens to reduce estrogen levels to near physiological range. ^{3}H-estradiol, ^{3}H-estrone, ^{3}H-estriol, or ^{3}H-DES was injected into the animals. One hour later the animals were sacrificed, and the

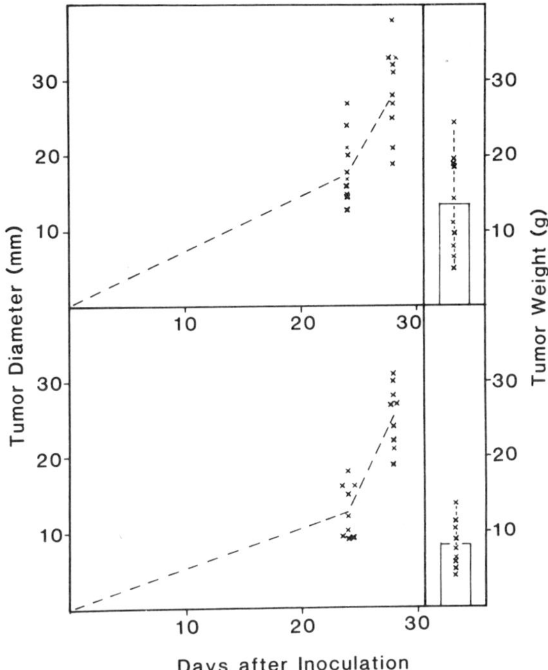

Figure 7. DES-dependent growth of H-301 tumor cells in hypophysectomized (lower panel) and normal (upper panel) male Syrian hamsters. Tumor sizes were measured on the twenty-fourth and twenty-eighth day after inoculation of each animal with 0.83×10^6 H-301 cells. After the last measurement the hamsters were killed, and the tumors were excised cleanly and weighed (right panels).

uptake of estrogens by H-301 cells and by various organs was measured. The highest concentrations of estrogens were found in liver and kidney, but not in tumor tissue. The estrogen concentration in H-301 cells was approximately that of muscle or brain or other nontarget tissue, whereas estrogen levels in liver or kidney were three to eight times that level. These results suggested that estrogen-stimulated H-301 cell tumor growth occurred via an indirect pathway and not via direct induction of H-301 mitogenesis (Sirbasku and Benson, 1979).

These experiments are in agreement with earlier evidence obtained (Sirbasku and Kirkland, 1976) with H-301 cells in culture. The usual growth medium for H-301 cells in culture was standard DME supplemented with 5% (vol/vol) fetal calf serum or 5% horse serum. The cell line showed a preference for a serum supplement from these two species and did not grow significantly when the DME growth medium was supplemented with sheep or calf serum. Despite the strong dependency of H-301 cells on estrogen for growth *in vivo,* addition of estradiol or estriol to the growth medium did not influence cell division. When the serum supplement was first charcoal extracted to remove most of the endogenous steroid hormones, H-301 cell growth studies still showed no influence of exogenous estrogens on growth. In cell culture, the growth rate of H-301 cells was not significantly influenced by addition of exogenous estradiol, of estradiol metabolites, such as estrone or estriol, or of antiestrogens [α-(4-pyrrolidin ethoxyl) phenyl-4-methoxy-2′-nitrostilbene or 1-2-[p(3,4-dihydro-6-methoxy-2-phenyl-1-naphthyl)phenoxyl] ethyl pyrrolidine hydrochloride]. Testosterone likewise had no effect on growth when added to cells in culture at various concentrations. These experiments indicated that estrogens were not directly mitogenic for H-301 cells, but that some indirect mechanism was operative. Similarly, the primary tumor, explanted to monolayer cell culture, failed to grow upon addition of various concentrations of DES (de Kernion and Fraley, 1971). The influence of nonsteroidal growth factors was suspected when heat treatment of fetal calf serum at 90°C for 45 min lowered the rapid rate of cell division found earlier. It was thus concluded that the heat-labile fraction of fetal calf serum contained the major growth-promoting factor (Sirbasku and Kirkland, 1976).

The elevated uptake of estrogen by kidney or liver, but not by H-301 cells *in vivo,* reported previously, suggested that macromolecular growth factors were induced in these or other organs and that these factors were responsible for H-301 mitogenesis (Leland *et al.,* 1981). Tissue extracts of liver, kidney, and other organs obtained from normal male or estrogen-treated male hamsters were prepared and used to supplement DME medium (Sirbasku, 1978). H-301 cell growth approximately doubled in medium supplemented with a liver extract derived from estradiol-treated hamsters as compared to an extract derived from normal hamster. Similar acceleration of growth was found when kidney extracts of estrogen-treated hamsters were used. The growth factor activity found in liver or kidney extracts of estrogen-treated hamsters was destroyed by heating the fluid to 100°C for 15 min. The growth factor activity was also shown to be specific for H-301 cell growth and the doubling of other rodent kidney cell lines.

The experiments described suggested a unique mechanism of estrogen-dependent growth of H-301 cells. An indirect rather than a direct mitogenic action of estrogen was indicated (Sirbasku and Benson, 1979).

4. HORMONE RECEPTOR STATUS OF THE HAMSTER KIDNEY TUMOR

Because of the unique role of estrogens in the induction and of other steroid hormones in the prevention of estrogen-induced renal carcinoma of Syrian hamsters, the steroid hormone receptor characteristics of the hamster kidney were investigated in several laboratories. Changes in receptor status during tumorigenesis and, finally, in tumor tissue itself were investigated for clues concerning mechanistic aspects. These studies will be described in this section.

4.1. Estrogen Receptor Status in Hamster Kidney

The estrogen receptor binding in hamster kidney was studied by Li (Li and Li, 1977; 1980; Li *et al.*, 1979) and also by Anderson *et al.* (1978). In both laboratories the total estrogen receptor content in normal hamster renal cytosols was found to be approximately 6.1–6.5 fmol/mg protein. With estrogen treatment the cytosolic estrogen receptor concentration increased. After 6 mo of treatment with estradiol, Anderson *et al.* (1978) found 9.2 fmol/mg protein and, in animals that had developed tumors, 13.7 fmol/mg protein. Li *et al.* (1979) measured an increase of kidney cytosolic estrogen receptors to 20.7 fmol/mg protein in DES-treated hamsters. The differences in these values have not been discussed but may be due to different estrogen absorption rates or other influences on the estrogen concentration in the animals. Both 8S and 4S estrogen-binding compounds were identified in sucrose-gradient analyses (Fig. 8). The 8S estrogen-binding component increased in the untransformed kidney with continuing hormone treatment.

From measurement of the estrogen receptor binding in cell nuclei isolated from hamster kidney, it was concluded that the cytosolic estrogen receptor complex was translocated to cell nuclei. Measurements were carried out in two different ways: Anderson *et al.* (1978) exposed kidney slices taken from untreated or estradiol-treated hamsters to tritiated estradiol. Cell nuclei were isolated from these kidney slices, and the nuclear binding of ^{3}H-estradiol was measured. For normal male hamsters the nuclear binding found by these authors was 32.8 fmol/mg protein or 3.2 fmol/mg DNA. In tumorous hamster kidneys nuclear estrogen receptor levels were 28.7 fmol/mg protein or 6.7 fmol/mg DNA. A similar large increase in nuclear binding levels as a function of estrogen treatment was found by Sufrin *et al.* (1981). These authors isolated kidney cell nuclei from normal male hamsters, from hamsters treated with DES for up to 6 mo, and from renal adenocarcinoma cells. The isolated nuclei were then exposed to tritiated estradiol in a buffer system. An increase in specific binding of ^{3}H-estradiol by kidney cell nuclei was found as a function of chronic DES treatment. In control animals approximately 0.2 fmol/mg DNA was found, whereas in animals treated with DES for 150 days, the specific nuclear binding was approximately 1.75 fmol/mg DNA. In tumor cell nuclei still higher binding levels were measured: 2.1 fmol/mg DNA.

4.2. Estrogen Receptor Status in Primary Renal Carcinoma

The first report on estrogen binding by Syrian hamster renal carcinoma tissue was published by Steggles and King (1972). They reported the presence of

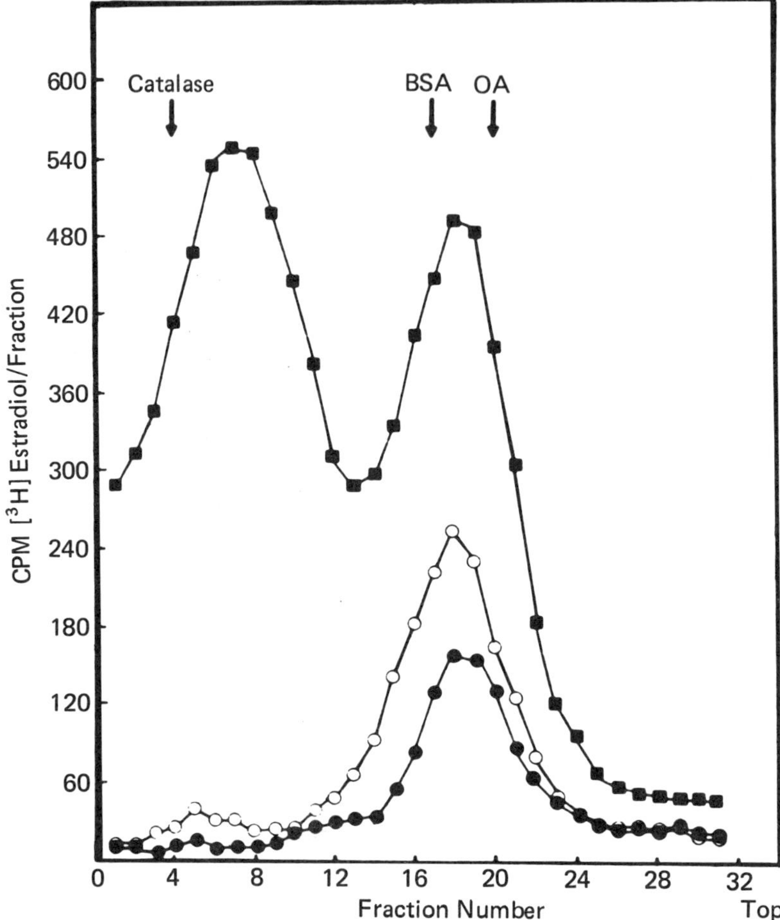

Figure 8. Representative sedimentation patterns of renal cytosol fractions from untreated castrate, DES-treated, and renal tumor-bearing male hamsters after *in vitro* incubation with 1 nM 17β-[³H]estradiol at 0°. Dextran–charcoal treatment (4 hr) was used to remove free steroid from cytosols of untreated control (●), 6.0-mo DES-treated (○), and primary renal tumor (■). Protein concentration was 13–15 mg/ml. Aliquots of 0.4 ml were layered on each gradient. Catalase, bovine serum albumin (BSA), and ovalbumin (OA) served as sedimentation coefficient standards (Li *et al.*, 1976).

cytoplasmic 4S or 8S + 4S and also nuclear 5S estrogen-binding complexes in estrogen-induced kidney tumor tissue. Li *et al.* (1979) expanded these studies and thoroughly investigated the estrogen receptor characteristics of tumor tissue and also of tissue taken from abdominal metastases. They showed that cytosols isolated from primary renal tumor tissue contained an 8S and variable amounts of

4S receptor proteins. Similar findings were obtained from abdominal metastases of the estrogen-dependent kidney tumor. In competitive kidney experiments, it was found that nonestrogenic steroids exhibited minimal competition for the cytosolic estrogen receptor.

The translocation of cytosolic estrogen receptor complexes into nuclei isolated from tumor tissue was studied by Li and Li (1977). Cytosols were prepared from renal tumor tissue, charged with tritiated estradiol, and then incubated with purified hamster kidney or tumor tissue nuclei. Extracts of these nuclei were examined by sucrose density gradient analysis. The nuclear 5S estrogen receptor complex was clearly distinguishable. Translocation of cytosolic estrogen receptor complex into nuclei was thus proved for renal clear-cell carcinoma tissue.

4.3. Receptor Binding of Other Hormones

One of the first physiological changes found upon estrogen treatment of male Syrian hamsters was the induction of a 4S progesterone binding component in renal cytosol (Li *et al.*, 1976, 1977b; Li and Li, 1977; Lin *et al.*, 1980). The amount of specific progesterone binding induced by estrogen in the hamster kidney was in the range of 40–50 fmol/mg protein. This amount represented an increase of approximately 30 times over progesterone binding levels in kidney of untreated hamsters. Measurements of progesterone receptor concentrations in renal carcinoma tissues were approximately 1050 fmol/mg protein, which was approximately 520-fold greater than binding levels in intact male hamsters. This dramatic increase in cytosolic progesterone receptor concentrations was taken as a possible explanation for the inhibitory effect of progesterone on renal tumorigenesis. Cytosolic progesterone receptor complexes were shown (Li and Li, 1977) to undergo translocation into nuclei purified from tumor tissue. In this way it was shown that progesterone receptor complexes translocate into nuclei in the same way estrogen receptor complexes do. Li and Li (1978) also demonstrated that antiestrogens, such as nafoxidine and enclomiphene, and also androgens effectively block the induction by estrogen of specific cytosolic progesterone binding in the hamster kidney.

The presence of a specific dihydrotestosterone 9S receptor complex has also been detected by Li *et al.* (1977a, 1979) in the renal cytosol of untreated, castrated hamsters. The amount of this receptor was increased with dihydrotestosterone treatment. Prolonged estrogen treatment, however, did not augment androgen receptor concentrations. Furthermore, nuclear translocation also occurred with cytosolic androgen receptor complexes and nuclei isolated from renal carcinoma. Salt-extractable 3.2S nuclear androgen receptors have also been identified.

Finally, the hamster kidney tumor was shown to contain cytosolic adrenocorticoid receptors (Li *et al.*, 1979). The concentrations of these were approximately 138 fmol/mg protein measured for dexamethasone and 40 fmol/mg protein for aldosterone. Although no cross competition was apparent for either estrogen or progesterone receptors, progesterone was found to compete for adrenocorticoid and also for androgen binding. Cytosolic corticosteroid receptor complexes were shown to translocate into purified tumor tissue nuclei.

The presence of five specific steroid receptors in this tumor tissue makes it a unique model system for the study of the various influences of hormones on this tumor and also for the study of hormonal interactions.

4.4. Hormone Receptors in Transplants

The receptor status of transplanted tumor tissue was studied by Lin *et al.* (1978) and Li and Li (1980) and compared with that of primary renal carcinoma. Li and Li (1980) characterized all five hormone receptors with regard to cross competition, sedimentation coefficients, and affinity constants. The receptors' properties were similar to those identified in primary renal carcinoma. However, the concentrations of both the estrogen and progesterone receptors were elevated with respect to those reported previously for primary tumor tissue (cytosolic estradiol receptor: 321 fmol/mg protein; cytosolic progesterone receptor: 2721 fmol/mg protein). Lin *et al.* (1978) did not find any significant difference in cytosolic progesterone concentration in the primary tumor as compared with the concentration in tumor tissue serially passaged beneath the renal capsule of Syrian hamsters. The hamster kidney tumor was observed to have a slower growth rate after the second serial passage beneath the renal capsule of hamsters. However, the decreased tumor growth rate after the second passage could not be correlated with any alterations of progesterone receptor levels in tumor tissue.

4.5. Estrogen Receptor Status of H-301 Cells

The H-301 cells growing in culture have been assayed for the presence of specific estradiol receptors by the usual methods used with the tumor tissue cited previously. By assays with ^{3}H-labeled estradiol, saturable binding was observed with these cells in culture (Fig. 9) at 10–20 nM estradiol. By Scatchard plot analysis of these data (Fig. 10) these sites were calculated at 11,000/cell with an apparent Kd for estradiol of 1.0 nM. These sites were similar in number and binding content to those of the primary tumors (D. Danielpour and D. A. Sirbasku, unpublished results).

5. CONCLUSION

The description of the properties of both primary estrogen-induced renal clear-cell carcinomas and cell lines derived from the tumor suggest that many

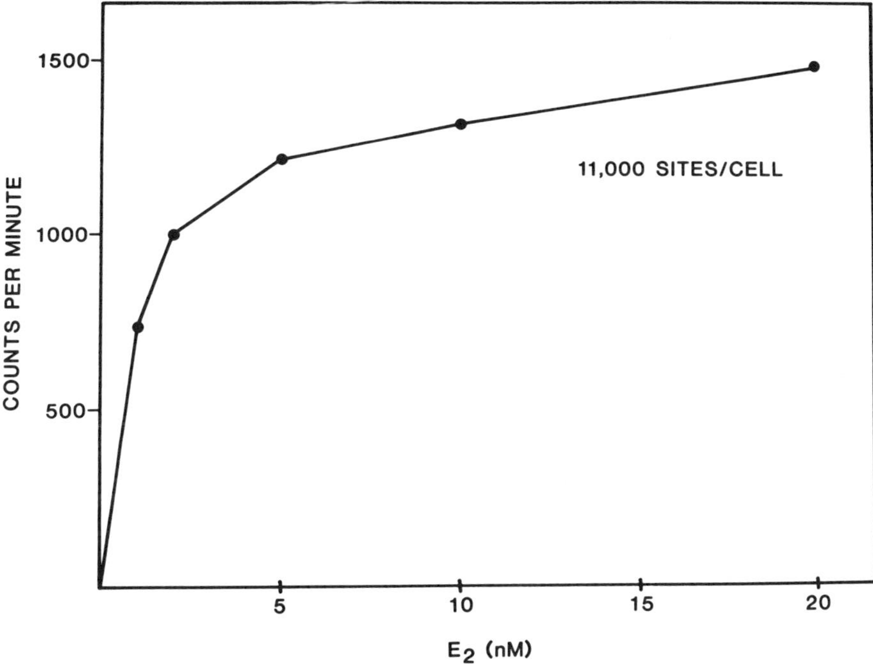

Figure 9. H-301 cells growing in culture were assayed for estrogen-specific binding using tritiated estradiol and parallel incubation with 100-fold molar excess of unlabeled estradiol. Specific binding was the difference between these two values.

areas are yet to be explained. At present the exact mechanism by which estrogens induce cancers is unknown. The system described here is unique in that a single molecular species is both initiator and promoter of tumor growth. Hence, a more detailed analysis of these two processes will prove to be a major advance in understanding the role of estrogen in carcinogenesis. Clearly, the continued study of tumor initiation must proceed with *in vivo* studies. However, the roles of estrogen as transforming agents of cells can now be studied in culture, and hence a combined *in vivo/in vitro* approach will be most productive. The process of estrogen-induced transformation of Syrian hamster kidney cells in culture can now be approached with normal kidney cell cultures described in this volume. Use of such an *in vitro* carcinogenesis assay system will eliminate many of the difficulties of identifying the active agents or metabolites and will provide a highly controlled environment for study of macromolecular changes occurring in carcinogenesis. The other study that remains to be completed is characterization of the factors and hormones other than estrogens that have direct effect in promotion of kidney tumor cell growth. These are now in the process of being studied in our laboratory, and new insights should be forthcoming soon.

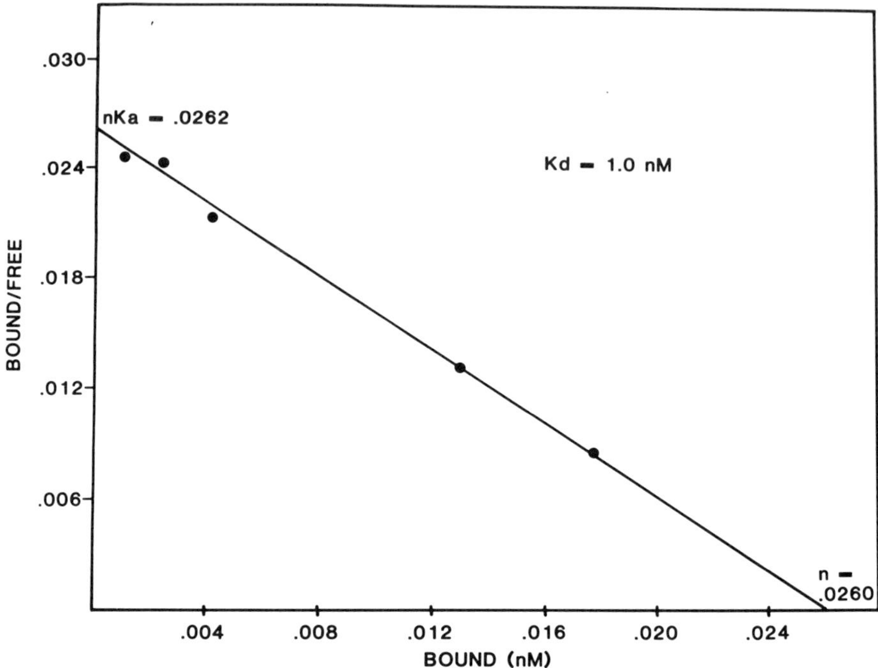

Figure 10. Scatchard plot analysis of the data presented in Fig. 9.

REFERENCES

Anderson, N. S., David, Y., and Fanestil, D. D., 1978, Estrogen receptor in hamster kidney during estrogen-induced renal tumorigenesis, *J. Steroid Biochem.* **10:**123–128.

Antonio, P., Gabaldon, M., Lacomba, T., and Juan, A., 1974, Effect of the antiestrogen nafoxidine on the occurrence of estrogen-dependent renal tumors in hamster, *Horm. Metab. Res.* **6:**522–524.

Bloom, H. J. G., and Wallace, D. M., 1964, Hormones and the kidney—possible therapeutic role of testosterone in a patient with regression of metastases from renal adenocarcinoma, *Brit. Med. J.* **2:**476–480.

Bloom, H. J. G., Baker, W. H., Dukes, C. E., and Ritchley, B. C. V., 1963a, Hormone-dependent tumors of the kidney. II. Effect of endocrine ablation procedures on the transplanted estrogen-induced renal tumour of the Syrian hamster, *Brit. J. Cancer* **17:**646–656.

Bloom, H. J. G., Dukes, C. E., and Ritchley, B. C. V., 1963b, Hormone-dependent tumors of the kidney. I. The oestrogen-induced renal tumour of the Syrian hamster—hormone treatment and possible relationship to carcinoma of the kidney in man, *Brit. J. Cancer* **17:**611–645.

Bloom, H. J. G., Roe, F. J. C., and Ritchley, B. C. V., 1967, Sex hormones and renal neoplasia—inhibition of tumor of hamster kidney by an estrogen antagonist, an agent of possible therapeutic value in man, *Cancer* **20:**2118–2124.

Clark, P. B., and Anderson, C. K., 1976, Tumors of the kidney and ureter, in: *Urology* (J. P. Blandy, ed.), Blackwell, Oxford, pp. 391–432.

Crabtree, C. E., 1941, The structure of Bowman's capsule in castrate and testosterone treated male mice as an index of hormonal effects on the renal cortex, *Endocrinology* **29**:197–203.

de Kernion, J. B., and Fraley, E. E., 1971, Growth characteristics of the stilbestrol-induced hamster kidney tumor, *J. Surg. Oncol.* **3**:507–515.

Farber, E., 1982, Sequential events in chemical carcinogenesis, in: *Cancer: A Comprehensive Treatise,* Volume 1: *Etiology: Chemical and Physical Carcinogenesis,* 2nd ed. (F. F. Becker, ed.), Plenum Press, New York, pp. 485–506.

Furth, J., 1982, Hormones as etiological agents in neoplasia, in: *Cancer: A Comprehensive Treatise,* Volume 1: *Etiology: Chemical and Physical Carcinogenesis,* 2nd ed. (F. F. Becker, ed.), Plenum Press, New York, pp. 89–134.

Gottschlich, R., and Metzler, M., 1980, Metabolic fate of diethylstilbestrol in the Syrian golden hamster, a susceptible species for diethylstilbestrol carcinogenicity, *Xenobiotica* **10**:317.

Grabstald, H., 1973, Experimental aspects or renal tumors, *J. Surg. Oncol.* **5**:509–516.

Heiman, D. F., Senderoff, J. Q., Katzenellenbogen, J. A., and Neeley, R. J., 1980, Estrogen receptor based imaging agents. I. Synthesis and receptor binding affinity of some aromatic and D-ring halogenated estrogens, *J. Med. Chem.* **23**:994–1002.

Horning, E. S., 1954, The influence of unilateral nephrectomy in the development of stilbestrol-induced renal tumours in the male hamster, *Brit. J. Cancer* **8**:627–634.

Horning, E. S., 1956a, Observations on hormone-dependent renal tumours in the golden hamster, *Brit. J. Cancer* **10**:678–681.

Horning, E. S., 1956b, Endocrine factors involved in the induction, prevention and transplantation of kidney tumours in the male golden hamster, *Zeitschr. Krebsforsch.* **61**:1–21.

Horning, E. S., and Whittick, J. W., 1957, The histogenesis of stilbestrol-induced renal tumours in the male golden hamster, *Brit. J. Cancer* **8**:451–457.

Kantor, A. F., 1977, Current concepts in the epidemiology and etiology of primary renal cell carcinoma, *J. Urol.* **117**:415–417.

Kirkman, H., 1951, Relation of sex hormones to the induction and control of renal tumors in the golden hamster, *Anal. Rec.* **109**:51.

Kirkman, H., 1957, Steroid tumorigenesis, *Cancer* **10**:757–764.

Kirkman, H., 1959a, Estrogen-induced tumors of the kidney in the Syrian hamster—III. Growth characteristics in the Syrian hamster, *Natl. Cancer Inst. Monogr. No. 1,* U.S. Department of Health, Education, and Welfare, United States Government Printing Office, Washington, D.C., pp. 1–57.

Kirkman, H., 1959b, Estrogen-induced tumors of the kidney in the Syrian hamster—IV. Incidence in female Syrian hamsters, *Natl. Cancer Inst. Monogr. No. 1,* U.S. Department of Health, Education, and Welfare, United States Government Printing Office, Washington, D.C., pp. 59–91.

Kirkman, H., 1974, Autonomous derivatives of estrogen-induced renal carcinomas and spontaneous renal tumors in the Syrian hamster, *Cancer Res.* **34**:2728–2744.

Kirkman, H., and Bacon, R. L., 1950, Malignant renal tumors in male hamsters (Cricetus auratus) treated with estrogen, *Cancer Res.* **10**:122–124.

Kirkman, H., and Bacon, R. L., 1952a, Estrogen-induced tumors of the kidney I. Incidence of renal tumors in intact and gonadectomized male golden hamsters treated with diethylstilbestrol, *J. Natl. Cancer Inst.* **13**:745–755.

Kirkman, H., and Bacon, R. L., 1952b, Estrogen-induced tumors of the kidney II. Effect of dose, administration, type of estrogen, and age on the induction of renal tumors in intact male golden hamsters, *J. Natl. Cancer Inst.* **13**:757–771.

Kirkman, H., and Robbins, M., 1959, Estrogen-induced tumors of the kidney in the Syrian hamster—V. Histology and histogenesis in the Syrian hamster, *Natl. Cancer Inst. Monogr. No. 1,*

U.S. Department of Health, Education, and Welfare, United States Government Printing Office, Washington, D.C., pp. 93–139.

Kirkman, H., and Wurster, D. H., 1957, Estrogen-induced kidney tumors in female Syrian hamsters, *Proc. Am. Assoc. Cancer Res.* **2**:221.

Krey, L. C., MacLusky, N. J., Pfeiffer, D. G., Parsons, B., Merriam, G. R., and Naftolin, F., 1983, Role of catechol estrogens in estrogen-induced lordosis behavior in the female rat, in: *Catechol Estrogens* (G. R. Merriam and M. R. Lipsett, eds.), Raven Press, New York, pp. 249–263.

Lee, K. Y., Toth, B., and Shubik, P., 1963, Carcinogenic response of the Syrian golden hamster treated at birth with 7,12-dimethylbenz(a) anthracene, *Proc. Soc. Exp. Biol. Med.* **114**:579–582.

Leland, F. E., Iio, M., and Sirbasku, D. A., 1981, Hormone-dependent cell lines, in: *Functionally Differentiated Cell Lines* (G. H. Sato, ed.), Liss, New York, pp. 1–46.

Li, J. J., Cuthbertson, T. L., and Li, S. A., 1977, Specific androgen binding in the kidney and estrogen-dependent renal carcinoma of the Syrian hamster, *Endocrinol.* **101**:1006–1015.

Li, J. J., Cuthbertson, T. L., and Li, S. A., 1980, Inhibition of estrogen tumorigenesis in the Syrian golden hamster kidney by antiestrogens, *J. Natl. Cancer Inst.* **64**:795–800.

Li, J. J., and Li, S. A., 1977, Translocation of specific steroid hormone receptors into purified nuclei in Syrian hamster tissues and estrogen-dependent renal tumor, in: *Research on Steroids,* Volume 7 (A. Vermeulen, P. Jungblut, A. Klopper, and I. Sciarra, eds.), North Holland, Amsterdam-Oxford, and American Elsevier, New York, pp. 325–344.

Li, J. J., and Li, S. A., 1980, High yield of primary serially transplanted hamster renal carcinomas: Steroid receptor and morphologic characteristics, *Eur. J. Cancer* **16**:1119–1125.

Li, J. J., and Li, S. A., 1981, Inhibition of estrogen-induced renal tumororigenesis in the Syrian golden hamster by BHA, β-, and α-naphthoflavone, *Proc. Am. Assoc. Cancer Res.* **22**:11.

Li, J. J., Li, S. A., and Cuthbertson, T. L., 1979, Nuclear retention of all steroid hormone receptor classes in the hamster renal carcinoma, *Cancer Res.* **39**:2647–2651.

Li, J. J., Li, S. A., Klicka, J. K., Parsons, J. A., and Lam, L. K. T., 1983, Relative carcinogenic activity of various synthetic and natural estrogens in the Syrian hamster kidney, *Cancer Res.* **43**:5200–5204.

Li, J. J., Talley, D. J., Li, S. A., and Villee, C. A., 1976, Receptor characteristics of specific estrogen binding in the renal adenocarcinoma of the golden hamster, *Cancer Res.* **36**:1127–1132.

Li, S. A., and Li, J. J., 1978, Estrogen-induced progesterone receptor in the Syrian hamster kidney. I. Modulation by antiestrogens and androgens, *Endocrinology* **103**:2119–2128,

Li, S. A., Li, J. J., and Villee, C. A., 1977, Significance of the progesterone receptor in the estrogen-induced and -dependent renal tumor of the Syrian golden hamster, *Ann. N.Y. Acad. Sci.* **186**:369–383.

Liehr, J. G., 1983, 2-Fluorestradiol: Separation of estrogenicity from carcinogenicity, *Molecular Pharmacol.* **23**:278–281.

Liehr, J. G., and Harden, S. L., 1981, Evidence for an indirect pathway of kidney tumor growth stimulation by estrogens in Syrian hamster, *Proc. Am. Assoc. Cancer Res.* **22**:11.

Liehr, J. G., and Wheeler, W. J., 1983, Inhibition of estrogen-induced renal carcinoma in Syrian hamsters by vitamin C, *Cancer Res.* **43**:4638–4642.

Liehr, J. G., Ballatore, A. M., and DaGue, B. B., 1984, Carcinogenic activity of hexestrol in male Syrian hamsters, submitted.

Liehr, J. G., Wheeler, W. J., and Ballatore, A. M., 1983a, Influence of vitamin C on estrogen-induced renal carcinogenesis in Syrian hamster, in: *Modulation and Mediation of Cancer by Vitamins* (F. L. Meyskens and K. N. Prasad, eds.), S. Karger, Basel, Switzerland, pp. 132–139.

Liehr, J. G., Ballatore, A. M., McLachlan, J. A., and Sirbasku, D. A., 1983b, Mechanism of diethylstilbestrol carcinogenicity as studied with the fluorinated analog E-3′,3″,5′,5″-tetra-fluorodiethylstilbestrol, *Cancer Res.* **43**:2678–2682.

Liehr, J. G., DaGue, B. B., Ballatore, A. M., and Henkin, J., 1983c, Diethylstilbestrol (DES) quinone: A reactive intermediate in DES metabolism, *Biochem. Pharmacol.* **32**:3711–3718.

Liehr, J. G., DaGue, B. B., Ballatore, A. M., and Sirbasku, D. A., 1982, Multiple roles of estrogen in estrogen-dependent renal clear-cell carcinoma of Syrian hamster, in: *Cold Spring Harbor Conferences on Cell Proliferation,* Volume 9: *Growth of Cells in Hormonally Defined Media* (G. H. Sato, A. B. Pardee, and D. A. Sirbasku, eds.), Cold Spring Harbor Laboratory, New York, pp. 445–457.

Lin, Y. C., Loring, J. M., and Villee, C. A., 1982, Permissive role of the pituitary in the induction and growth of estrogen-dependent renal tumors, *Cancer Res.* **42**:1015–1019.

Lin, Y. C., Talley, D. J., and Villee, C. A., 1978, Progesterone receptor levels in estrogen-induced renal carcinomas after serial passage beneath the renal capsule of Syrian hamsters, *Cancer Res.* **38**:1286–1290.

Lin, Y. C., Talley, D. J., and Villee, C. A., 1980, Dynamics of progesterone binding in nuclei and cytosol of estrogen-induced adenocarcinoma cells in primary culture, *J. Steroid Biochem.* **13**:29–37.

Llombart-Bosch, A., and Beydro, A., 1975, Morphological histochemical and ultrastructural observations of diethylstilbestrol-induced kidney tumors in the Syrian golden hamster, *Eur. J. Cancer* **11**:403–412.

Lutz, W. K., 1979, *In vivo* covalent binding of organic chemicals to DNA as a quantitative indicator in the process of chemical carcinogenesis, *Mutat. Res.* **65**:289–356.

Marks, F., Berry, D. L., Bertsch, S., Furstenberger, G., and Richter, H., 1982, On the relationship between epidermal hyperproliferation and skin tumor promoter, in: *Carcinogenesis,* Volume 7: *Cocarcinogenesis and Biological Effects of Tumor Promoters* (E. Hecker, N. E. Fusenig, W. Kunz, F. Marks, and H. W. Thielmann, eds.), Raven Press, New York, pp. 331–346.

Martucci, C., and Fishman, J., 1976, Uterine estrogen receptor binding of cortical estrogens and of estrol (1,3,5(10)-Estaratriene-3,15α,16α,17β-tetrol), *Steroids* **27**:325–332.

Matthews, V. S., Kirkman, H., and Bacon, R. L., 1947, Kidney damage in the golden hamster following chronic administration of diethylstilbestrol and sesame oil, *Proc. Soc. Exp. Biol. Med.* **66**:195–196.

McGregor, R. F., Putch, J. D., and Ward, D. N., 1960, Estrogen-induced kidney tumors in the golden hamster. I. Biochemical composition during tumorigenesis, *J. Natl. Cancer Inst.* **24**:1057–1066.

McLachlan, J. A., Metzler, M., and Lamb, J. C., 1978, Possible role of peroxidase in the di-ethylstilbestrol-induced lesions of the Syrian hamster kidney, *Life Sciences* **23**:2521–2524.

Metzler, M., 1981, The metabolism of diethylstilbestrol, *CRC Critical Rev. Biochem.* **10**:171–212.

Metzler, M., and McLachlan, J. A., 1978, Oxidative metabolism of diethylstilbestrol and steroidal estrogens as a potential factor in their fetotoxicity, in: *Role of Pharmacokinetics in Prenatal and Perinatal Toxicology* (D. Neubert, H. J. Merker, H. Nau, and J. Langman, eds.), Georg Thieme, Stuttgart, Germany, pp. 157–163.

Murphy, G. P., Johnston, G. S., and Melby, E. C., 1967, Comparative aspects of experimentally induced and spontaneously observed renal tumors, *J. Urol.* **97**:965–972.

Nawata, H., Yamamoto, R. S., and Poirier, L. A., 1981, Elevated levels of ornithine decarboxylase and polyamines in the kidneys of estradiol-treated male hamsters, *Carcinogenesis* **2**:1207–1211.

Robison, A. K., Sirbasku, D. A., and Stancel, G. M., 1983, op′-DDT and estrogen-dependent tumor growth, *Toxicologist* **3**:143.

Sempronj, A., and Morelli, E., 1939, Carcinoma of the kidney in rats treated with β-anthra-quinoline, *Am. J. Cancer* **35**:534.

Sirbasku, D. A., 1978, Estrogen induction of growth factors specific for hormone-responsive mammary, pituitary, and kidney tumor cells, *Proc. Natl. Acad. Sci. USA* **75**:3786–3790.

Sirbasku, D. A., and Benson, R. H., 1979, Estrogen-inducible growth factors that may act as mediators (estromedins) of estrogen-promoted tumor cell growth, in: *Cold Spring Harbor Conferences on Cell Proliferation,* Volume 6: *Hormones and Cell Culture* (G. Sato and R. Ross, eds.), Cold Spring Harbor Laboratory, New York, pp. 477–497.

Sirbasku, D. A., and Kirkland, W. L., 1976, Control of cell growth. IV. Growth properties of a new cell line established from an estrogen-dependent kidney tumor of the Syrian hamster, *Endocrinology* **98**:1260–1272.

Steggles, A. W., and King, R. J. B., 1967, The uptake and localization of [6,7-^{3}H] oestradiol-17β in tissues of male Syrian hamsters, *J. Endocrinol.* **38**:25–32.

Steggles, A. W., and King, R. J. B., 1968, The uptake of [6,7-^{3}H] oestradiol by oestrogen dependent and independent hamster kidney tumors, *Eur. J. Cancer* **4**:395–401.

Steggles, A. W., and King, R. J. B., 1972, Oestrogen receptors in hamster tumors, *Eur. J. Cancer* **8**:323–334.

Sufrin, G., Alvarez, J., and Swaneck, G. E., 1981, Role of nuclear estrogen-binding in experimental renal carcinoma, *Surg. Forum* **32**:589–591.

Tomatis, L., Magee, P. N., and Shubilk, P., 1964, Induction of liver tumors in the Syrian hamster by feeding dimethylnitrosamine, *J. Natl. Cancer Inst.* **46**:81–93.

Toth, B., 1971, Tumor indication by repeated injections of urethan in newborn and adult hamsters. II. Age Influence. *J. Natl. Cancer Inst.* **46**:81–93.

Toth, B., Tomatis, L., and Shubilk, P., 1961, Multipotential carcinogenesis with urethran in the Syrian golden hamster, *Cancer Res.* **21**:1537–1541.

Tsutsui, T., Maizumi, H., McLachlan, J. A., and Barrett, J. C., 1983, Aneuploidy induction and cell transformation by diethylstilbestrol: A possible chromosomal mechanism in carcinogenesis, *Cancer Res.* **43**:3814–3821.

Tsutsui, T., Degen, G. H., Schiffmann, D., Wong, A., Maizumi, H., McLachlan, J. A., and Barrett, J. C., 1984, Dependence on exogenous metabolic activation for induction and unscheduled DNA synthesis in Syrian hamster embryo cells by diethylstilbestrol and related compounds, *Cancer Res.* **44**:184–189.

Vasquez–Lopez, E., 1944, The reaction of the pituitary gland and related hypothalamic centers in the hamster to prolonged treatment with oestrogens, *J. Pathol. Bacteriol.* **56**:1–13.

Ward, D. N., Putch, J. D., McGregor, R. F., and Chang, J. P., 1964, Estrogen-induced kidney tumors in the golden hamster. II. Diethylstilbestrol absorption and distribution in tissues, *Cancer Res.* **24**:319–326.

Williams, P. D., and Murphy, G. P., 1971, Dose related effects of cycasin induced renal and hepatic tumors, *Res. Commun. Chem. Pathol. Pharmacol.* **2(445)**:627–632.

12

Hormonally Defined, Serum-Free Media for Epithelial Cells in Culture

DAVID BARNES

1. INTRODUCTION

Through most of the 75-year history of *in vitro* culture of mammalian cells and tissues, investigators have found it necessary for maintenance and growth of the cells to add to basal nutrient culture media some type of undefined and often inconsistent biological fluid such as lymph, milk, plasma, or serum, amniotic or spinal fluid (Harrison, 1907; Carrel, 1913; Temin *et al.,* 1972; Brooks, 1975). Most of the studies in the last 50 years have used serum-supplemented media for the growth of both fibroblastic and epithelial cell types in culture. Although the disadvantages of the use of undefined supplements such as serum in culture media were recognized from the beginning, progress toward replacing serum with defined components of the media was slow, and the most successful early attempts at cell culture in defined media were those in which the approach was to "adapt" cells to a predetermined serum-free medium formulation, a procedure that in most cases probably selected for a small subpopulation within the cells used to initiate the cultures (Higuchi, 1973).

Often such a subpopulation may prove to be of limited value for studies of the cell biology of the cultures, since the cells are likely to be variant from the original population in that they may require less or none of some of the growth-stimulatory components supplied by serum or may have acquired the ability to synthesize some of the protein or lipid components of serum or other molecules

DAVID BARNES • Department of Biological Sciences, University of Pittsburgh, Pittsburgh, Pennsylvania 15260 that can replace these components. In this respect, the cells may not accurately reflect the properties of the original cells from which they were derived. Furthermore, such studies almost entirely have been carried out with established cell lines derived by culture in serum-containing media, and these lines may be themselves already changed in many respects from the cells of the tissue of origin. Almost no success has been reported at establishing primary cultures directly from sources *in vivo* by "adaptation" to serum-free media.

In the last 10–15 years, several approaches have proved successful at developing serum-free culture media that allow the maintenance and growth of the entire cell population to be studied, including serum-free primary cell culture. The major contributions toward the development of these media have come from the laboratories of Dr. Richard Ham (Ham and McKeehan, 1979; Ham, 1981, 1982) and Dr. Gordon Sato (Sato, 1975; Hayashi and Sato, 1976; Reid and Sato, 1978; Bottenstein *et al.*, 1979a; Barnes and Sato, 1980a,b; Rizzino *et al.*, 1979).

The approach of the laboratory of Ham has been primarily to develop improved nutritional formulations of the basal culture media to allow the clonal growth of a number of different cell types in media containing little or no serum as a supplement. This approach has been quite successful because many of the protein components of serum that are active at promoting growth of cells in culture may do so by mechanisms related to availability of nutrients to the cells. Such serum components might include transferrin, ceruloplasmin, lipoproteins, and albumin with associated nutritional components. In addition, complicated interactions are evident between the availability of some nutritional components and some hormonal components supplied by serum (McKeehan, 1982; Mather *et al.*, 1982a), suggesting that the requirement of some hormones can be replaced by a properly balanced nutritional environment. Furthermore, this approach also has been successful at reducing or replacing serum in culture medium partially because serum itself contributes important nutritional components to the medium under conventional cell culture conditions (Ham and McKeehan, 1978, 1979; Ham, 1981, 1982).

The approach pioneered by the laboratory of Sato involves replacing the components of serum that are active in supporting the maintenance and growth of cells in culture with purified components capable of replacing the functions carried out by serum. These factors include hormones, supplementary nutrients, binding proteins that modulate the action of hormones and nutrients, attachment factors, and extracellular matrix components (Barnes and Sato, 1980a,b) and, in a few cases, extracellular enzymes (Cherington *et al.*, 1979; Simonian *et al.*, 1982). Some of these factors are themselves purified serum components, such as the iron-binding protein transferrin, albumin used as a lipid-binding protein, the copper-binding protein ceruloplasmin, the substratum components fibronectin and serum-spreading factor, and the growth-stimulatory enzyme thrombin. Also, many of the hormones used in this approach to serum-free cell culture, such as insulin, glucocorticoids, sex steroids, triiodothyronine (T_3), glucagon, and some growth factors, are found in serum at concentrations that may affect cell growth in serum-containing media. Some of the supplements in these serum-free media, however, are probably not found in serum at concentrations high enough to expect that they might be relevant to effects of the serum supplement in culture media. These include the hormonelike hypothalamic-releasing factors and some substratum molecules such as laminin or collagen.

Because these approaches are not limited in regard to the components that might be added to serum-free culture medium in order to allow maintenance and growth of cells in culture, they allow the development not only of media that support the growth of cell types that are conventionally grown in serum-containing media, but also of media that will support the growth and maintenance in a differentiated state of cell types in primary or multipassage culture that cannot be maintained by serum supplementation of a basal nutrient mixture (Mather and Sato, 1979a; Ambesi–Impiombato *et al.*, 1980; Orly *et al.*, 1980). The approach of Sato has been applied more extensively to epithelial cell culture, and it is culture systems developed by that approach that will be reviewed in this chapter, although some studies (Barnes and Sato, 1980a,b; Ham, 1982; McKeehan, 1982) have combined both approaches, such as using the hormonal supplements identified in the approach of Sato and co-workers with basal nutritional media developed by Ham *et al.* This combination of the advantages of both methods is likely to yield most interesting results in the future. In fact, the basal nutrient media used most commonly in the studies to be reviewed here is a one-to-one mixture (Mather and Sato, 1979a,b) of Dulbecco–modified Eagle's medium (DME) and F12, a medium formulation developed in Ham's laboratory.

The use of serum-free culture media is advantageous from the point of view of experimental design for a number of reasons, primarily related to removal of undefined serum components that may act on the cells of the culture in ways that make experimental design and interpretation of data for some kinds of experiments difficult or impossible (Barnes and Sato, 1980a,b). The problems encountered in conventional animal cell culture methods and the advantages of serum-free cell culture for experimental design of some types of experiments are discussed in the concluding section of this chapter. An additional advantage of the serum-free culture approach is its use in the development of culture systems for cell types that previously could not be grown *in vitro*. These include culture of differentiated rat thyroid, human mammary, and rat ovarian epithelia discussed in the following sections. Although the techniques involved in the work reviewed in this chapter still are fairly new, the field already has expanded to the point that a comprehensive treatment, even of only the area of epithelial cells in serum-free media, is virtually impossible in a single chapter, and only a limited number of published investigations in the area will be reviewed here. More comprehensive descriptions of the concepts and techniques of serum-free, hormone-supplemented cell culture are available (Sato *et al.*, 1982; Barnes *et al.*, 1984; Mather, 1984).

2. EPITHELIAL CELLS OF THE ENDOCRINE SYSTEM

The GH$_3$ pituitary carcinoma cell line (Yasumura *et al.*, 1966) was derived from a transplantable rat pituitary tumor established in Jacob Furth's laboratory (Takemoto *et al.*, 1962) and expresses a number of interesting differentiated functions, including secretion of prolactin and growth hormone. This line was the first for which serum-free, hormone-supplemented media allowing the continuous growth of the cells was developed by the methods subsequently applied extensively by Sato and co-workers to other epitheloid cell types (Hayashi and Sato, 1976). The basal nutrient medium used is Ham's F12, to which are added the following supplements: T$_3$, thyrotropin-releasing hormone (TRH), transferrin, parathyroid hormone (PTH), insulin, fibroblast growth factor (FGF), and somatomedin C. All the supplements are added at a final concentration in the culture medium of a few nanograms per milliliter or less except insulin and transferrin, which are added at 5 μg/ml. This concentration of transferrin, a protein present in serum at about 3 mg/ml, is not unreasonable, but the optimum concentration of insulin for these cells and most other cell types in serum-free media is superphysiological. The requirement of cells in serum-free media for high concentrations of insulin for optimal growth is not well understood, but probably relates at least partly to rapid inactivation of the hormone in serum-free culture media (Hayashi *et al.*, 1978). The medium developed for GH$_3$, like several of the other serum-free media developed for various differentiated established cell lines, including those of tumor origin, also has been used to support short-term primary cultures of normal rat pituitary.

Another epithelial cell type for which studies led to some of the early advances in the development of serum-free cell culture are Sertoli cells from the testes (Mather and Sato, 1979a; Mather, 1980; Mather and Haour, 1981; Mather *et al.*, 1982b). These cells require for optimal growth in serum-free culture the following supplements: insulin, transferrin, epidermal growth factor (EGF), follicle-stimulating hormone (FSH), somatomedin C, and growth hormone. All these components are effective at concentrations less than 1 μg/ml except for insulin and transferrin, which are effective in the range of 5 μg/ml. The approach also has been extended to primary cultures of Sertoli cells (Mather and Sato, 1979a; Mather and Haour, 1981; Mather *et al.*, 1982b), and similar steps have been taken toward the development of serum-free media for primary culture of Leydig cells from mouse and pig testes (Mather *et al.*, 1981, 1982a).

The first line for which the importance of the proper substratum in serum-free culture was demonstrated was the RF1 rat ovarian cell line (Orly and Sato,

1979). This diploid, nontumorigenic line (Bottenstein *et al.*, 1979a) can be grown in a farily simple serum-free formulation, consisting of a one-to-one mixture of F12 and DME supplemented with insulin, transferrin, hydrocortisone, and fibronectin (Orly and Sato, 1979). The observation by Orly and Sato that an isolated attachment factor (fibronectin) was important for growth of these cells eventually was extended to many other cell types of both fibroblastic and epithelial origin (Barnes and Sato, 1980a,b) and led to the use of other attachment factors such as a serum-spreading factor (Barnes *et al.*, 1980, 1982, 1983a) as routine supplements in serum-free media formulations. The medium developed for the established rat cell line is also applicable to primary culture of rat ovarian granulosa cells (Orly *et al.*, 1980), and it has been shown that the serum-free medium allows maintenance of an FSH-responsive differentiated state for these cells in culture under conditions in which the use of conventional, serum-containing media results in loss of differentiated function.

Another example of a situation in which the use of serum-free medium allows the maintenance of functional, differentiated epithelia in primary culture and, in this case, in subsequent transfers is the culture of rat thyroid follicle cells (Ambesi–Impiombato *et al.*, 1980). Although culture of rat thyroid cells in serum-containing media results in cell death, fibroblastic overgrowth, or outgrowth of tumorigenic, heteroploid epithelioid cells that do not exhibit differentiative properties of the cells of the tissue of origin, the culture of rat thyroid cells in a serum-free medium formulation results in maintenance of thyroid epithelia that are diploid, nontumorigenic, and differentiated. These cells in serum-free culture secrete thyroglobulin, concentrate iodide, and show a mitogenic response to thyroid-stimulating hormone (TSH) (Ambesi–Impiombato *et al.*, 1980).

Another epithelial cell type responsive to hormones from the pituitary are cells from the adrenal cortex. In serum-containing medium, it is possible to maintain bovine adrenocortical cells for about 60 cell divisions and to demonstrate that the steroidogenic pathway for these cells is functional and inducible and stimulated by adrenocorticotropin (ACTH) (Gill *et al.*, 1979). Similar properties are maintained by these cells in a serum-free formulation consisting of a mixture of F12 and DME as the basal nutrient medium, further supplemented with additional nutritional components (low-density lipoprotein, ascorbic acid, alpha-tocopherol, and selenium added as sodium selenite or selenous acid), hormones (insulin, FGF), binding proteins (transferrin, bovine serum albumin), and an extracellular enzyme (thrombin) (Simonian *et al.*, 1982).

Serum-free media also have been developed for normal and transformed rodent and human mammary epithelial cells in culture. Work from the laboratory of Kano–Sueoka on the 64-24 rat mammary cell line (Kano–Seuoka and Hsieh, 1973), established from a prolactin and estrogen-dependent transplantable mammary carcinoma, identified ethanolamine and phosphoethanolamine as important nutritional supplements to serum-free medium for this (and subsequently other) cell types (Kano-Sueoka *et al.*, 1979a, 1979b; Kano-Sueoka and Errick, 1981; Murakami *et al.*, 1982). A serum-free medium has been developed for the line, consisting of an F12/DME mixture supplemented with insulin, hydrocortisone, triiodothyronine, estradiol, transferrin, prolactin, and albumin as a source of fatty acids (Kano–Sueoka *et al.*, 1979b; Kano–Sueoka and Errick, 1982). Cholera toxin and epidermal growth factor (EGF) are also stimulatory for these cells. Although no prolactin stimulation of cell growth can be demonstrated for this line in conventional serum-containing culture medium, a mitogenic effect of prolactin on the cells in serum-free medium is seen.

Serum-free culture conditions for normal and tumorigenic human and mouse mammary epithelial cells grown in collagen gels also have been developed (Yang *et al.*, 1980a,b, 1982; Imagawa *et al.*, 1982). As in most cases, the basal nutrient medium chosen is a F12/DME mixture, and supplements include insulin and transferrin. Also included for rodent cells are cholera toxin, EGF, and albumin, again probably acting as a source of fatty acids. Putrescine also was found to be stimulatory for growth of the cells. Rodent mammary cells grown under these conditions synthesize casein and produce epithelial mammary gland outgrowths when transplanted into fat pads of syngeneic hosts. Some of the tumorigenic cultures no longer require EGF for optimal growth, a phenomenon reported previously for transformed fibroblastic cells (Cherington *et al.*, 1979).

A serum-free medium for normal rat mammary epithelia in monolayer culture has been developed, in this case using a somewhat different basal nutrient medium ("improved Eagle's medium") (Kidwell *et al.*, 1982, 1984). Stimulatory supplements included are insulin, transferrin, dexamethasone, EGF, ascorbic acid, and fetuin. The active component in fetuin for these cells appears to be a molecule isolated by these investigators and termed "embryonin" because it also stimulates the growth of embryonal carcinoma cells in serum-free culture (Kidwell *et al.*, 1984; Salomon *et al.*, 1984). The factor is similar in many respects to alpha-2-macroglobulin. All the stimulatory supplements in the serum-free medium formulation except insulin stimulate the accumulation of extracellular type IV (basement membrane) collagen by these cells in culture, although the mechanism of this affect is different for some of the factors compared to others (Salomon *et al.*, 1981; Wicha *et al.*, 1979, 1982). For instance, EGF stimulates collagen synthesis by the cells, whereas dexamethasone inhibits the production by the cells of an active collagenase, thus preventing breakdown of the collagenous matrix (Salomon *et al.*, 1981). Other factors found to be stimulatory for growth of rat mammary epithelial cells in serum-free medium include multiplication-stimulating activity (somatomedin C), dibutryl cyclic AMP, and prostaglandin E_1. All these supplements also stimulate collagen production in the cultures. Trophic hormones affecting mammary epithelium *in vivo* such as prolactin and estradiol are without effect in this system (Kidwell *et al.*, 1984).

Human mammary tumor lines have been grown in serum-free medium, particularly the ZR-75-1 line and the MCF-7 line (Allegra and Lipman, 1978; Barnes and Sato, 1979; Barnes, 1980; Barnes *et al.*, 1981; Lippman, 1983). Both lines were established initially from effusions of mammary carcinoma metastases. The ZR-75-1 line in serum-free medium is responsive to estrogen (Allegra and Lippman, 1978; Lippman, 1984). The basal medium for the serum-free formulation developed for ZR-75-1 is "improved minimal essential medium" supplemented with glutamine. Factors added to this medium are insulin, transferrin, T_3, estradiol, dexamethasone, and, at subculture, FGF and additional nutrients. MCF-7 cells are grown in serum-free medium using F12/DME supplemented with insulin, transferrin, prostaglandin F_2-alpha, fibronectin, EGF, and serum-spreading factor (Barnes and Sato, 1979; Barnes, 1980; Barnes *et al.*, 1981). An estrogen-dependent increase in cell number for these cells in serum-free medium also can be observed, but only if cells are plated under conditions suboptimal for growth (i.e., omission of other stimulatory supplements) (Barnes and Sato, 1979; Barnes, 1980). Cultures of normal human mammary epithelia also have been grown in monolayer in a hormone-supplemented medium containing a low level of serum supplementation and some serum-containing conditioned media, and most recently, in serum-free medium composed of Ham's MCDB 170 supplemented with insulin, hydrocortisone, EGF, ethanolamine, phosphoethanolamine, transferrin, and bovine pituitary extract (Stampfer, 1984; Stampfer *et al.*, 1980; Ham, 1982). Th active component in the pituitary extract has not yet been identified, but the low level required for stimulation *in vitro* (70 μg/ml) suggests that the factor is a hormone active at low levels (Stampfer, 1984).

Normal or benign human prostatic epithelial cells also have been grown in serum-free medium (Chaproniere–Rickenberg and Webber, 1982). The basal nutrient medium used in the studies was RPMI 1640, and the supplements were insulin, transferrin, dexamethasone, and zinc chloride. These cells in serum-free culture are organized into bilayer structures similar to the organization observed *in vivo* in the prostatic epithelia and also form hemicysts or "domes" and desmosomes. Enzymatic markers of prostatic epithelium, such as prostatic acid phosphatase and plaminogen activator, are also present.

3. EPITHELIA OF OTHER ORGANS

Another of the earliest cell types for which serum-free, hormone-supplemented medium was developed is the HeLa human cervical carcinoma cell line (Hutchings and Sato, 1978). The medium devised for these cells calls for Ham's F12 as the basal nutrient formulation, supplemented with a mixture of trace elements also devised by Ham. To this medium is added insulin, transferrin,

EGF, FGF, and hydrocortisone. Critical among these supplements for survival and growth of the cells are hydrocortisone and EGF. Hydrocortisone can be replaced in this system by aldosterone (Hutchings and Sato, 1978). If another recently developed and more complicated basal nutrient medium is used, the MCDB 105 formulation of Ham (Ham and McKeehan, 1979), several of the hormonal supplements to the medium are no longer required for serum-free HeLa cell growth (Wu and Sato, 1978). The MCDB 105 medium was originally developed for the growth in low serum of human fibroblasts at clonal density. Studies of HeLa cell membrane components indicate that some differences exist when HeLa cells grown in serum-free medium are compared to those grown in serum-containing medium (Wu and Sato, 1978). In addition, studies of EGF binding to HeLa cells in serum-free and serum-containing media show that the time course of binding of radiolabeled EGF and subsequent loss of radioactivity is different for HeLa cells in the two culture systems (Wolfe *et al.*, 1980).

Transplantable human colon tumor lines, carried in athymic mice, have been used as models to develop serum-free media for colon carcinoma cells (Murakami and Masui, 1980; Van der Bosch *et al.*, 1981; Van der Bosch, 1984). The initial line studied was found to grow in an F12/DME mixture supplemented with insulin, transferrin, EGF, hydrocortisone, T_3, glucagon, and ascorbate (Murakami and Masui, 1980). The requirement for addition of ascorate to the medium for optimal growth may represent a dependence of these cells on the presence of a collagen substratum, similar to the collagen substratum dependence observed for mammary epithelia in serum-free medium. Plating efficiency of the colon tumor cells in serum-free medium is improved by the use of a collagen gel substratum. These cells also exhibit a mitogenic response to gastrin under conditions otherwise suboptimal for growth, but inclusion of gastrin in the serum-free medium developed for optimal growth of the cells has no effect (Murakami and Masui, 1980). When cells from transplantable tumors of this line are placed in primary culture in the serum-free medium developed for these cells in culture, the primary cultures grow considerably faster in the serum-free medium than in serum-containing medium, and fibroblastic overgrowth also is prevented in the serum-free medium. With time in serum-free culture these cells form three-dimensional structures somewhat similar to "inside-out" intestinal villi and appear to secrete a mucinlike material into the centers of these structures (Murakami and Masui, 1980). The serum-free medium developed for the first colon tumor line studied in this way has proved useful, with some adaptations, for the growth of a number of other transplantable human colon tumor lines, as well as for the growth of some primary tumor specimens directly from biopsy (Van der Bosch *et al.*, 1981; Van der Bosch, 1984).

Initial work in the development of serum-free media for kidney epithelia was carried out on the well-characterized MDCK canine kidney cell line (Taub *et al.*, 1979, 1981, 1983). Supplementation of an F12/DME mixture with insulin, transferrin, prostaglandin E_1, T_3, selenium, and hydrocortisone allows long-term growth of these cells in mass culture as well as colony formation with good efficiency. Cultures of MDCK in serum-free medium form hemicysts under the influence of hydrocortisone and prostaglandin E_1 (Taub *et al.*, 1979; Taub and Sato, 1979). Although this medium is effective at establishing and maintaining functional primary cultures of kidney epithelia from a number of species (Taub and Sato, 1979, 1980; Taub *et al.*, 1981), it does not allow optimal growth of some other established kidney cell lines, such as BSC-1 and epitheloid NRK (Taub and Sato, 1979). Other factors that are stimulatory for growth of MDCK cells in serum-free medium under suboptimal conditions, but are not included in the final serum-free medium formulation, include glucagon, EGF, FGF, and norepinephrine. Several of these components apparently act like prostaglandin in the serum-free formulation to increase cyclic AMP levels in the cell. Dibutyryl cyclic AMP and phosphodiesterase inhibitors are also effective at promoting MDCK cell growth and will replace prostaglandin E_1 when added to serum-free medium.

Whereas the MDCK line in culture is a system exhibiting properties similar in some respects to epithelia of the kidney collecting duct or distal tubule, another line, isolated from procine kidney and designated $LLC-PK_1$, is similar in many respects to epithelial cells of the proximal kidney tubule with the exception that this cell line responds to arginine vasopressin, a distal tubule marker. This line will not grow long term in the serum-free medium developed for MDCK, but will grow in a somewhat similar medium formulation (Chuman *et al.*, 1982). Using the same basal nutrient medium as was used to develop the serum-free medium formulation for MDCK, the $LLC-PK_1$ line can be grown by adding the following supplements: insulin, transferrin, selenium, hydrocortisone, T_3, vasopressin, and cholesterol. Although both prostaglandin E_1 and vasopressin stimulate adenyl cyclase, vasopressin but not prostaglandin E_1 is stimulatory for growth of $LLC-PK_1$ in serum-free medium (Taub and Livingston, 1981). Some growth stimulation of $LLC-PK_1$ also is observed with dibutyryl cyclic AMP or with phosphodiesterase inhibitors. Cholesterol is growth stimulatory for $LLC-PK_1$, but not for MDCK. EGF also is mitogenic for $LLC-PK_1$ under suboptimal serum-free conditions, but is not stimulatory if added to the serum-free medium formulation given above. Although MDCK cells in the serum-free medium formulation developed for MDCK grow as well as cells in serum-containing medium, $LLC-PK_1$ cells in the serum-free medium formulation developed for them do not grow as well as $LLC-PK_1$ cells in serum-containing medium (Chuman *et al.*, 1982).

$LLC-PK_1$ cells in the serum-free medium developed for them will form hemicysts and appear similar in morphology to $LLC-PK_1$ cells in serum-containing medium although some differences appear in the morphology of microvilli of the brush border when cells grown in serum-free and serum-containing media are compared by electron microscopy. MDCK cells will grow in the serum-free medium developed for LLC–PK$_1$ but will not form domes, since dome formation for MDCK in serum-free medium is dependent on the addition of prostaglandin E$_1$ to the medium as a supplement. The medium developed for the growth of LLC–PK$_1$ in the absence of serum can be used for serum-free primary culture of epithelial cells isolated from distal tubule of the kidney of rabbits, although this medium is not a selective medium for that cell type only. Epithelial kidney cells from other areas of the kidney also will grow in primary culture in this formulation, and complete differentiation of the epithelial cells in the LLC–PK$_1$ serum-free medium in primary culture does not occur (Saier, 1984). Both the medium developed for MDCK and the medium developed for LLC–PK$_1$ prevent fibroblastic overgrowth in serum-free primary culture of kidney epithelia. An even simpler serum-free medium for rabbit proximal tubule cells by primary culture has been developed, and this medium supports maintenance in the cells *in vitro* of a number of differentiated functions of proximal tubule epithelia observed *in vivo* (Chung *et al.*, 1982). For further details of the primary kidney cells and this medium refer to Chapter 13 by Taub.

An approach toward serum-free cell culture of human bronchogenic epidermoid carcinoma cells similar to that taken in developing serum-free culture conditions for human colon carcinoma cells has led to serum-free growth and differentiation of this cell type *in vitro* (Miyazaki *et al.*, 1982). Primary cultures of these cells, isolated from transplantable tumors carried in athymic mice, will grow slowly in culture in a mixture of F12 and DME with no further supplementation, but improved growth, particularly at lower cell densities, is best achieved by supplementation of the medium with insulin, glucagon, selenium, and retinoic acid. The presence of selenium is particularly important for the serum-free growth of these cells: a proper substratum is also required and can be supplied by addition of fibronectin to the system. Retinoic acid can be replaced with T$_3$, but no further stimulation of growth is seen if T$_3$ is added in addition to retinoic acid in the complete serum-free medium. Concentrations of retinoic acid greater than 10 nM result in a decrease in cell number from the optimal achieved in serum-free medium. Inhibitory effects also are seen with hydrocortisone, transferrin, and EGF. Deletion of T$_3$ or retinoic acid from the serum-free medium results in eventual keratinization and stratification. In addition to retinoic acid and T$_3$, retinol, retinal, retinyl acetate, retinyl palmitate, and thyroxine also prevent or delay differentiation of these cells *in vitro* (Miyazaki *et al.*, 1982; Miyazaki and Sato, 1984). If primary cultures are maintained in the serum-free formulation containing one of the substances that prevents keratinization, the cells can be subcultured several times, at least. A serum-free medium also has been developed to support the growth of human small-cell lung cancer *in vitro* (Simms *et al.*, 1980; Carney *et al.*, 1981; Minna *et al.*, 1982). This medium contains as supplements hydrocortisone, insulin, transferrin, estradiol, and selenium and allows the growth of cells of this type directly from biopsy material in the absence of fibroblastic contamination.

Normal human bronchial epithelial cells also have been grown in serum-free medium (Lechner *et al.*, 1982). In this case, the basal nutrient medium used is a modification of Ham's MCDB 150 series (Tsao *et al.*, 1982; Peehl and Ham, 1980), a medium developed for human epidermal keratinocytes, and to this the following supplements are added: EGF, insulin, transferrin, hydrocortisone, phosphoethanolamine, ethanolamine, and trace elements. A growth-promoting substratum is provided by treatment of the culture dish with a mixture of fibronectin, soluble collagen, and albumin. Also stimulatory for growth of these cells in serum-free medium are pituitary and brain extracts, although the identity of the mitogenic components in these extracts is not clear. Both serum-free, multipassage culture of cells at high density and clonogenic culture of cells at low density have been achieved with this medium formulation. Cholera toxin, stimulatory for a number of epithelial cell types, including epidermal keratinocytes, was not stimulatory for bronchial epithelial cells in this medium.

A medium has been developed for the A431 human epidermoid carcinoma cell line that allows the long-term, multipassage growth of these cells in the absence of serum (Barnes, 1982a,b). The formulation is a mixture of F12 and DME supplemented with insulin, transferrin, ethanolamine, and fibronectin. Serum-spreading factor can substitute for fibronectin for these cells in serum-free medium, as it can for many other cell types (Barnes and Sato, 1980b; Barnes *et al.*, 1980). Also stimulatory for A431 cell growth to some extent in serum-free medium are glycyl-histidyl-lysine, parathyroid hormone, and somatostatin. EGF in serum-free medium at concentrations that are stimulatory for growth of many cell types is inhibitory for the growth of A431 cells.

4. CELLS OF NEUROEPITHELIAL ORIGIN

A serum-free medium devised for the growth of the B104 rat neuroblastoma cells has been developed (Bottenstein *et al.*, 1979b; Bottenstein and Sato, 1979). The basal nutrient medium is supplemented with insulin, transferrin at a concentration that is rather high compared to that needed by most cell types in serum-free culture, progesterone, putrescine, and selenium. The plastic tissue culture substratum is also modified for serum-free growth of these cells by treatment with polylysine and fibronectin (Bottenstein and Sato, 1980). Slight modifications of the formulation developed for the B104 line allow the serum-free growth of established human neuroblastoma and rat pheochromocytoma lines and the maintenance of postmitotic neurons in primary, serum-free culture (Bottenstein *et al.*, 1979b, 1980; Bottenstein and Sato, 1979; Bottenstein, 1980, 1981). Some of the modifications involve the addition of NGF, and, in the case of one clonal isolate of a rat pheochromocytoma (PC-G2), the addition of EGF, which is inhibitory for the growth of the B104 neuroblastoma line in serum-free culture (Goodman *et al.*, 1979; Goodman, 1984).

A modification of the medium developed for neuroblastoma also has been used for the serum-free primary culture of fetal mouse hypothalamic cells (Faivre-Bauman *et al.*, 1981; Puymirat *et al.*, 1982; Loudes *et al.*, 1984). This serum-free formulation contains, in addition to the components of the medium designed for the B104 line, estradiol, arachidonic acid, and docosahexaenoic acid. Estradiol improves the attachment and morphological appearance of the cells in the cultures but does not affect cell number. T_3 was found to increase neurite elongation and survival of the cultures in the serum-free medium (Puymirat *et al.*, 1983). Plating the cells on a laminin-treated substratum also promotes growth for these cells and for embryonal carcinoma cells that differentiate into neurons (Darmon, 1983). The critical components of the serum-free medium for the maintenance of primary fetal mouse hypothalmic cell cultures were transferrin and selenium.

5. CONCLUSION—THE ADVANTAGES OF SERUM-FREE CELL CULTURE

In addition to the advantages of the use of serum-free media for the establishment of cultures of some cell types that cannot be maintained *in vitro* in serum-containing media, a number of advantages from the point of view of experimental design also exist in the use of serum-free cell culture (Barnes and Sato, 1980a,b). For instance, effects of hormones or growth factors can be studied in the absence of unknown serum factors that may mimic or act synergistically or antagonistically with the factor to be studied and in the absence of a background of endogenous levels of hormones or growth factors in the serum-containing medium. Experiments also can be designed to control effects of serum factors on the nature and quantity of particular cell membrane components and the functionality of those components. Subtle effects of this type have been found to be exerted by serum components in studies in which cell membrane behavior and composition in serum-containing and serum-free media have been compared (Wu and Sato, 1978; Wolfe *et al.*, 1980).

Experiments also can be carried out under conditions that allow control of effects of serum factors on cellular metabolism (Hayashi *et al.*, 1978; Bottenstein *et al.*, 1979b). In addition to hormone and growth factor effects, such serum-mediated changes include those related to actions of serum transport and binding proteins such as transferrin, ceruloplasmin, lipid-binding albumin, and lipoproteins (Mather and Sato, 1979b; Mather, 1983; Rockwell *et al.*, 1980; Simonian *et al.*, 1982; Gospodarowicz, 1984). Furthermore, the absolute control of the nutritional environment of cells in culture, including control of the nature and concentration of binding proteins modulating the action of nutrients, allows studies in which the nutritional composition of the medium can be precisely defined. Such studies are virtually impossible to carry out in culture in serum-containing medium (Ham and McKeehan, 1978; Ham, 1981, Rizzino *et al.*, 1979; McKeehan, 1982).

Problems avoided in experiments designed to examine effects of hormones or growth factors in serum-free media are also avoided when this approach is taken in the study of drug effects on cells in culture (Lippman, 1984); considerably less attention need be given in serum-free cultures to environmental factors that may interfere with drug action, potentiate drug action, or enzymatically alter the compound of interest. Finally, because serum-free media can be adjusted so that growth or survival of cells is critically dependent on a particular set of environmental conditions, it is possible to use manipulations of serum-free media to select for rare cells, mutant or variant from the parent population in response to a particular hormone, drug, growth factor, transport protein, or nutrients (Taub *et al.*, 1981, 1983; Shimizu, 1984). All these advantages have been used to some extent in recent years, allowing new approaches to questions that investigators have found difficult to address with the use of conventional cell culture methods, and all may be applied in various ways to the particular problems associated with studies of the mechanisms and regulation of transport processes in epithelial cells. Progess made thus far in this regard indicates that considerable improvements in the understanding of epithelial transport mechanisms and the factors that affect them are likely to result.

REFERENCES

Allegra, J. C., and Lippman, M. E., 1978, Growth of a human breast cancer cell line in serum-free hormone-supplemented medium, *Cancer Res.* **38**:3823–3829.

Ambesi–Impiombato, F. S., Parks, L. A. M., and Coon, H. G., 1980, Culture of hormone dependent functional epithelial cells from rat thyroids, *Proc. Natl. Acad. Sci. USA* **77**:3455–3459.

Barnes, D., 1980, Factors that stimulate proliferation of breast cancer cells in vitro in serum-free medium, in: *Cell Biology of Breast Cancer* (C. McGrath, M. Brennan, and M. Rich, eds.), Academic Press, New York, pp. 277–287.

Barnes, D. W., 1982a, Epidermal growth factor inhibits growth of A431 human epidermoid carcinoma in serum-free cell culture, *J. Cell Biol.* **93**:1–4.

Barnes, D. W., 1982b, Growth of A-431 human epidermoid carcinoma in serum-free cell culture: Inhibition by epidermal growth factor, in: *Cold Spring Harbor Conferences on Cell Proliferation*, Volume 9 (G. H. Sato, A. B. Pardee, and D. A. Sirbasku, eds.), Cold Spring Harbor Press, Cold Spring Harbor, New York, pp. 937–941.

Barnes, D., and Sato, G., 1979, Growth of a human mammary tumor cell line in a serum-free medium, *Nature* **281**:388–389.

Barnes, D., and Sato, G., 1980a, Methods for growth of cultured cells in serum-free medium, *Anal. Biochem.* **102:** 255–270.

Barnes, D., and Sato, G., 1980b, Serum-free cell culture; a unifying approach, *Cell* **22:**649–655.

Barnes, D., Wolfe, R., Serrero, G., McClure, D., and Sato, G., 1980, Effects of a serum spreading factor on growth and morphology of cells in serum-free medium, *J. Supramol. Struct.* **14:**47–63.

Barnes, D. W., van der Bosch, J.,Masui, H., Miyazaki, K., and Sato, G., 1981, The culture of human tumor cells in serum-free medium, in: *Methods in Enzymology,* Volume 79: *Interferons, Part B* (S. Pestka, ed.), Academic Press, New York, pp. 368–391.

Barnes, D. W., Darmon, M., and Orly, J., 1982, Serum-spreading factor: Effects on RF1 rat ovary cells and 1003 mouse embryonal carcinoma cells in serum-free media, in: *Cold Spring Harbor Conferences on Cell Proliferation,* Volume 9 (G. H. Sato, A. B. Pardee, and D. A. Sirbasku, eds.), Cold Spring Harbor Press, Cold Spring Harbor, New York, pp. 155–167.

Barnes, D. W., Silnutzer, J., See, C., and Shaffer, M., 1983a, Characterization of human serum spreading factor with monoclonal antibody, *Proc. Natl. Acad. Sci. USA* **80:**1362–1366.

Barnes, D., Sirbasku, D., and Sato, G. (eds.), 1984, *Cell Culture Methods for Molecular and Cell Biology,* Volumes 1–4 (D. Barnes, D. Sirbasku, and G. Sato, eds.) Liss, New York.

Bottenstein, J., 1980, Serum-free culture of neuroblastoma cells, in: *Advances in Neuroblastoma Research* (A. Evans, ed.), Raven Press, New York, pp. 161–170.

Bottenstein, J., 1981, Differentiated properties of neuronal cell lines, in: *Functionally Differentiated Cell Lines* (G. Sato, ed.), Liss, New York, pp. 155–184.

Bottenstein, J. E., and Sato, G. H., 1979, Growth of a rat neuroblastoma cell line in serum-free medium, *Proc. Natl. Acad. Sci. USA* **76:**514–517.

Bottenstein, J., and Sato, G., 1980, Fibronectin and polylysine requirement for proliferation of neuroblastoma cells in defined medium, *Exp. Cell Res.* **129:**361–366.

Bottenstein, J., Hayashi, I., Hutchings, S., Masui, H., Mather, J., McClure, D. B., Ohasa, S., Rizzino, A., Sato, G., Serrero, G., Wolfe, R., and Wu, R., 1979a, The growth of cells in serum-free hormone-supplemented media, *Methods Enzymol.* **58:**94–109.

Bottenstein, J. E., Sato, G. H., and Mather, J. P., 1979b, Growth of neuroepithelial-derived cell lines in serum-free hormone-supplemented media, in: *Cold Spring Harbor Conferences on Cell Proliferation: Hormones in Cell Culture,* Volume 6 (G. H. Sato and R. Ross, eds.), Cold Spring Harbor Press, Cold Spring Harbor, New York, pp. 531–544.

Bottenstein, J. E., Skaper, S. D., Varon, S. S., and Sato, G. H., 1980, Selective survival of neurons from chick embryo sensory ganglionic dissociates utilizing serum-free supplemented medium, *Exp. Cell Res.* **125:**183–190.

Brooks, R. F., 1975, Growth regulation in vitro and the role of serum, in: *Structure and Function of Plamsa Proteins* (A. C. Allison, ed.), Plenum Press, New York, pp. 1–112.

Carney, D. N., Bunn, P. A., Gazdar, A. F., Pagan, J. A., and Minna, J. D., 1981, Selective growth in serum-free hormone supplemented medium of tumor cells obtained by biopsy from patients with small cell carcinoma of the lung, *Proc. Natl. Acad. Sci. USA* **78:**3185–3189.

Carrel, A., 1913, Artificial activation of the growth in vitro of connective tissue, *J. Exp. Med.* **17:**14–19.

Chaproniere–Rickenberg, D. M., and Webber, M. M., 1982, A chemically defined medium for the growth of adult human prostatic epithelium, in: *Growth of Cells in Hormonally Defined Media,* Volume 9, Book B (G. H. Sato, A. B. Pardee, and D. A. Sirbasku, eds.), Cold Spring Harbor Press, Cold Spring Harbor, New York, pp. 1109–1115.

Cherington, P. V., Smith, B. L., and Pardee, A. B., 1979, Loss of epidermal growth factor requirement and malignant transformation, *Proc. Natl. Acad. Sci. USA* **76:**3937–3941.

Chuman, L., Fine, L. G., Cohen, A. I., and Saier, M. H., 1982, Continuous growth of proximal tubular kidney cells in hormone-supplemented serum-free medium, *J. Cell Biol.* **94:**506–510.

Chung, S. D., Alavi, N., Livingston, D., Hiller, S., and Taub, M., 1982, Characterization of primary rabbit kidney cultures that express proximal tubule functions in hormonally defined medium, *J. Cell Biol.* **95:**118–126.

Darmon, M., 1983, Laminin provides a better substratum than fibronectin for attachment, growth and differentiation of 1003 embryonal carcinoma cells, *In Vitro* **18:**997–1003.

Faivre-Bauman, A., Rosenbaum, E., Puymirat, J., Grouselle, D., and Tixier-Vidal, A., 1981, Differentiation of fetal mouse hypothalmic cells in serum-free medium, *Dev. Neurosci.* **4:**118–125.

Gill, G. N., Hornsby, P. J., and Simonian, M. H., 1979, Regulation of growth and differentiated function of bovine adrenocortical cells, in: *Cold Spring Harbor Conferences on Cell Proliferation; Hormones in Cell Culture,* Volume 6 (G. H. Sato and R. Ross, eds.), Cold Spring Harbor Press, Cold Spring Harbor, New York, pp. 701–715.

Goodman, R., 1984, Growth and differentiation of pheochromocytoma cells in chemically defined medium, in: *Cell Culture Methods for Molecular and Cell Biology,* Volume 4 (D. Barnes, D. Sirbasku, and G. Sato, eds.), Liss, New York, pp. 23–36.

Goodman, R., Chandler, C., and Herschman, H. R., 1979, Pheochromocytoma cell lines as models of neuronal differentiation, in: *Cold Spring Harbor Conferences on Cell Proliferation, Hormones in Cell Culture,* Volume 6 (G. H. Sato and R. Ross, eds.), Cold Spring Harbor Press, Cold Spring Harbor, New York, pp. 653–669.

Gospodarowicz, D., 1984, Preparations and uses of lipoproteins to culture normal diploid and tumor cells under serum-free conditions, in: *Cell Culture Methods for Molecular and Cell Biology,* Volume 1 (D. Barnes, D. Sirbasku, and G. Sato, eds.), Liss, New York, pp. 69–88.

Ham, R. G., 1981, Survival and growth requirements of nontransformed cells, in: *Handbook of Experimental Pharmacology* (I. N. Baserga, ed.), Springer, New York, pp. 13–38.

Ham, R. G., 1982, Importance of the basal nutrient medium in the design of hormonally defined media, in: *Cold Spring Harbor Conferences on Cell Proliferation,* Volume 9 (G. H. Sato, A. B. Pardee, and D. A. Sirbasku, eds.), Cold Spring Harbor Press, Cold Spring Harbor, New York, pp. 39–60.

Ham, R. G., and McKeehan, W. L., 1978, Nutritional requirements for clonal growth of nontransformed cells, in: *Nutritional Requirements of Cultured Cells* (I. N. Katsuta, ed.), Japan Scientific Society Press, Tokyo, pp. 361–115.

Ham, R. G., and McKeehan, W. L., 1979, Media and growth requirements, *Methods Enzymol.* **58:**44–95.

Harrison, R. G., 1907, Observations on the living developing nerve fiber, *Proc. Soc. Exp. Biol. Med.* **4:**140–143.

Hayashi, I., and Sato, G. H., 1976, Replacement of serum by hormones permits growth of cells in a defined medium, *Nature (London)* **259:**132–134.

Hayashi, I., Sato, G. H., and Larner, J., 1978, Hormonal growth control of cells in culture, *In Vitro* **14:**23–30.

Higuchi, K., 1973, Cultivation of animal cells in chemically defined media, a review, *Adv. Appl. Microbiol.* **16:**111–160.

Hutchings, S. E., and Sato, G. H., 1978, Growth and maintenance of HeLa cells in serum-free medium supplemented with hormones, *Proc. Natl. Acad. Sci. USA* **75:**901–904.

Imagawa, W., Tomooka, Y., and Nandi, S., 1982, Serum-free growth of normal and tumor mouse mammary epithelial cells in primary culture, *Proc. Natl. Acad. Sci. USA* **71:**4074–4077.

Kano–Sueoka, T., and Errick, J. E., 1981, Effects of phosphoethanolamine and ethanolamine on growth of mammary carcinoma cells in culture, *J. Exp. Cell Res.* **136:**137–145.

Kano–Sueoka, T., and Errick, J. E., 1982, Roles of phosphoethanolamine, ethanolamine, and prolactin on mammary cell growth, in: *Cold Spring Harbor Conferences on Cell Proliferation,*

Volume 9 (G. H. Sato, A. B. Pardee, and D. A. Sirbasku, eds.), Cold Spring Harbor Press, Cold Spring Harbor, New York, pp. 727–740.

Kano–Sueoka, T., and Hsieh, P., 1973, A rat mammary carcinoma in vivo and in vitro: Establishment of clonal lines of the tumor, *Proc. Natl. Acad. Sci. USA* **70:**1922–1926.

Kano–Sueoka, T., Cohen, D. M., Yamaizumi, Z., Nishimura, S., Mori, M., and Fujiki, H., 1979a, Phosphoethanolamine as a growth factor of a mammary carcinoma cell line of rat, *Proc. Natl. Acad. Sci. USA* **76:**5741–5744.

Kano–Sueoka, T., Errick, J. E., and Cohen, D. M., 1979b, Effects of hormones and a novel mammary growth factor on a rat mammary carcinoma in culture, in: *Cold Spring Harbor Conferences on Cell Proliferation; Hormones in Cell Culture,* Volume 6 (G. H. Sato and R. Ross, eds.), Cold Spring Harbor Press, Cold Spring Harbor, New York, pp. 499–512.

Kidwell, W. R., Salomon, D. S., Liotta, L. A., Zweibel, J. A., and Bano, M., 1982, Effects of growth factors on mammary epithelial cell proliferation and basement-membrane synthesis, in: *Growth of Cells in Hormonally Defined Media* (G. H. Sato, A. B. Pardee, and D. A. Sirbasku, eds.) Cold Spring Harbor Press, Cold Spring Harbor, New York, pp. 807–818.

Kidwell, W. R., Banc, M., and Salomon, D. S., 1984, The growth of normal mammary epithelium on collagen in serum-free medium, in: *Cell Culture Methods for Molecular and Cell Biology,* Volume 2 (D. Barnes, D. Sirbasku, and G. Sato, eds.), Liss, New York, in press.

Lechner, J. F., Haugen, A., McClendon, I. A., and Pettis, E. W., 1982, Clonal growth of normal adult human bronchial epithelial cells in a serum-free medium, *In Vitro* **18:**633–642.

Lippman, M. E., 1984, Definition of hormones and growth factors required for optimal proliferation and expression of phenotypic response in human breast cancer cells, in: *Cell Culture Methods for Molecular and Cell Biology,* Volume 2 (D. Barnes, D. Sirbasku, and G. Sato, eds.), Liss, New York, pp. 183–200.

Loudes, C., Faivre–Bauman, A., Barret, A., Grouselle, D., Puymirat, A., and Tixier–Vidal, A., 1984, Functional maturation of TRH immunoreactive neurons in serum-free cultures of mouse hypothalmic cells, *Develop. Brain Res.,* in press.

Mather, J. P., 1980, The establishment and characterization of two distinct mouse testicular epithelial cell lines, *Biol. Reprod.* **23:**243–250.

Mather, J. P. (ed.), 1984, *Mammalian Cell Culture: The Use of Serum-Free Hormone-Supplemented Media,* Plenum Press, New York.

Mather, J. P., and Haour, F., 1981, Hormone response of testicular cells in culture: Established cell lines and primary cultures, in: *Functionally Differentiated Cell Lines* (G. Sato, ed.), Liss, New York, pp. 93–108.

Mather, J. P., and Sato, G. H., 1979a, The use of hormone-supplemented serum-free media in primary cultures, *Exp. Cell Res.* **124:**215–221.

Mather, J. P., and Sato, G. H., 1979b, The growth of mouse melanoma cells in hormone-supplemented serum-free medium, *Exp. Cell Res.* **12:**191–200.

Mather, J. P., Saez, J. M., and Haour, F., 1981, Primary cultures of Leydig cells for rat, mouse and pig: Advantages of porcine cells for the study of gonadotropin regulation of Leydig cell function, *Steroids* **38:**35–44.

Mather, J. P., Saez, J. M., Dray, F., and Haour, F., 1982a, Hormone–hormone and hormone–vitamin interactions in the control of growth and function of Leydig cells in vitro, in: *Cold Spring Harbor Conferences on Cell Proliferation,* Volume 9 (G. H. Sato, A. B. Pardee, and D. A. Sirbasku, eds.), Cold Spring Harbor Press, Cold Spring Harbor, New York, pp. 1117–1128.

Mather, J. P., Zhuang, L. Z., Perez–Infante, V., and Phillips, D. M., 1982b, Culture of testicular cells in hormone-supplemented serum-free medium, *Ann. N.Y. Acad. Sci.* **383:**44–68.

McKeehan, W. L., 1982, Growth-factor–nutrient interrelationships in control of normal and transformed cell proliferation, in: *Cold Spring Harbor Conferences on Cell Proliferation,* Volume 9

(G. H. Sato, A. B. Pardee, and D. Sirbasku, eds.), Cold Spring Harbor Press, Cold Spring Harbor, New York, pp. 65–74.

Minna, J. D., Carney, D. N., Oie, H., Bunn, P. A., and Gazadai, A. F., 1982, Growth of human small-cell lung cancer in defined medium, in: *Growth of Cells in Hormonally Defined Medium* (D. Sirbasku, A. Pardee, and G. Sato, eds.), Cold Spring Harbor Press, Cold Spring Harbor, New York, pp. 627–639.

Miyazaki, K., and Sato, G. H., 1984, Methods for growth and differentiation of human bronchogenic epidermoid carcinoma cells in serum-free media, in: *Cell Culture Methods for Molecular and Cell Biology*, Volume 3 (D. Barnes, D. Sirbasku, and G. Sato, eds.), Liss, New York, in press.

Miyazaki, K., Masui, H., and Sato, G., 1982, Control factors for keratinization of human bronchoogenic epidermoid carcinoma cells, in: *Growth of Cells in Hormonally Defined Media*, Volume 9, Book B (G. H. Sato, A. B. Pardee, and D. A. Sirbasku, eds.), Cold Spring Harbor Press, Cold Spring Harbor, New York, pp. 657–661.

Murakami, H., and Masui, H., 1980, Hormonal control of human colon carcinoma cell growth in serum-free medium, *Proc. Natl. Acad. Sci. USA* **77:**3464–3468.

Murakami, H., Masui, H., Sato, G. H., Sueoka, N., Chow, T. P., and Kano–Sueoka, T., 1982, Growth of hybridoma cells in serum-free medium: Ethanolamine is an essential component, *Proc. Natl. Acad. Sci. USA* **79:**1158–1162.

Orly, J., and Sato, G., 1979, Fibronectin mediates cytokinesis and growth of rat follicular cells in serum-free medium, *Cell* **17:**295–305.

Orly, J., Sato, G., and Erickson, G. F., 1980, Serum suppresses the expression of hormonally induced functions in cultured granulosa cells, *Cell* **20:**817–827.

Peehl, D. M., and Ham, R. G., 1980, Clonal growth of human keratinocytes with small amounts of dialyzed serum, *In Vitro* **16:**526–538.

Puymirat, J., Loudes, C., Faivre–Bauman, A., Bourre, J. M., and Tixier–Vidal, A., 1982, Expression of neuronal functions by mouse fetal hypothalamic cells cultured in hormonally defined medium, in: *Growth of Cells in Hormonally Defined Medium*, Volume 9 (D. Sirbasku, A. Pardee, and G. Sato, eds.), Cold Spring Harbor Press, Cold Spring Harbor, New York, pp. 1033–1051.

Puymirat, J., Barrett, A., Picart, R., Vigny, A., Loudes, C., Faivre–Bauman, A., and Tixier–Vidal, A., 1984, Triiodothyronine enhances the morphological maturation of dopaminergic neurons in serum-free medium cultures of fetal mouse hypothalmic cells, *Neuroscience*, in press.

Reid, L., and Sato, G., 1978, Replacement of serum in cell culture by hormones, in: *Biochemistry and Mode of Action of Hormones* 11 (H. V. Rickenberg, ed.), University Park Press, Baltimore, pp. 219–251.

Rizzino, A., Rizzino, H., and Sato, G., 1979, Defined media and the determination of nutritional and hormonal requirements of mammalian cells in culture, *Nutr. Rev.* **37:**369–378.

Rockwell, G. A., McClure, D., and Sato, G. H., 1980, The growth requirements of SV40 virus-transformed Balb/C-3T3 cells in serum-free monolayer culture, *J. Cell Physiol.* **103:**323–331.

Saier, M. H., Jr., 1984, Hormonally-defined, serum-free medium for proximal tubular kidney epithelial cell line, LLC–PK$_1$, in: *Cell Culture Methods for Molecular and Cell Biology*, Volume 3 (D Barnes, D. Sirbasku, and G. Sato, eds.), Liss, New York, in press.

Salomon, D. S., Liotta, L. A., and Kidwell, W. R., 1981, Differential response to growth factor by rat mammary epithelium plated on different collagen substrata in serum-free medium, *Proc. Natl. Acad. Sci. USA* **78:**382–386.

Salomon, D. S., Smith, K. B., Losonczy, I., Bano, M., Kidwell, W. R., Allesandri, G., and Gullino, P. M., 1984, Alpha$_2$-macroglobulin, a contaminant of commercially-prepared Pedersen fetuin: Isolation, characterization and biological activity, in: *Cell Culture Methods for*

Molecular and Cell Biology, Volume 3 (D. Barnes, D. Sirbasku, and G. Sato, eds.), Liss, New York, in press.

Sato, G. H., 1975, The role of serum in cell culture, in: *Biochemical Actions of Hormones* (G. Litwack, ed.), Academic Press, New York, pp. 391–396.

Sato, G., Pardee, A. B., and Sirbasku, D. A. (eds.), 1982, *Cold Spring Harbor Conferences on Cell Proliferation,* Volume 9, Cold Spring Harbor Press, Cold Spring Harbor, New York.

Shimizu, N., 1984, The use of hormone-toxin conjugates and serum-free media for the isolation and study of cell variants in hormone responses, in: *Cell Culture Methods For Molecular and Cell Biology,* Volume 3 (D. Barnes, D. Sirbasku, and G. Sato, eds.), Liss, New York, in press.

Simms, E., Gazdar, A. F., Abrams, P. G., and Minna, J. D., 1980, Growth of human small cell (oat cell) carcinoma of the lung in serum-free growth factor-supplemented medium, *Cancer Res.* **40:**4356–4363.

Simonian, M. H., White, M. L., and Gill, G. N., 1982, Growth and function of cultured bovine adrenocortical cells in a serum-free defined medium, *Endocrinology* **111:**919–927.

Stampfer, M., 1984, Methods for growth of human mammary epithelial cells in monolayer culture, in: *Cell Culture Methods for Molecular and Cell Biology,* Volume 2 (D. Barnes, D. Sirbasku, and G. Sato, eds.), Liss, New York, pp. 171–182.

Stampfer, M., Hallowes, R. C., and Hackett, A. D., 1980, Growth of normal human mammary cells in culture, *In Vitro* **16:**415–423.

Takemoto, H., Yokoro, K., Furth, J., and Cohen, A. J., 1962, Adrenotropic activity of mammosomatotropic tumors in rats and mice, *Cancer Res.* **22:**917–924.

Taub, M., and Livingston, D., 1981, The development of serum-free hormone-supplemented media for primary kidney cultures and their use in examining renal functions, *Ann. New York Acad. Sci.* **372:**406–421.

Taub, M., and Sato, G., 1979, Growth of kidney epithelial cells in hormone-supplemented, serum-free medium, *J. Supramol. Struc.* **11:**207–216.

Taub, M., and Sato, G., 1980, Growth of functional primary cultures of kidney epithelial cells in defined medium, *J. Cell. Physiol.* **105:**369–378.

Taub, M., Chuman, L., Saier, M. H., and Sato, G., 1979, Growth of Madin Darby canine kidney epithelial cell (MDCK) line in hormone-supplemented serum-free medium, *Proc. Natl. Acad. Sci. USA* **76:**3338–3342.

Taub, M. U. B., Chuman, L., Rindler, M. J., Saier, M. H., Jr., and Sato, G., 1981, Alterations in growth requirements of kidney epithelial cells in defined medium associated with malignant transformation, *J. Supramol. Struct.* **15:**63–72.

Taub, M., Saier, M. H., Jr., Chuman, L., and Hiller, S., 1983, Loss of the PGE_1 requirement for MDCK cell growth associated with a defect in cyclic AMP phosphodiesterase, *J. Cell. Physiol.* **114:**153–161.

Temin, H. M., Pierson, R. W., and Dulak, N. C., 1972, The role of serum in control of multiplication of avian and mammalian cells in culture, in: *Growth, Nutrition and Metabolism of Cells in Culture,* Volume 1 (G. Rothblat and V. J. Cristafalo, eds.), Academic Press, New York, pp. 50–81.

Tsao, M. C., Walthall, B. J., and Ham, R. G., 1982, Clonal growth of normal human epidermal keratinocytes in a defined medium, *J. Cell Physiol.* **110:**219–229.

Van der Bosch, J., 1984, Tissue culture of human colon carcinomas. Therapy experiments in vitro, in: *Proceedings of the First European Conference on Serum-Free Cell Culture* (G. Fischer, ed.), Springer, Heidelberg, in press.

Van der Bosch, J., 1984, Primary tissue cultures of human colon carcinomas in serum-free medium: An in vitro system for tumor analysis and therapy experiments, in: *Cell Culture Methods for Molecular and Cell Biology,* Volume 3 (D. Barnes, D. Sirbasku, and G. Sato, eds.), Liss, New York, in press.

Van der Bosch, J., Masui, H., and Sato, G., 1981, Growth characteristics of primary tissue cultures from heterotransplanted human colorectal carcinomas in serum-free medium, *Cancer Res.* **41**:611–618.

Wicha, M. S., Liotta, L. A., Garbisz, S., and Kidwell, W. R., 1979, Basement membrane collagen requirements for attachment and growth of mammary epithelium, *Exp. Cell Res.* **124**:181–190.

Wicha, M. S., Lowrie, G., Kohn, E., Baganandos, P., and Mahn, T., 1982, Extracellular matrix promotes mammary epithelial growth and differentiation in vitro, *Proc. Natl. Acad. Sci. USA* **79**:3213–3217.

Wolfe, R. A., Wu, R., and Sato, G., 1980, EGF-induced down regulation of receptor does not occur in HeLa cells grown in defined medium, *Proc. Natl. Acad. Sci. USA* **77**:2735–2739.

Wu, R., and Sato, G. H., 1978, Replacement of serum in cell culture by hormones: A study of hormonal regulation of cell growth and specific gene expression, *J. Tox. Env. Health* **4**:427–448.

Yang, J., Guzman, R., Richards, J., Jintott, V., DeVault, M. R., Wellings, S. R., and Nandi, S., 1980a, Epithelial cells embedded in collagen gels, *J. Natl. Cancer Inst.* **65**:337–341.

Yang, J., Richards, J., Guzman, R., Imagawa, W., and Nandi, S., 1980b, Sustained growth in primary culture of normal mammary epithelial cells imbedded in collagen gels, *Proc. Natl. Acad. Sci. USA* **77**:2088–2092.

Yang, J., Larson, L., Flynn, D., Elias, J., and Nandi, S., 1982, Serum-free primary culture of human normal mammary epithelial cells in collagen gel matrix, *Cell Biol. Int. Rep.* **6**:969–975.

Yasumura, Y., Tashjian, A. H., Jr., and Sato, G., 1966, Establishment of four functional clonal strains of animal cells in culture, *Science* **154**:1186–1189.

13

Importance of Hormonally Defined, Serum-Free Medium for in Vitro Studies Concerning Epithelial Transport

MARY TAUB

1. INTRODUCTION

Hormonally defined, serum-free media have been developed for a number of different types of epithelial cell lines. Some of these cell lines, such as the MDCK and the LLC–PK$_1$ pig kidney cell line, have been demonstrated to possess the capacity for transepithelial solute transport *in vitro*. MDCK and LLC–PK$_1$ cells are of particular interest to study in this regard, as their primary function *in vivo* is to reabsorb solutes. However, the transport properties of many other epithelial cell lines are also of interest. Although many of the epithelial cells in such cultures possess differentiated functions that are not primarily reabsorptive in nature, these cells often must transport particular solutes across the cell layer, in order to carry out their primary differentiated functions. For example, thyroid and testicular epithelial cells possess the capacity for transepithelial transport of select solutes *in vivo*. Such transepithelial transport ultimately permits the secretion of thyroid hormone in the thyroid and facilitates germ cell maturation in the testes.

Epithelial cell cultures have particular advantages in studies concerned with transport across epithelial cells. The transport systems involved in transepithelial

---

MARY TAUB • Department of Biochemistry, State University of New York at Buffalo, School of Medicine, Buffalo, New York 14214.   Work funded by National Institutes of Health Grant 1 R01 CA28111-04 and National Institutes of Health Research Career Development Award 1 K04 CA/AM 00888-01.

solute transport can be more precisely defined when using cultured cell mono-layers than when using slices. Furthermore, the factors that regulate these transport systems may be readily identified *in vitro*. Subsequently, the mechanisms by which such factors regulate transport can be defined.

The serum in the tissue culture medium has limited *in vitro* studies concerning the effects of hormones and other regulatory factors on transport. The serum is a very complex supplement, which contains numerous hormones and undefined factors. These undefined serum factors have prevented the use of primary epithelial cell cultures for transport studies. Such serum factors are cytotoxic to the epithelial cells in the primaries, yet generally permit fibroblast overgrowth to occur. As a consequence, the number of established epithelial cell lines is limited. Transport studies with these epithelial cell lines have also been limited by the use of serum. Hormones and other factors in the serum have masked the effects of hormones on transport. As a consequence, few observations concerning hormonal regulation of transport *in vitro* have been reported.

Hormonally defined, serum-free medium can alleviate these problems to a significant extent. When the serum is removed and replaced with appropriate defined supplements, many additional types of epithelial cells can be maintained *in vitro*. The transport systems in such epithelial cell cultures can then be studied. Previous *in vitro* investigations concerning the effects of hormones on these transport systems were impeded by the serum in the tissue culture medium. When the serum is removed and hormonally defined, serum-free conditions are used, these hormonal effects can then be observed. The concentrations of the hormone supplements in the serum-free medium can be readily manipulated, and thus, the effects of hormones on transport can be precisely defined. In this chapter the use of hormonally defined medium in the study of transport in cultured kidney epithelia is discussed in detail. The use of defined medium in the study of transport in other types of cultured epithelial cells is also described.

2. TRANSEPITHELIAL SOLUTE TRANSPORT BY CULTURED EPITHELIAL CELLS

Epithelial cells in a number of tissues in the body possess the capacity for transepithelial solute transport. Such transporting epithelial cells generally possess a polarized morphology (Fig. 1). Solutes present in a lumenal space may enter the cells through the mucosal surface (which faces the lumen) and leave the cells through the serosal surface (which faces the blood). Such vectorial fluid transport permits epithelial cells to translocate particular solutes from one tissue compartment to another.

The initial reports that suggested that epithelial cell cultures possess the capacity for transepithelial solute transport were based on the microscopic obser-

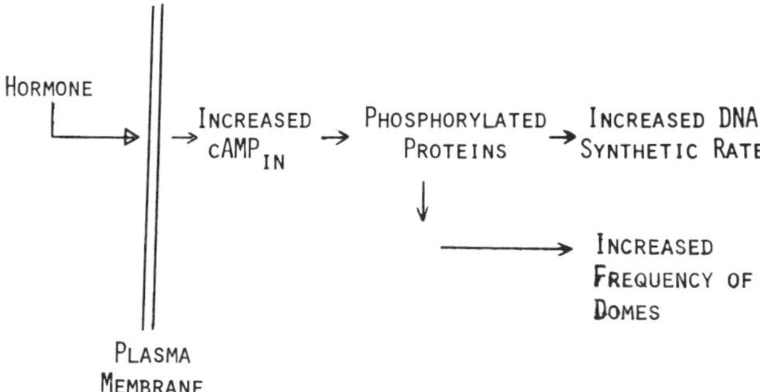

Figure 1. MDCK as a model of transporting kidney epithelial cells.

vation of domes or hemicysts in confluent epithelial cell cultures (Leighton *et al.*, 1969). When focusing on the majority of the cells in the monolayer, groups of cells (domes) were found to be out of focus, indicating that the cells in the domes were slightly elevated from the tissue culture dish surface. Dome formation has been observed in a number of established epithelial cell lines, as well as primary epithelial cell cultures (Table I).

The phenomenon of dome formation was first examined in detail in the MDCK cell line by Dr. Joseph Leighton and his associates (Leighton *et al.*, 1969, 1970). Leighton observed that MDCK cells attached to the tissue culture dish in a polarized manner. The mucosal surface of the cells uniformly faced the tissue culture medium, whereas the serosal surface faced the plastic. Adjacent cells formed tight junctions such that at confluency a contiguous epithelial sheet was formed. The domes or hemicysts in such monolayers were examined by time lapse photography. The time lapse studies indicated that dome formation was a

Table I. Cell Cultures Reported to Exhibit Domes

Tissue of origin	Cell culture	
Bladder	Rat bladder carcinoma	Toyoshima *et al.*, 1976
Cervix	Human cervical carcinoma	Auersperg, 1969
Choroid plexus	Primary rat choroid plexus	Cameron, 1953
Colon	Human colon carcinoma	McCombs *et al.*, 1976
Kidney	MDCK (dog kidney)	Leighton *et al.*, 1969
	LLC–PK$_1$ (pig kidney)	Mullin *et al.*, 1980
Lung	Primary pulmonary alveolar type II cells	Mason *et al.*, 1982
Mammary	Primary neoplastic mouse mammary	McGrath 1975
	RAMA 25 (rat mammary)	Lever, 1979b
Oral	Primary pig oral epithelia	Birek *et al.*, 1982
Stomach	Primary rat glandular stomach	Huh *et al.*, 1977

dynamic process (domes constantly formed and burst within MDCK mono-layers). Dome formation was stimulated by dibutyryl cyclic AMP (Valentich *et al.* 1979; Lever, 1979a) and was inhibited by ouabain (Abaza *et al.*, 1974).

Leighton explained these observations by postulating that salt and water was transported across the MDCK cell layer. Fluid entered the cells via the mucosal surface (which faced the medium) and left the cells via the serosal surface (which faced the dish). The transported salt and water, being caught between the mono-layer and the dish, caused the development of hydrostatic pressure and conse-quently dome formation. Thus, these studies indicated that MDCK cells had retained the morphological and functional polarity of kidney tubule cells, after their establishment in culture.

More direct evidence for transepithelial solute transport by MDCK cells was obtained in the studies of Misfeldt *et al.* (1976) and of Cereijido *et al.* (1978). MDCK cells were grown to confluency on collagen-coated semisolid supports. Their electrophysiological properties were then examined in Ussing flux cham-bers. A transepithelial potential of 1.0 mV and a resistance of 84 ohm. cm^2 were found, and water transport was measured. The transepithelial potential was re-duced by the addition of the pyrazine diuretic amiloride to the mucosal bathing solution (Cereijido *et al.*, 1978). Ouabain bound primarily to the basolateral surface and had an inhibitory effect on sodium efflux (Cereijido *et al.*, 1980). These studies are consistent with model for transepithelial salt transport in MDCK cells illustrated in Fig. 1.

The properties of the amiloride-sensitive component of $^{22}Na^+$ uptake by MDCK cells have been examined in detail (Rindler *et al.*, 1979). The uptake of $^{22}Na^+$ into MDCK cells is a saturable process which depends on the extra-cellular Na^+ concentration. Amiloride inhibits $^{22}Na^+$ uptake into MDCK cells with a K_i of 5×10^{-4} M. An inhibitory effect of amiloride on dome formation was similarly observed (the K_i was similarly 5×10^{-4} M). The $^{22}Na^+$ uptake studies indicated that the amiloride-sensitive Na^+ transport system was capable of catalyzing both Na^+/proton antiport and Na^+/Na^+ exchange. Variants re-sistant to killing by the cytotoxic effects of amiloride have been isolated from MDCK cell cultures. These amiloride-resistant variants have a reduced rate of amiloride-sensitive $^{22}Na^+$ uptake (Taub and Saier, 1981).

Several other established epithelial cell lines and primary epithelial cell cultures have also been shown to exhibit transepithelial solute transport (Table II). A transepithelial resistance of 211 ohm·cm^2 and a potential of 2.8 mV was measured across monolayers of LLC–PK_1 pig kidney cells (Misfeldt and Sand-ers, 1981). The transepithelial resistance across the LLC–PK_1 cell layer was dependent on the presence of glucose in the mucosal bath, and the potential was sensitive to inhibition by phlorezin. Similarly, a transepithelial resistance and potential have been measured across such cell types as primary mammary epi-thelial cells and primary pulmonary alveolar cells (Table II).

Table II.   Epithelial Cell Cultures that Exhibit a Transepithelial Resistance and Potential

Tissue of origin	Cell culture	
Bladder	TB-M (toad urinary bladder)	Handler *et al.*, 1979
	TB-bc (toad urinary bladder)	Handler *et al.*, 1979
Kidney	MDCK	Misfeldt *et al.*, 1976; Cereijido *et al.*, 1980
	LLC–PK$_1$	Misfeldt and Sanders, 1981
Lung	Primary rat pulmonary alveolar type II cells	Mason *et al.*, 1982
Mammary	Primary mouse mammary	Bisbee *et al.*, 1979
Stomach	Primary fetal rabbit gastric cells	Logsdon *et al.*, 1982

3. IMPORTANCE OF HORMONALLY DEFINED, SERUM-FREE MEDIUM IN THE STUDY OF HORMONAL REGULATION OF EPITHELIAL TRANSPORT

The mechanisms by which hormones regulate transepithelial solute transport in such epithelial cells is of interest. However, hormonal regulation of transport has only been reported in several types of cultured epithelial cells. Handler *et al.* (1979) have observed that aldosterone has a stimulatory effect on the I_{sc} of an established toad urinary bladder epithelial cell line (these cells were maintained with serum). Dibutyryl cyclic AMP also was observed to have a significant effect on the transepithelial resistance and potential of these cells. The possibility was examined that vasopressin, which increases cyclic AMP levels in these cells under physiological conditions, also affects the resistance and potential. However, a significant effect of vasopressin was not observed.

The transport properties of primary mouse mammary epithelial cells in serum-supplemented medium were observed to be regulated by a physiologically significant hormone, prolactin (Bisbee *et al.*, 1979). Prolactin increased the transepithelial resistance across the mammary cell cultures twofold and also increased the net absorption of sodium into the mammary cells threefold. However, a very high dosage of prolactin was needed in these studies (10 µg/ml), perhaps owing to the presence of serum in the tissue culture medium.

Effects of hormones on either the transepithelial resistance or the potential across MDCK cells have not been reported. However, calcium has been observed to regulate the activity of the Na^+ channel in MDCK cells (Taub and Sato, 1979). Extracellular calcium has an inhibitory effect on Na^+ uptake, whereas intracellular calcium has a stimulatory effect on the rate of Na^+ uptake. The stimulatory effect of intracellular calcium on Na^+ uptake cannot be simply explained by means of Na^+/calcium exchange. Instead intracellular calcium, or an intracellular complex of calcium and calmodulin, may directly interact with a regulatory site on the Na^+ channel, so as to increase transport rates. The involvement of intracellular calcium and calmodulin in the regulation of membrane transport systems has been similarly indicated by the observations of Rephaeli and Parsons (1982) and of Villereal (1981) with other types of animal cells.

Presumably, the intracellular effect of calcium on Na^+ transport is hormonally controlled. However, hormones that regulate the rate of Na^+ uptake in MDCK cells have not been identified. The use of serum in the tissue culture medium has hampered attempts at determining the effects of hormones on Na^+ uptake in MDCK.

4. IMPORTANCE OF HORMONALLY DEFINED MEDIUM IN THE STUDY OF EPITHELIAL TRANSPORT *IN VITRO*

In general, the use of serum either prior to or during such transport studies has been a limiting factor in the investigations. The serum is not only very complex, but is also undefined. Unidentified factors in the serum may not only synergistically interact with the hormones of interest (also present in serum) to increase the rate of transport, but may also have inhibitory effects on transport. Serum factors may cause the stimulatory effects of hormones on transport to be lost, by causing the down regulation of the hormone receptors involved in mediating the transport response.

A number of investigators have observed that the serum must be removed in order for the effects of hormones on transport to be detected. These investigators first grew their epithelial cell cultures in serum-supplemented medium. The medium was then changed to serum-free medium at the time of the transport study. After this procedure, Allen *et al.* (1981) observed stimulatory effects of insulin, epidermal growth factor (EGF), and fibroblast growth factor (FGF) on 3-O-methyl-D-glucose uptake in primary ovarian granulosa cells. Similarly, Koch and Leffert (1979) observed stimulatory effects of insulin, glucagon, and EGF on Na^+ uptake in primary rat heptaocytes. These investigators postulated that the increase in the rate of Na^+ uptake acted as a signal to initiate an increase in the rate of hepatocyte proliferation.

Even though these transport studies were done serum free, they are subject to several limitations. First, the cells used in these studies were previously grown with serum. As a consequence, the serum components may continue to affect the cells' phenotype for a number of generations after the serum is removed. For example, steroids such as estradiol may remain in the nucleus of mammalian cells for several generations after the cells are switched to serum-free medium lacking steroids (Rudland *et al.*, 1979). Thus, the effects of such steroids on transport may be difficult to study. Second, when the serum is removed, and simply replaced with a serum-free basal medium, the cells are no longer under a condition that supports their long-term survival. The normal metabolic processes of the cells are disrupted. As a consequence, the physiologically relevant effects of hormones on transport may not be observed. Instead, effects of hormones on transport may be observed that are simply a result of the hormones' effects on cell viability. If physiologically relevant effects of hormones on transport are to be observed, the cells must be kept in a viable steady-state condition, which closely resembles the *in vivo* situation.

In vivo transporting epithelial cells are in a complex hormonal milieu. The cells are exposed to solutes present in a lumenal space, as well as to solutes present in a second fluid compartment. For example, kidney tubule epithelial cells are affected by molecules present within the tubule (the mucosal surface of the cells faces this lumenal fluid). In addition, the kidney cells are affected by serum components that have diffused through the blood capillaries (the serosal surface of the cells is initially exposed to these serum factors). Some of these serum factors are involved in regulating the growth of the renal cells, whereas other factors regulate their transport functions. These serum factors may include not only known hormones, but also urinary proteins that have not yet been identified. Renal cell cultures may be used as a means to identify and purify such factors.

First, however, the optimal conditions for the growth of the epithelial cells in the absence of serum must be defined. In 1976, Hayashi and Sato demonstrated that the GH_3 rat pituitary cell line could grow in serum-free medium, supplemented with seven hormone supplements. Subsequently, Sato and his associates demonstrated that a number of differentiated animal cell lines could grow serum free, provided that an appropriate group of supplements was added (Sato and Reid, 1978). These supplements included hormones, growth factors, transport-binding proteins, and cell attachment factors. The group of supplements that was added *in lieu* of serum differed according to the cell type of interest. *In vivo* these growth supplements may similarly be required for the cells to grow. Thus, these *in vitro* studies may be a means of identifying the *in vivo* growth requirements of differentiated animal cells.

The regulatory effects of hormones on transport can be readily observed in such hormonally defined growth media. Thus, new regulatory factors for transport systems may be identified under these serum-free conditions. Their mechanisms of action may then be studied serum free. The stimulatory effects of any particular hormone on transport in serum-free medium may be due to the direct effects of the hormone on a transport system or may be the consequence of the hormone's effects on cell growth. These possibilities can be readily delineated by using culture media which not only permit optimal growth, but which are also hormonally defined. Hormones may be identified that affect transport rather than growth. Their actions may then be studied while the animal cells are in a viable steady-state condition. Other hormones may be identified which not only regulate the activity of transport systems, but which are also required for optimal growth serum free. Such hormones may affect transport systems as a consequence of their stimulatory effects on epithelial cell growth. Alternatively, the growth-stimulatory effects of such hormone supplements may be the consequence of the hormone's effects on transport systems. Hormonally defined, serum-free media permits these alternative mechanisms of hormone action to be identified and studied.

5. USE OF THE MDCK CELL LINE IN THE STUDY OF HORMONAL REGULATION OF TRANSPORT

The MDCK cell line is an excellent system for *in vitro* studies concerning hormonal regulation of transport. Several transport systems involved in the translocation of ions across the cells have been characterized in detail. The effects of hormones on adenylate cyclase have been studied (Rindler *et al.*, 1979), and several hormone receptors have been characterized (Ludens *et al.*, 1978; Lin *et al.*, 1982). Thus, the effects of specific hormones on ion transport systems in MDCK cells may now be defined. Several hormones may affect the rate of ion transport by modulating intracellular calcium concentrations (intracellular calcium has been shown to regulate the rate of Na^+ transport into MDCK cells). The observation of such hormonal effects on intracellular calcium concentrations, and ion flux may be facilitated by a hormonally defined, serum-free medium (medium K-1), which has been developed for the MDCK cell line (Taub *et al.*, 1979).

Medium K-1, the hormonally defined medium for MDCK cells, consists of a serum-free basal medium supplemented with five factors, insulin, transferrin, triiodothyronine (T_3), prostaglandin E_1 (PGE_1), and hydrocortisone. When MDCK cells are maintained in medium K-1, the cells grow over the long term and at the same rate as in serum-supplemented medium. However, when any of the five supplements in medium K-1 is individually deleted from K-1, MDCK cell growth is reduced. The individual deletion of either PGE_1 or transferrin results in a more substantial decrease in growth than the individual deletion of the other three supplements. Thus, although all five factors are required for optimal growth, PGE_1 and transferrin are more important than the other three factors in medium K-1 for MDCK cells.

Medium K-1 is not only an optimal growth medium for MDCK cells, but also permits MDCK cells to express their differentiated functions. Confluent monolayers of MDCK cells form domes in medium K-1. Dome formation is dependent on the presence of PGE_1 and is increased in frequency after the further addition of either T_3 or hydrocortisone to the culture medium (Taub *et al.*, 1979).

These supplements may increase the frequency of dome formation as a direct consequence of their affects on transport systems involved in transepithelial sodium transport. Indeed PGE_1, T_3, and hydrocortisone have all been reported to affect salt reabsorption in the kidney (Iino and Imai, 1978; Edelman, 1974; Marver *et al.*, 1981). Alternatively, the increased frequency of dome formation caused by PGE_1, T_3, and hydrocortisone may be secondary effects of these factors, their primary effects being on cell viability.

The mechanisms by which these factors affect transport may be delineated by means of the genetic approach. This use of the genetic approach is most clearly demonstrated in studies concerning the effects of PGE_1 on MDCK cells. PGE_1 increases intracellular cyclic AMP levels in MDCK cells (Rindler *et al.*, 1979), which as a consequence activates cyclic-AMP-dependent protein kinase activity. According to the model presented in Fig. 2, the increase in intracellular cyclic AMP levels activates two different forms of cyclic-AMP-dependent protein kinase. One form of kinase then causes protein phosphorylations resulting in the growth response to PGE_1, whereas the other form of kinase causes the protein phosphorylations involved in the dome response to PGE_1.

Consistent with this model, variants of MDCK cells have been isolated (PGE_1-independent variants), which have lost the growth response to PGE_1, but which retain the dome response to PGE_1 (Taub *et al.*, 1984). Another class of MDCK variant cells, dibutyryl cyclic-AMP-resistant variants, has lost the dome response to PGE_1 but retains the growth response to PGE_1 (Devis *et al.*, 1982). This latter class of variants has decreased cyclic-AMP-dependent protein kinase activity. An analysis of the defective kinase activity may lead to the identification of the kinases involved in the growth response and the dome response to PGE_1.

These variant cells may also be used to identify the molecular sites affected by PGE_1 which are critical in the induction of dome formation. Included among these molecular sites are the transport systems involved in determining the rate

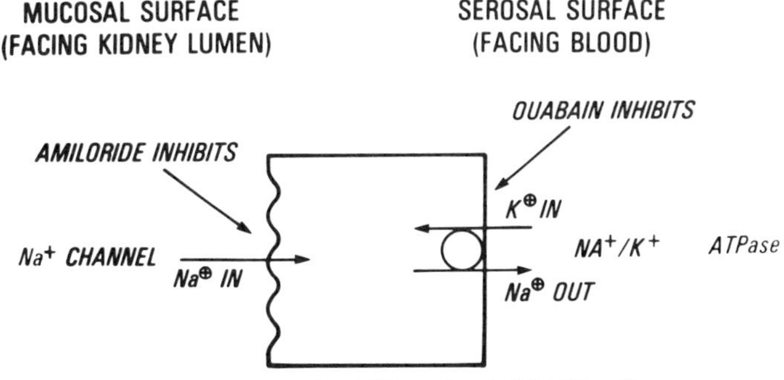

Figure 2. Model for PGE_1 action in MDCK cells.

vectorial salt transport, as well as the tight junctions. The possibility that agents such as PGE_1 may affect the junctions is indicated by the work of Duffey *et al.* (1981). These investigators observed that the permeability of the tight junctions is regulated by intracellular cyclic AMP levels. The adhesion of MDCK cells to their substratum is another parameter that has been reported as being regulated by cyclic AMP (Rabito *et al.*, 1980). PGE_1 may increase the frequency of dome formation in MDCK cells as a consequence of such cyclic-AMP-mediated events.

The effects of other hormones on MDCK cells are also of interest. Hydrocortisone increases the frequency of dome formation in MDCK cells in defined medium (Taub *et al.*, 1979). A similar observation has been made with aldosterone. These steroids have also been shown to increase the rate of vectorial Na^+ transport in kidney tubule cells. Such increases in the rate of Na^+ transport in the kidney have been proposed to be mediated by mineralocorticoid receptors. However, in MDCK cells the increase in the frequency of dome formation caused by aldosterone and hydrocortisone is probably not mediated by such mineralocorticoid receptors. Ludens *et al.* (1978) did not observe the binding of steroids to mineralocorticoid receptors, but did not obtain evidence for the binding of steroids to DOC receptors in MDCK cells. As a consequence, hydrocortisone may affect dome formation in MDCK cells as a result of its interaction with such DOC receptors. The involvement of glucocorticoid receptors is also possible, as the existence of glucocorticoid receptors in MDCK was not excluded by these studies.

The mechanisms by which hormones such as hydrocortisone and T_3 affect ion transport systems have been extensively studied. Hydrocortisone and T_3 have both been observed to complex with receptors which then are translocated to the nucleus. The rate of synthesis of particular classes of mRNA and proteins is then increased. Subsequently, increased rates of transepithelial ion transport are observed. Steroids such as aldosterone and hydrocortisone have been shown to increase the rate of transepithelial ion transport as a consequence of their effects on the activity of the Na^+ channel (Edelman, 1983, personal communication; Marver *et al.*, 1981). In contrast, T_3 has been shown to increase the rate of transepithelial solute transport as a consequence of its effects on the activity of the Na^+/K^+ ATPase (Edelman, 1974). These hormones may similarly affect the ion transport systems in MDCK cells and, as a consequence, increase the frequency of dome formation.

The use of hormonally defined, serum-free medium is critical in examining the effects of hormones such as T_3 and hydrocortisone on transport in MDCK. The effects of these hormones can be observed at physiological concentrations, unlike the case with serum-supplemented medium. The concentrations of hormones can be readily manipulated, permitting synergistic interactions between hormones to be studied. Of particular importance in studies concerning hormonal regulation with MDCK is the fact that MDCK cells can be maintained in a viable steady-state condition over long time intervals in medium K-1. As a consequence, the long-term effects (hours, days) of steroids and prostaglandins on transport and dome formation can readily be studied. The routine use of hormonally defined medium furthermore avoids the problem of the "carry-over" of serum factors, which occurs when cells are first grown with serum and then switched to a serum-free medium. Under such conditions, factors from the serum such as steroids may remain in the cell nucleus for many generations thereafter. When a hormonally defined medium is used, the effects of steroids on animal cells can be more clearly defined.

6. USE OF HORMONALLY DEFINED, SERUM-FREE MEDIUM IN THE STUDY OF TRANSPORT IN OTHER ESTABLISHED EPITHELIAL CELL LINES

The studies described previously with MDCK cells illustrate the use of hormonally defined, serum-free medium in the study of hormonal regulation of transepithelial solute transport. Similar studies may be done with other established kidney epithelial cell lines, as well as with established epithelial cell lines that originate from other tissues.

A number of established kidney epithelial cell lines can be grown in hormonally defined, serum-free medium. Several of these kidney cell lines can be maintained in medium K-1, the hormonally defined, serum-free medium for MDCK cells (Table III). However, the transport properties of the kidney cell lines listed in Table III have not been examined, and their hormonal responses have not been defined. A hormonally defined medium has been developed for the LLC–PK$_1$ cell line, using medium K-1 as a starting point (Chuman *et al.*, 1982). LLC–PK$_1$ cells, like MDCK cells, are well characterized. LLC–PK$_1$ cells possess the capacity for transepithelial glucose transport *in vitro* (Sanders *et al.*, 1983), a characteristic of the proximal tubule of the kidney. Polarized amino acid transport has also been reported (Rabito and Karish, 1982). The effects of hormones on adenylate cyclase activity have been examined in LLC–PK$_1$ cells. These pig kidney cells possess an arginine vasopressin-sensitive adenylate cyclase typical of the distal tubule (Goldring *et al.*, 1978), rather than a parathyroid hormone (PTH)-sensitive adenylate cyclase typical of the proximal tubule (Morel, 1981). Consequently, the physiological significance of studies concerning the regulation of transport in LLC–PK$_1$ by hormones is unclear.

Established epithelial cell lines derived from other tissues can similarly be maintained in hormonally defined medium and studied with regard to their transport properties. Hormonally defined, serum-free media have been designed specifically for the established epithelial cell lines summarized in Table IV. Many of

Table III.   Growth of Kidney Cell Lines in Medium K-1[a]

Cell type, cell line	Species of origin	Growth rate (doublings/day)	
		Serum-free medium + 10% FCS	Medium K-1
LLC–MK$_2$	Monkey	1.6	1.6
MDCK	Bovine	0.8	1.0
BSC-1	African green monkey	0.7	0.4
NRK	Normal rat kidney	1.3	0.1
RAG	Rat adenocarcinoma	1.6	1.2
MDCK	Canine	1.5	1.5

[a]Cells were inoculated at 2.5×10^4 cells/60-mm dish into dishes containing serum-free DME/F12 (50:50 mix) supplemented with 10% fetal calf serum. The next day the medium was removed by aspiration, and the cells were washed three times with serum-free DME/F12 supplemented with 10% fetal calf serum. The growth rate was estimated from daily cell counts in duplicate dishes over a 5-day period. (Reprinted with permission from Taub and Sato, 1979.)

these epithelial cell lines were derived from common human tumors. For example, the HC845 cell line and the MCF-7 cell line originate from a human colon tumor and a human mammary tumor, respectively. The normal tissue from which these tumors are derived contains transporting epithelial cells. In the colon transepithelial ion transport by these cells is regulated by vasopressin, whereas in normal mammary tissue vectorial transport is regulated by prolactin. The possibility that these human tumor cells retain the transport properties and the hormone responses of the normal tissue has not been examined. However, studies concerning the differentiated functions of the cultured human tumor cells may be done in defined medium. Such studies may aide in the analysis of the etiology of the disease state in patients with such tumors, as well as in cancer chemotherapy.

In order to evaluate the resemblance of such established human tumor cell lines to normal cells and to understand the alterations that occur upon malignant transformation, good *in vitro* model systems of normal human epithelial cells are needed. Primary cultures of normal human cells have the potential of being important model systems. However, the use of primary cultures of normal human epithelial cells has been limited, not only by the lack of human tissue, but also by the serum in the tissue culture medium.

7. IMPORTANCE OF HORMONALLY DEFINED, SERUM-FREE MEDIUM IN THE STUDY OF PRIMARY EPITHELIAL CELL CULTURES

The serum in the tissue culture medium has prevented the development of many primary epithelial cell culture systems. Serum contains many different

Table IV. Established Epithelial Cell Lines for Which Serum-Free Media Have Been Developed

Cell line	Growth supplements[a]						
	Transferrin (μg/ml)	Insulin (μg/ml)	Other hormones	Growth factors (ng/ml)	Nutrients	Undefined components	Other
Cervix							
HeLa: human cervical carcinoma (Hutchings and Sato, 1978)	5	5	50 nM HC	EGF (10)			
Colon							
T84: human colon carcinoma (Murakami and Masui, 1980)	2	2	0.2 μg/ml glucagon, 50 nM HC	EGF (1)	10 μg/ml ascorbate		
Embryo							
F9: mouse embryonal EM2: carcinoma (Rizzino and Crowley, 1980; Rizzino and Sato, 1978): EM1	5 5	1 10				500 μg/ml Pedersen	5 μg/ml CIg 10 μM 2-mercapto ethanol
C17-S clone 1003: mouse embryonal carcinoma (Darmon *et al.*, 1981)	10	5					5 μg/ml CIg
Kidney							
MDCK: dog kidney (Taub *et al.*, 1979)	5	5	50 nM HC, 5 pMT$_3$, 25 ng/ml PGE$_1$				
LLC–PK$_1$: pig kidney (Chuman *et al.*, 1982)	5	5	200 nm HC, 5pM T$_3$, 10 μ U/ml vasopressin		10^{-8} M cholesterol		
Lung							
A431: human long epidermal carcinoma (Barnes, 1982)	10	10			0.5 mM ethenolamine		5 μg/ml CIg

(*continued*)

Table IV. *(Continued)*

Growth supplements[a]

Cell line	Transferrin (μg/ml)	Insulin (μg/ml)	Other hormones	Growth factors (ng/ml)	Nutrients	Undefined components	Other
Mammary							
MCF7: human mammary carcinoma (Barnes and Sato, 1979)	25	0.1	100 ng/ml $PGF_2{}^\alpha$	EGF (100)		α-1 Spreading protein	
2R-75-1: human mammary carcinoma (Allegra and Lippman, 1978)	1	0.5 nM	10nM T_3, 10nM DEX, 10 nM 17-β estradiol				
Melanoma							
M2R: mouse (Mather and Sato, 1979)	1	5	10 nM testosterone, 0.05 ug/ml FSH, 3 ng/ml NGF, 1 ng/ml LHRH				
Ovarian							
RF-1 (Orly *et al.*, 1980)	5	2	10 nM HC				8 μg/ml CIg
Pituitary							
GH$_3$: rat pituitary carcinoma (Hayashi *et al.*, 1978)	5	5	100 nM T_3, 0.5 ng/ml PTH, 1 ng/ml TRH, 1 ng/ml SM-C				
Testicular							
TM4: mouse	5	5	0.5 μg/ml FSH, 1 ng/ml GH, 1 ng/ml SM-C,	EGF (3000)	50 ng/ml retinoic acid		

[a]HC, hydrocortisone; T_3, triiodothyronine; CIg, cold insoluble globulin; EGF, epidermal growth factor; FGF, fibroblast growth factor; DEX, dexamethasone; FSH, follicle-stimulating hormone; NGF, nerve growth factor; LHRH, luteinizing hormone-releasing hormone; PTH, parathyroid hormone; TRH, thyroid-releasing hormone; SM-C, somatomedin-C; FSH, follicle-stimulating hormone; GH, growth hormone; $PGF_2{}^\alpha$, prostaglandin $F_2{}^\alpha$; PGE_1, prostaglandin E_1.

components that permit extensive fibroblast overgrowth of epithelial cell cultures. In addition, serum contains components that are deleterious to the growth of the epithelial cells of interest. As a consequence, the number of established epithelial cell lines available for study is limited. Those epithelial cell lines which have become established in culture may have had to undergo a number of genetic changes, in order to survive indefinitely *in vitro*. As a consequence, primary epithelial cell culture systems are needed as a means to check the physiological significance of studies done with established epithelial cell lines.

Previous problems encountered with primary epithelial cell cultures can be alleviated to a significant extent by the use of hormonally defined, serum-free medium. Two major approaches have been taken. First, the hormonally defined, serum-free media developed for established epithelial cell lines have been used to maintain primary cultures of epithelial cells derived from the same organ, without fibroblast overgrowth (Taub and Sato, 1980). Second, hormonally defined, serum-free media have been developed directly for primary epithelial cell cultures.

An example of the first approach toward growing primaries is exemplified with baby mouse kidney epithelial cells. Primary cultures of epithelial cells derived from the kidneys of 10-day-old Balb/c mice were maintained in medium K-1 (the defined medium for MDCK cells) without fibroblast overgrowth. In contrast, in cultures maintained in medium supplemented with 10% fetal calf serum, the epithelial cells were overgrown with fibroblasts (Taub and Sato, 1980).

The growth properties of the baby mouse kidney primaries were examined in detail. When plated at high densities (greater than 5×10^3 cells/cm^2), the majority of the cells that attached were epithelial in morphology, both in defined and in serum-supplemented medium. The primary baby mouse kidney epithelial cells subsequently grew at 0.6 doublings/day in these media. At lower plating densities, baby mouse kidney epithelial colonies formed. The cells in the colonies had an epithelial morphology in K-1 but were fibroblastic in serum-supplemented medium. A hormone deletion study indicated that in the case of both baby mouse kidney epithelial cells and MDCK cells, (1) each of the five supplements in medium K-1 was required to attain the maximal frequency of colonies, and (2) PGE$_1$ and transferrin had the most dramatic effects on colony formation (Taub and Sato, 1980). Thus, MDCK cells possess growth properties similar to those of epithelial cells in the mouse kidney.

The baby mouse kidney primaries exhibited multicellular domes in medium K-1, which suggested that the monolayers contained transporting epithelial cells. Dome formation was not observed, however, in serum-supplemented medium. Amiloride-sensitive Na$^+$ uptake, a distal tubule marker, was observed in the baby mouse kidney primary cultures. However, Na$^+$-dependent α-methylglucoside uptake (a proximal tubule marker) was not detected. These studies illustrate the use of a hormonally defined, serum-free medium developed for an established kidney epithelial cell line to culture primary kidney cells with similar growth and functional properties.

Primary kidney cell cultures that express other functional properties would also be of interest to study serum free. In order to develop a primary kidney culture system with other functional properties, the second approach described previously was taken. Optimal serum-free growth conditions were determined for primary rabbit kidney proximal tubule cells. Proximal tubule cells were chosen for study as their hormone responses and transport systems are distinct from those of MDCK cells.

The proximal tubules were purified from the rabbit kidney, and the cells were cultured as illustrated in Fig. 3 (Chung *et al.*, 1982). The proximal tubules attached to dishes containing serum-free medium supplemented with insulin, transferrin, and hydrocortisone. Epithelial cells migrated from the tubules and after several weeks formed confluent monolayers, without fibroblast overgrowth. Figure 4 illustrates the epithelial cells in confluent monolayers of proximal tubule cell cultures.

The functional properties of 2-wk-old primary rabbit proximal tubule cells have been examined. The rabbit kidney proximal tubule cultures exhibit Na^+-dependent sugar transport and Na^+-dependent phosphate transport, which is typical of cells in the proximal tubule. In addition, the primaries have hormone responses typical of the cells in the proximal tubule. As observed in the proximal tubule, the primary cells synthesize cyclic AMP in response to parathyroid

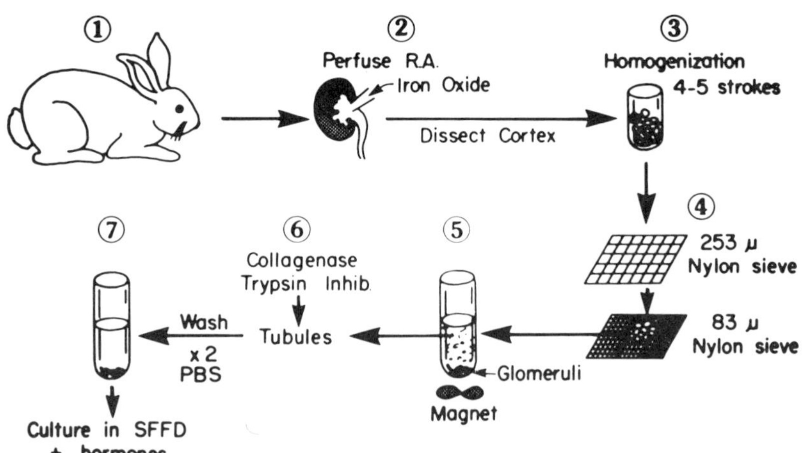

Figure 3. Preparation of primary cultures of rabbit kidney proximal tubules. Primary rabbit kidney cultures were prepared as diagrammed. Details are described in Chung *et al.* (1982).

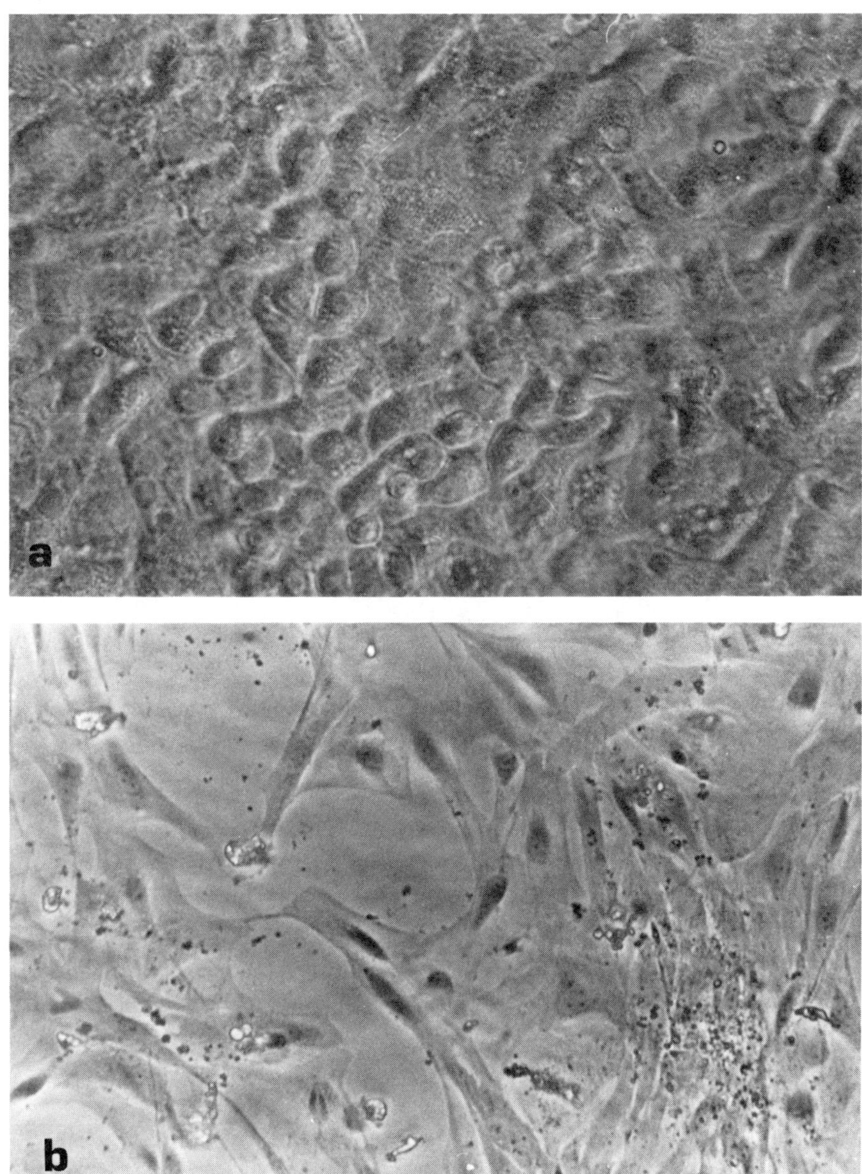

Figure 4. Primary cultures of rabbit kidney proximal tubule cells after maintenance for 2 wk in serum-free DME/F12 (50:50 mix) supplemented with either (a) 5 μg/ml transferrin, 5 μg/ml insulin, and 4×10^{-8} M hydrocortisone, or (b) 10% fetal calf serum.

Table V. Primary Cultures of Epithelial Cells Maintained in Hormonally Defined Media

Primary cells	Growth supplements[a]						
	Transferrin	Insulin	Other hormones	Growth factors	Nutrients	Undefined components	Other
Adrenal: adult bovine zona fasciculata (Simonian *et al.*, 1982)	—	0.2 nM		44 nM FGF	10 μg/ml LDL		2 μg/cm^2 CIg, 1.5 nM thrombin
Embryo: mouse blastocyst (Rizzino and Sherman, 1979)	5	10				0.3% BSA, 0.05% Pedersen fetuin	
Kidney: baby mouse (Taub and Sato, 1980)	5	5	5 pMT$_3$, 50 nM HC, 25 ng/ml PGE$_1$				
adult proximal tubule (Taub *et al.*, 1982)	5	5	50 nM HC				
Liver: adult rat (Koch and Leffert, 1979)		10 ng/ml		10 ng/ml EGF			
fetal mouse (Salas–Prato, 1982)	—	—	100 ng/ml glucagon		0.25 nM, selenium, 10 nM ethanolamine		2 ng/ml cholera toxin
Mammary: mouse (Imagawa *et al.*, 1982)	10	10		EGF (100)		5 mg/ml BSA	10 ng/ml cholera toxin

rat (Salomon *et al.*, 1981)	5	5	5 nM DEX	EGF (10) FGF (100)		1 mg/ml Pedersen fetuin
human (Stampfer *et al.*, 1981)	—	10	0.1 µg/ml HC, 10 nM T_3, 10 nM estradiol	EGF (5)		0.5% FCS; conditioned medium
Ovarian: rat granulosa (Orly *et al.*, 1980)	5	2	10 nM HC			8 µg/ml CIg
Prostate: mouse (Waymouth *et al.*, 1982)	5	8	2 µg/ml HC, 0.5 µg/ml, 5 dihydro-testosterone, 25 IU L ovine prolactin		0.1 µg/ml spermine	
Testicular: mouse (Perez-Infante and Mather, 1982)	5	10		EGF (10)		
Tracheal: rabbit (Wu and Smith, 1982)	5	5		10 ng/ml EGF		0.3% FCS
Thyroid: rat (Ambesi–Impiombato *et al.*, 1980)	5	10	10 ng/ml somatostatin, 10 mU/ml thyrotropin			10 ng/ml glycl-L-histidyl-L-lysine acetate

[a]T_3, triiodothyronine; HC, hydrocortisone; PGE_1, prostaglandins E_1; FGF, fibroblast growth factor; EGF, epidermal growth factor; LDL, low-density lipoprotein; BSA, bovine serum albumin; FCS, fetal calf serum; CIg, cold insoluble globulin; DEX, dexamethasone.

hormone, whereas a significant response to calcitonin and arginine vasopressin was not observed (Chung *et al.*, 1982). Thus, the use of hormonally defined medium has resulted in the development of a cell culture system that more closely resembles the proximal tubule cells *in vivo* than the other available system, LLC–PK$_1$.

Hormonally defined, serum-free medium has also been used as a means to develop primary epithelial cell culture systems from other tissues. The different types of primary epithelial cells that have been maintained in appropriate hormonally defined, serum-free media are summarized in Table V. *In vivo* many of these epithelial cells possess the capacity for transepithelial solute transport. For example, tracheal epithelial cells possess the capacity for chloride transport across the cell layer. Such vectorial ion transport by the tracheal cells is important in mucociliary clearance from the airways. Secretogogues such as PGE$_2$ and epinephrine regulate vectorial transport in tracheal epithelial cells *in vivo* by affecting the activity of a Na$^+$-dependent chloride transport system. The Sertoli cells in the testes are similarly involved in transepithelial ion movements. Such ion transport affects the ionic composition of the seminiferous tubules and as a consequence the maturation of germ cells. Even hepatocytes possess the capacity for transepithelial salt movements. Bile is transported from the caniculi (the lumenal space of hepatocytes) through the cells and then into the blood. Such bile transport is regulated by cyclic AMP (Boyer, 1980).

The transport systems involved in such transepithelial transport may be retained in the primary epithelial cell cultures that have been maintained in hormonally defined medium. Scharschmidt and Stephens (1981) have observed Na$^+$-dependent bile transport tin primary rat hepatocyte cultures. The regulation of bile transport by cyclic AMP may be examined using hormonally defined serum-free medium. The primary tracheal epithelial cells of Wu and Smith (1982) and the primary Sertoli cells of Mather *et al.* (1982) may similarly be studied with regard to their ion transport properties.

The investigations of Sinback and Coon (1982) have clearly demonstrated the importance of examining the transport properties of the primary epithelial cell cultures maintained in serum-free medium. These investigators examined the transport properties of rat thyroid cell strains that were developed in serum-free medium. Primary thyroid epithelial cell cultures were initiated from rat thyroid tissue by the use of serum-free medium supplemented with 10 μg/ml insulin, 10 nM hydrocortisone, 5 μg/ml transferrin, 10 ng/ml glycyl-L-histidyl-L-lysine acetate, 2.5 mU/ml thyroid-stimulating hormone and, 10 mU/ml somatostatin (Ambesi–Impiombato *et al.*, 1980). The cultures were then subcultured for more than 2 yr in this hormonally defined medium. The thyroid cells were observed to possess the capacity to concentrate iodide and to secrete thyroglobulin, over this time period.

Similarly, the epithelial cells that form the thyroid follicles possess these functional properties. Both iodide and thyroglobulin are concentrated in the follicles by the thyroid cells. In order to concentrate iodide into the follicles, iodide outside the follicle is transported into the thyroid cells from their serosal surface (which faces the outside of the follicle) and then out of the cells' mucosal surface (which faces the inside of the follicle). Thyroid hormone synthesis can then occur, using both the concentrated iodide and the thyroglobulin. The iodide transport systems in the thyroid cells, which are required for thyroid hormone synthesis, have not been well characterized. However, the regulation of these transport systems by hormones is clearly an important factor in modulating thyroid function.

Sinback and Coon (1982) have made several novel observations concerning iodide transport in the thyroid, by studying their thyroid cell cultures. The electrophysiological properties of the thyroid cells were determined by means of microelectrodes. The average resting membrane potential of individual thyroid cells was observed to be between -70 mV and -80 mV. Slow transitions in the membrane potential from this high potential state to a lower potential state (between -25 mV and -40 mV) occurred spontaneously. Noradrenaline elicited both a fast depolarization, and a biphasic response (a fast depolarization of 35 mV followed by a slow depolarization of 22 mV).

Iodide uptake may be affected as a consequence of two different types of changes that occur after such depolarization. First, the depolarization causes a decrease in the membrane potential, which as a consequence results in increased iodide uptake (the electrical potential had opposed iodide uptake). Second, the results of Sinback and Coon indicated that the depolarization causes an increase in chloride conductance through chloride conductance channels. Such chloride channels have been observed to admit iodide and chloride equally well (Take-huchi, 1977). Thus, the rate of iodide transport into chloride channels may also increase dramatically as a consequence of the depolarization.

8. SUMMARY

The transport properties of epithelial cells derived from other tissues may similarly be studied *in vitro*. Primary cultures from many different tissues may now be maintained in hormonally defined, serum-free medium. The use of hormonally defined, serum-free medium is important for studies with primary cultures, for several reasons. First, when primary epithelial cell cultures are maintained in hormone-supplemented, serum-free medium, fibroblast overgrowth is not observed. When primary epithelial cell cultures are maintained with serum, however, fibroblast overgrowth is observed and has prevented reproducible results from being obtained. Second, when primary epithelial cell cultures are maintained in serum-free medium, the cells retain differentiated functions that are not observed when the same primaries are maintained with serum (Orly *et al.*, 1980). Finally, the use of hormonally defined medium has permitted new types of epithelial cells to be cultured, which could not previously be studied with serum.

In vitro studies concerning the factors that regulate the vectorial movements of solutes across epithelial cells are of interest. Kidney tubule cell cultures are particularly good model systems for such studies, as *in vivo* their primary function is to reabsorb solutes. Several established kidney epithelial cell lines have been shown to retain the capacity for transepithelial solute transport. Transport systems involved in transepithelial solute transport have been characterized in detail by using these cell lines. The effects of hormones on these transport systems can now be studied, using appropriate hormonally defined, serum-free media. Similar studies can also be done with established epithelial cell lines and primary cultures derived from a number of tissues, by the use of hormonally defined medium.

ACKNOWLEDGMENTS.   The author acknowledges Angela Dudek for preparation of the manuscript.

REFERENCES

Abaza, N. A., Leighton, J., and Schultz, S., 1974, Effects of ouabain on the function and structure of a cell line (MDCK) derived from canine kidney, *In Vitro* **10**:172–183.

Allegra, J. C., and Lippman, M. E., 1978, Growth of a human breast cancer cell line in serum-free hormone-supplemented medium, *Cancer Res.* **38**:3823–3829.

Allen, W. R., Nilsen–Hamilton, M., and Hamilton, R. T., 1981, Insulin and growth factors stimulate rapid posttranslational changes in glucose transport in ovarian granulosa cells, *J. Cell Physiol.* **108**:15–24.

Ambesi–Impiombato, F. S., Parks, L. A. M., and Coon, H. G., 1980, Culture of hormone-dependent functional epithelial cells from rat thyroids, *Proc. Natl. Acad. Sci. USA* **77**:3455–3459.

Auersperg, N., 1969, Histogenetic behavior of tumors. I. Morphologic variation *in vitro* of two related human carcinoma cell lines, *J. Natl. Cancer Inst.* **43**:151–173.

Barnes, D. W., 1982, Epidermal growth factor inhibits growth of A431 human epidermoid carcinoma in serum-free cell cultures, *J. Cell Biol.* **93**:7–4.

Barnes, D., and Sato, G., 1979, Growth of a human mammary tumor cell line in a serum free medium, *Nature* **281**:388–389.

Birek, C., Aubin, J. E., Bhargava, U., Brunette, D. M., and Melcher, A. H., 1982, Dome formation by oral epithelia *in vitro*, *In Vitro* **18**:382–392.

Bisbee, C. A., Machen, T. E., and Bern, H. A., 1979, Mouse mammary epithelial cells in floating collagen gels: Transepithelial ion transport and effects of prolactin, *Proc. Natl. Acad. Sci. USA* **76**:536–540.

Boyer, J. L., 1980, New concepts of mechanisms of hepatocyte bile formation, *Physiol. Rev.* **60**:303–326.

Cameron, G., 1953, Secretory activity of the choroid plexus in tissue culture, *Anat. Rec.* **117**:115–125.

Cereijido, M., Robbins, E. S., Dolan, W. J., Rotunno, C. A., and Sabatini, D. D., 1978, Polarized monolayers formed by epithelial cells on a permeable and translucent support, *J. Cell Biol.* **77:**853–880.

Cereijido, M., Ehrenfeld, J., Meza, I., and Martinez-Palomo, 1980, Structural and functional membrane polarity in cultured monolayers of MDCK cells, *J. Membr. Biol.* **52:**147–159.

Chuman, L., Fine, L. G., Cohen, A. H., and Saier, M. H., 1982, Continuous growth of proximal tubular kidney epithelial cells in hormone-supplemented serum-free medium, *J. Cell Biol.* **94:**506–510.

Chung, S. D., Alavi, N., Livingston, D., Hiller, S., and Taub, M., 1982, Characterization of primary rabbit kidney cultures that express proximal tubule functions in a hormonally defined medium, *J. Cell Biol.* **95:**118–126.

Darmon, M., Bottenstein, J., and Sato, G., 1981, Neural differentiation following culture of embryonal carcinoma cells in a serum free defined medium, *Dev. Biol.* **85:**463–473.

Devis, P., Hiller, S., and Taub, M., 1982, Prostaglandin E_1 regulates the growth and function of Madin Darby canine kidney (MDCK) cells in a hormonally defined medium, *J. Cell Biol.* **95:**188a.

Duffey, M. E., Hainau, B., Ho, S., and Bentzel, C. J., 1981, Regulation of epithelial tight junction permeability by cyclic AMP, *Nature* **294:**451–453.

Edelman, I. S., 1974, Thyroid thermogenesis, *N. Engl. J. Med.* **290:**1303–1308.

Goldring, S. R., Dayer, J. M., Ausiello, D. A., and Krane, S. M., 1978, A cell strain cultured from porcine kidney increases cyclic AMP content upon exposure to calcitonin or vasopressin, *Biochem. Biophys. Res. Com.* **83:**434–440.

Handler, J. S., Steel, R. E., Sahib, M. K., Wade, J. B., Preston, A. S., Lawson, N. L., and Johnson, J. P., 1979, Toad urinary bladder epithelial cells in culture: Maintenance of epithelial structure, sodium transport and response to hormones, *Proc. Natl. Acad. Sci. USA* **76:**4151–4155.

Hayashi, I., and Sato, G. H., 1976, Replacement of serum by hormones permits growth of cells in defined medium, *Nature* **256:**132–134.

Hayashi, I., Larner, J., and Sato, G., 1978, Hormonal growth control of cells in culture, *In Vitro* **14:**23–30.

Huh, N., Takaoka, T., and Katsuka, H., 1977, Establishment of epithelial cell lines from rat glandular stomachs, *Jpn. J. Exp. Med.* **47:**413–424.

Hutchings, S. E., and Sato, G. H., 1978, Growth and maintenance of HeLa cells in serum-free medium supplemented with hormones, *Proc. Natl. Acad. Sci. USA* **75:**901–904.

Iino, Y., and Imai, M., 1978, Effects of prostaglandins on Na transport in isolated collecting tubules, *Pflugers Arch.* **373:**125–132.

Imagawa, W., Tommoka, Y., and Nandi, S., 1982, Serum-free growth of normal and tumor mouse mammary epithelial cells in primary cultures, *Proc. Natl. Acad. Sci. USA* **79:**4074–4077.

Koch, K. S., and Leffert, H. L., 1979, Increased sodium ion influx is necessary to initiate rat hepatocyte proliferation, *Cell* **18:**153–163.

Leighton, J., Brada, Z., Estes, L. W., and Justh, G., 1969, Secretory activity and oncogenicity of a cell line (MDCK) derived from canine kidney, *Science* **158:**472–473.

Leighton, J., Estes, L. W., Mansukhani, S., and Brada, Z., 1970, A cell line derived from normal dog kidney (MDCK) exhibiting qualities of papillary adenocarcinoma and or renal tubular epithelium, *Cancer* **26:**1022–1028.

Lever, J. E., 1979a, Inducers of mammalian differentiation stimulate dome formation in a differentiated kidney epithelial cell line (MDCK), *Proc. Natl. Acad. Sci. USA* **76:**1323–1327.

Lever, J. E., 1979b, Cyclic AMP and inducers of mammalian cell differentiation stimulate dome formation in mammary and renal epithelial cell cultures, in: *Hormones and Cell Culture,* Book

B, Cold Spring Harbor Conferences on Cell Proliferation, Volume 6, Cold Spring Harbor Laboratory, Cold Spring Harbor, New York, pp. 727–738.

Lin, M. C., Koh, S. W. M., Dykman, D. D., Beckner, S. K., and Shih, T. Y., 1982, Loss and restoration of glucagon receptors and responsiveness in a transformed kidney cell line, *Exp. Cell Res.* **142:**181–189.

Logsdon, C. D., Bisbee, C. A., Rutter, M. J., and Machen, T. E., 1982, Fetal rabbit gastric epithelial cells cultures on floating collagen gels, *In Vitro* **18:**233–242.

Ludens, J. H., Vaughn, D. A., Mawe, R. C., and Fanestil, D. D., 1978, Specific binding of deoxycorticosterone by canine kidney cells in culture, *J. Steroid Biochem.* **9:**17–21.

Marver, D., Schwartz, M. J., and Kokko, J. P., 1981, Direct effects of corticoid hormones on the mammalian nephron, *Ann. N.Y. Acad. Sci.* **372:**39–55.

Mason, J. R., Williams, M. C., Widdicombe, J. H., Sanders, M. J., Misfeldt, D. S., and Berry, L. C., Jr., 1982, Transepithelial transport by pulmonary alveolar type II cells in primary culture, *Proc. Natl. Acad. Sci. USA* **79:**6033–6037.

Mather, J. P., and Sato, G. H., 1979, The growth of mouse melanoma cells in hormone-supplemented serum-free medium, *Exp. Cell Res.* **120:**191–200.

Mather, J. P., Zhuang, L. Z., Perez–Infante, V., and Phillips, D. M., 1982, Culture of testicular cells in hormone-supplmented serum-free medium, *Ann. N.Y. Acad. Sci.* **383:**44–68.

McCombs, W. B., III, Leibovitz, A., McCoy, C. E., Stinson, J. C., and Berlin, J. D., 1976, Morphologic and immunologic studies of a human colon tumor cell line (SW-48), *Cancer* **38:**2316–2327.

McGrath, C. M., 1975, Cell organization and responsiveness to hormone *in vitro:* Genesis of domes in mammary cell cultures, *Amer. Zool.* **15:**231–236.

Misfeldt, D. S., and Sanders, M. J., 1981, Transepithelial transport in cell culture, D-glucose transport by a pig kidney cell line (LLC–PK$_1$), *J. Mem. Biol.* **59:**13–18.

Misfeldt, D. S., Hamamoto, S. T., and Pitelka, D. R., 1976, Transepithelial transport in cell culture, *Proc. Natl. Acad. Sci. USA* **73:**1212–1216.

Morel, F., 1981, Sites of hormone action in the mammalian nephron, *Am. J. Physiol.* **240:**F159–F164.

Mullin, J. M., Weibel, J., Diamond, L., and Kleinzeller, A., 1980, Sugar transport in the LLC–PK$_1$ renal epithelial cell line: Similarity to mammalian kidney and the influence of cell density, *J. Cell Physiol.* **104:**375–389.

Murakami, H., and Masui, H., 1980, Hormonal control of human colon carcinoma cell growth in serum-free medium, *Proc. Natl. Acad. Sci. USA* **77:**3464–3468.

Orly, J., Sato, G., and Erickson, G. F., 1980, Serum suppresses the experssion of hormonally induced functions in cultured granulosa cells, *Cell* **20:**817–827.

Perez–Infante, V., and Mather, J. P., 1982, The role of transferrin in the growth of testicular cell lines in serum-free medium, *Exp. Cell Res.* **142:**325–332.

Rabito, C. A., and Karish, M. V., 1982, Polarized amino acid transport by an epithelial cell line of renal origin (LLC–PK$_1$), *J. Biol. Chem.* **257:**6802–6808.

Rabito, C. A., Tchao, R., Valantich, J., and Leighton, J., 1980, Effect of cell–substratum interaction on hemicyst formation by MDCK cells, *In Vitro* **16:**461–468.

Rephaeli, A., and Parsons, S. M., 1982, Calmodulin stimulation of $^{45}Ca^{2+}$ transport and protein phosphorylation in cholinergic synaptic vesicles, *Proc. Natl. Acad. Sci. USA* **79:**5783–5787.

Rindler, M. J., Chuman, L. M., Shaffer, L., and Saier, M. H., 1979, Retention of differentiated properties in an established dog kidney epithelial cell line (MDCK), *J. Cell Biol.* **81:**635–648.

Rizzino, A., and Crowley, C., 1980, Growth and differentiation of embryonal carcinoma cell line F9 in defined media, *Proc. Natl. Acad. Sci. USA* **77:**457–461.

Rizzino, A., and Sato, G., 1978, Growth of embryonal carcinoma cells in serum-free medium, *Proc. Natl. Acad. Sci. USA* **75:**1844–1848.

Rizzino, A., and Sherman, M. I., 1979, Development and differentiation of mouse blastocysts in serum-free medium, *Exp. Cell Res.* **121**:221–233.

Rudland, P. S., Bennett, D. C., and Warburton, M. D., 1979, Hormonal control of growth and differentiation of cultured rat mammary gland epithelial cells, in: *Hormones and Cell Culture,* Book B, Cold Spring Harbor Conference on Cell Proliferation, Volume 6, Cold Spring Harbor Laboratory, Cold Spring Harbor, New York, pp. 677–700.

Salas–Prato, M., 1982, Growth of fetal mouse liver cells in hormone-supplemented serum-free medium, in: *Growth of Cells in Hormonally Defined Media,* Book A, Cold Spring Harbor Conferences in Cell Proliferation, Volume 9, Cold Spring Harbor Laboratory, Cold Spring Harbor, New York, pp. 615–624.

Salomon, D. S., Liotta, L. A., and Kidwell, W. R., 1981, Differential response to growth factor by rat mammary epithelium plated on different collagen substrates in serum-free medium, *Proc. Natl. Acad. Sci. USA* **78**:382–386.

Sanders, M. J., Simon, L. M., and Misfeldt, D. S., 1983, Transepithelial transport in cell culture: Bioenergetics of Na, D-glucose-coupled transport, *J. Cell Physiol.* **114**:263–266.

Sato, G., and Reid, L., 1978, Replacement of serum in cell culture by hormones, in: *International Review of Biochemistry,* Volume 20, *Biochemistry and Mode of Action of Hormones* II (H. V. Rickenberg, ed.), University Park Press, Baltimore, Maryland, pp. 219–251.

Scharschmidt, B. F., and Stephens, J. E., 1981, Transport of sodium, chloride, and taurocholate by cultured rat hepatocytes, *Proc. Natl. Acad. Sci. USA* **78**:986–990.

Simonian, M. H., White, M. L., and Gill, G. N., 1982, Growth and function of cultured bovine adrenocortical cells in a serum-free, defined medium, *Endocrinology* **111**:919–927.

Sinback, D. N., and Coon, H. G., 1982, Electrophysiological and pharmacological properties of cultured rat thyroid cells, *J. Cell Physiol.* **112**:391–402.

Stampfer, M. R., Hackett, A. J., Smith, H. S., Hancock, M. C., Leung, J. P., and Edgington, T. S., 1981, Growth of human mammary epithelium in culture and expression of tumor-specific properties, in: *Growth of Cells in Hormonally Defined Media,* Book B, Cold Spring Harbor Conferences on Cell Proliferation, Volume 9, Cold Spring Harbor Laboratory, Cold Spring Harbor, New York, pp. 819–830.

Takehuchi, A., 1977, Functional transmission, I. Postsynaptic mechanisms, in: *Handbook of Physiology,* Volume 1 (E. R. Kandel, ed.), American Physiological Society, Bethesda, Maryland, Section 1, pp. 295–327.

Taub, M., and Saier, M. H., Jr., 1979, Regulation of ^{22}Na$^+$ uptake by calcium in an established kidney epithelial cell line, *J. Biol. Chem.* **254**:11440–11444.

Taub, M., and Saier, M. H., Jr., 1981, Amiloride-resistant Madin–Darby canine kidney (MDCK) cells exhibit decreased cation transport, *J. Cell Physiol.* **106**:191–199.

Taub, M., and Sato, G. H., 1979, Growth of kidney epithelial cells in hormone-supplemented serum-free medium, *J. Supramol. Struct.* **11**:207–216.

Taub, M., and Sato, G., 1980, Growth of functional primary cultures at kidney epithelial cells in defined medium, *J. Cell. Physiol.* **105**:369–378.

Taub, M., Chuman, L., Saier, M. H., and Sato, G., 1979, Growth of Madin–Darby canine kidney epithelial cell (MDCK) line in hormone-supplemented serum-free medium, *Proc. Natl. Acad. Sci. USA* **76**:3338–3342.

Taub, M., Devis, P., and Hiller, S., 1984, Madin Darby canine kidney variant cells have altered cyclic AMP metabolism and altered responsiveness to PGE$_1$ in a hormonally defined medium, *J. Cell. Biochem.* submitted for publication.

Toyoshima, K., Valentich, J. D., Tchao, R., and Leighton, J., 1976, Conditions of cultivation required for the formation of hemicysts *in vitro* by rat bladder carcinoma R-4909, *Cancer Res.* **36**:2800–2806.

Valentich, J. D., Tchao, R., and Leighton, J., 1979, Hemicyst formation stimulated by cyclic AMP in dog kidney cell line MDCK, *J. Cell Physiol.* **100:**291–304.

Villereal, M. L., 1981, Sodium fluxes in human fibroblasts: Effects of serum Ca^{2+} and amiloride, *J. Cell Physiol.* **107:**359–369.

Waymouth, C., Ward, P. F., and Blake, S. L., 1982, Mouse prostatic epithelial cells in defined cultured media, in: *Growth of Cells in Hormonally Defined Media,* Book B, Cold Spring Harbor Conferences on Cell Proliferation, Volume 9, Cold Spring Harbor Laboratory, Cold Spring Harbor, New York, pp. 1097–1108.

Wu, R., and Smith, D., 1982, Continuous multiplication of rabbit tracheal epithelial cells in a defined, hormone-supplemented medium, *In Vitro* **18:**800–812.

Index